STEP PROJECT MANAGEMENT

Guide for Science, Technology, and Engineering Projects

Industrial Innovation Series

Series Editor
Adedeji B. Badiru
Department of Systems and Engineering Management
Air Force Institute of Technology (AFIT) – Dayton, Ohio

PUBLISHED TITLES

Computational Economic Analysis for Engineering and Industry
Adedeji B. Badiru & Olufemi A. Omitaomu

Handbook of Industrial and Systems Engineering
Adedeji B. Badiru

Handbook of Military Industrial Engineering
Adedeji B.Badiru & Marlin U. Thomas

Industrial Project Management: Concepts, Tools, and Techniques
Adedeji B. Badiru, Abidemi Badiru, and Adetokunboh Badiru

Knowledge Discovery from Sensor Data
Auroop R. Ganguly, João Gama, Olufemi A. Omitaomu, Mohamed Medhat Gaber, and Ranga Raju Vatsavai

STEP Project Management: Guide for Science, Technology, and Engineering Projects
Adedeji B. Badiru

Systems Thinking: Coping with 21st Century Problems
John Turner Boardman & Brian J. Sauser

Techonomics: The Theory of Industrial Evolution
H. Lee Martin

Triple C Model of Project Management: Communication, Cooperation, Coordination
Adedeji B. Badiru

FORTHCOMING TITLES

Beyond Lean: Elements of a Successful Implementation
Rupy (Rapinder) Sawhney

Handbook of Industrial Engineerng Calculations and Practice
Adedeji B. Badiru & Olufemi A. Omitaomu

Industrial Control Systems: Mathematical and Statistical Models and Techniques
Adedeji B. Badiru, Oye Ibidapo-Obe, & Babatunde J. Ayeni

Modern Construction: Productive and Lean Practices
Lincoln Harding Forbes

Project Management: Systems, Principles, and Applications
Adedeji B. Badiru

Research Project Management
Adedeji B. Badiru

Statistical Techniques for Project Control
Adedeji B. Badiru

Technology Transfer and Commercialization of Environmental Remediation Technology
Mark N. Goltz

STEP PROJECT MANAGEMENT

Guide for Science, Technology, and Engineering Projects

WITHDRAWN

Adedeji B. Badiru

CRC Press
Taylor & Francis Group
Boca Raton   London   New York

CRC Press is an imprint of the
Taylor & Francis Group, an **informa** business

CRC Press
Taylor & Francis Group
6000 Broken Sound Parkway NW, Suite 300
Boca Raton, FL 33487-2742

© 2009 by Taylor & Francis Group, LLC
CRC Press is an imprint of Taylor & Francis Group, an Informa business

No claim to original U.S. Government works
Printed in the United States of America on acid-free paper
10 9 8 7 6 5 4 3 2 1

International Standard Book Number-13: 978-1-4200-7235-8 (Hardcover)

This book contains information obtained from authentic and highly regarded sources. Reasonable efforts have been made to publish reliable data and information, but the author and publisher cannot assume responsibility for the validity of all materials or the consequences of their use. The authors and publishers have attempted to trace the copyright holders of all material reproduced in this publication and apologize to copyright holders if permission to publish in this form has not been obtained. If any copyright material has not been acknowledged please write and let us know so we may rectify in any future reprint.

Except as permitted under U.S. Copyright Law, no part of this book may be reprinted, reproduced, transmitted, or utilized in any form by any electronic, mechanical, or other means, now known or hereafter invented, including photocopying, microfilming, and recording, or in any information storage or retrieval system, without written permission from the publishers.

For permission to photocopy or use material electronically from this work, please access www.copyright.com (http://www.copyright.com/) or contact the Copyright Clearance Center, Inc. (CCC), 222 Rosewood Drive, Danvers, MA 01923, 978-750-8400. CCC is a not-for-profit organization that provides licenses and registration for a variety of users. For organizations that have been granted a photocopy license by the CCC, a separate system of payment has been arranged.

Trademark Notice: Product or corporate names may be trademarks or registered trademarks, and are used only for identification and explanation without intent to infringe.

Library of Congress Cataloging-in-Publication Data

Badiru, Adedeji Bodunde, 1952-
 STEP project management : guide for science, technology, and engineering projects / Adedeji B. Badiru.
  p. cm. -- (Industrial innovation series)
 Includes bibliographical references and index.
 ISBN 978-1-4200-7235-8 (hardcover : alk. paper)
  1. Project management. 2. Science--Managment. 3. Technology--Management. 4. Engineering--Management. I. Title. II. Series.

T56.8.B334 2009
658.4'04--dc22
                                                            2009006945

Visit the Taylor & Francis Web site at
http://www.taylorandfrancis.com

and the CRC Press Web site at
http://www.crcpress.com

Dedication

To Abi, Ade, and Tunji who, along with their progeny, are charged with carrying on the torch of science, technology, and engineering.

Contents

Preface xv
Acknowledgment xvii
Author xix

Chapter 1 Science, Technology, and Engineering
Project Methodology 1

Introduction to STEP Methodology 2
Importance of General Project Management Knowledge 2
    Large Projects (Large Budget: Has High Visibility
    and External Consequences) 3
    Midsize Projects (Medium Budget: Has Average Visibility) 3
    Small Projects (No Formal Budget: Focused
    on an Internal Goal) 4
Systems Engineering and Program Management 4
Systems Architecture for Project Management 5
Guide to Using This Book 6
Building Blocks for STEP Methodology 7
Engineering Challenges for Twenty-First Century 8
STEM Initiatives and Project Conceptions 8
Emergence of Project Management for STEP 10
Project Definitions 12
Project Management Knowledge Areas 14
Components of the Knowledge Areas 14
Project Management Processes 16
Projects and Operations 20
Factors of STEP Success or Failure 21
Work Breakdown Structure 22
Project Organization Structures 23
Traditional Formal Organization Structures 24
    Functional Organization 25
Projectized Organization 26
    Matrix Organization Structure 28
Elements of a Project Plan 30
General Applicability of Project Management 31
Documenting Project Lessons Learned 32
Marvels of STEPs 33
Advancement of Society on the Back of STE 35
Levels of Project Execution 35
Categories of Project Outputs 37
Unique Aspects of STEPs 37

Integrated Systems Approach to STEPs 38
STEP Project Management Steps 39
Managing Project Requirements 39

Chapter 2   STEP Integration 43

Project Integration: Step-by-Step Implementation 44
    Step 1: Developing Project Charter 47
        Definition of Inputs to Step 1 52
        Definition of Tools and Techniques for Step 1 53
        Definition of Output of Step 1 54
    Step 2: Develop Preliminary Project Scope Statement 54
        Definition of Inputs to Step 2 54
        Definition of Tools and Techniques for Step 2 54
        Definition of Output of Step 2 55
    Step 3: Develop Project Management Plan 55
        Definition of Inputs to Step 3 56
        Definition of Tools and Techniques for Step 3 56
        Definition of Output of Step 3 56
    Step 4: Direct and Manage Project Execution 57
        Definition of Inputs to Step 4 57
        Definition of Tools and Techniques for Step 4 58
        Definition of Outputs of Step 4 58
    Step 5: Monitor and Control Project Work 58
        Definition of Inputs to Step 5 58
        Definition of Tools and Techniques for Step 5 59
        Definition of Outputs of Step 5 59
    Step 6: Integrated Change Control 59
        Definition of Inputs to Step 6 60
        Definition of Tools and Techniques for Step 6 60
        Definition of Outputs of Step 6 61
    Step 7: Close Project 61
        Definition of Inputs to Step 7 61
        Definition of Tools and Techniques of Step 7 61
        Definition of Outputs of Step 7 62
Application of CMMI and Carpé Futurum 62
Project OPR 64
Project Sustainability 64

Chapter 3   STEP Scope Management 67

Scope Definitions 67
Scope Management: Step-by-Step Implementation 68
Use of DMAIC for Scope Management 72
Use of SIPOC Diagram for Scope Management 74
Scope Feasibility Analysis 76
Dimensions of Scope Feasibility 77

Industry Conversion for Project Scoping 78
Assessment of Local Resources and Work Force 79
Developing Scope-Based Project Proposal 79
Proposal Preparation Scope 81
    Technical Section of Project Proposal 81
    Management Section of Project Proposal 82
Scope Budget Planning 82
    Scoping Top-Down 83
    Scoping Bottom-Up 83
    Zero-Base Scoping 84
Project Scoping with WBS 84
    Project Scope Selection Criteria 84
    Criteria for Project Review 85
    Hierarchy of Selection 85
    Sizing of Projects 86
    Planning Levels 86
    Hammersmith's Project Alert Scale: Red, Yellow, Green Convention 86
    Product Assurance Concept for Industrial Projects 86

Chapter 4   STEP Time Management 89

Time Management: Step-by-Step Implementation 89
CPM Network Scheduling 95
Working with Activity Precedence Relationships 100
Example of CPM Analysis 101
CPM Forward Pass 102
CPM Backward Pass 103
Determination of Critical Activities 104
Subcritical Paths 106
Schedule Templates 107
Gantt Charts 107
Project Crashing 109
Critical Chain Analysis 112

Chapter 5   STEP Cost Management 115

Cost Management: Step-by-Step Implementation 115
STEP Portfolio Management 116
Project Cost Elements 118
    Basic Cash Flow Analysis 121
        Time Value of Money Calculations 121
        Calculations with Compound Amount Factor 121
        Calculations with Present Value Factor 122
        Calculations with Uniform Series Present Worth Factor 122
        Calculations with Uniform Series Capital Recovery Factor 123
        Calculations with Uniform Series Compound Amount Factor 124
        Calculations with Uniform Series Sinking Fund Factor 125

        Calculations with Capitalized Cost Formula 125
        Arithmetic Gradient Series 126
        Internal Rate of Return 127
    Benefit–Cost Ratio Analysis 128
        Simple Payback Period 129
        Discounted Payback Period 129
        Time Required to Double Investment 130
        Effects of Inflation on Project Costing 131
    Breakeven Analysis 136
        Profit Ratio Analysis 138
Project Cost Estimation 141
    Optimistic and Pessimistic Cost Estimates 142
    Project Budget Allocation 142
    Top-Down Budgeting 142
    Bottom-Up Budgeting 143
    Budgeting and Risk Allocation for Types of Contract 144
    Cost Monitoring 145
Project Balance Technique 146
    Cost and Schedule Control Systems Criteria 146
    Elements of Cost Control 150
    Contemporary Earned Value Technique 150
    Activity-Based Costing 153

Chapter 6   STEP Quality Management 157

Quality Management: Step-by-Step Implementation 157
Six Sigma and Quality Management 157
Taguchi Loss Function 160
Identification and Elimination of Sources of Defects 161
Roles and Responsibilities for Six Sigma 162
Statistical Techniques for Six Sigma 162
    Control Charts 163
        Types of Data for Control Charts 163
        X-Bar and Range Charts 164
        Data Collection Strategies 164
        Subgroup Sample Size 164
        Frequency of Sampling 165
        Stable Process 165
        Out-of-Control Patterns 166
        Calculation of Control Limits 166
        Plotting Control Charts for Range and Average Charts 168
        Plotting Control Charts for Moving Range and Individual
            Control Charts 169
        Case Example: Plotting of Control Chart 169
        Calculations 170
        Trend Analysis 172

Process Capability Analysis for Six Sigma 177
    Capable Process (C_p) 177
    Capability Index (C_{pk}) 178
    Possible Applications of Process Capability Index 181
    Potential Abuse of C_p and C_{pk} 181
Lean Principles and Applications 182
Applying Kaizen to a Process 182
Lean Task Value Rating System 183
Lean–Six Sigma within Project Management 185

Chapter 7   STEP Human Resource Management 187

Aging Workforce in Science, Technology, and Engineering 188
Knowledge Workers in STE Work Environment 189
Elements of Human Resource Management 189
    Human Resource Management: Step-by-Step Implementation 198
Managing Human Resource Performance 199
Quantitative Modeling of Worker Assignment 202
Resource Work Rate Analysis 204
Model for Technical Human Resource Training 208
A Conceptual Approach 210
Extended Training Model 210

Chapter 8   STEP Communications Management 219

Communications Management: Step-by-Step Implementation 219
Complexity of Multiperson Communication 221
Communicating through Triple C Model 224
Typical Triple C Questions 226
Triple C Communication 227
SMART Communication 230
Triple C Cooperation 232
Triple C Coordination 234
Conflict Resolution Using Triple C Approach 235
Application of Triple C to STEPS 236
DMAIC and Triple C 237

Chapter 9   STEP Risk Management 239

Risk Definition 239
Risk Management: Step-by-Step Implementation 241
Project Decisions under Risk and Uncertainty 242
Cost Uncertainties 244
Schedule Uncertainties 245
Performance Uncertainties 246
Risk and Decision Trees 246

Chapter 10 STEP Procurement Management 253
Procurement Management: Step-by-Step Implementation 254
Completion and Term Contracts 260
Procurement Management Plan 260
Contractor Statement of Work 261
Organization Process Assets 261
Contract Feasibility Analysis 261
Contents of Project Proposal 263
    Technical Section of Project Proposal 264
    Management Section of Project Proposal 264
Contract Teamwork and Cooperation 265
Vendor Rating System 266
Rating Procedure 267
    Requirements 267
    Computation Steps 267
Multicriteria Vendor Selection Technique 268
    Wadhwa–Ravindran Vendor Selection Technique 269
    Weighted Objective Method 271
Goal Programming 271
Compromise Programming 273
Inventory Analysis and Procurement 274
Economic Order Quantity Model 274
Quantity Discount 276
Calculation of Total Relevant Cost 277
Evaluation of the Discount Option 277
Sensitivity Analysis 279

Chapter 11 STEP Case Study: Space Shuttle *Challenger* 281
Case Background 282
Foundation for Lessons Learned 283
Space Shuttle Mission Background and Authorization 283
Space Shuttle Design Decisions 284
Shuttle Development Process 285
Chronology of Developments 285
The *Challenger* Mission 287
    Launch Delays 287
    Launch Problems 287
    The Accident 288
    The Investigation 289
    Flaws in the Decision Process 289
Testimonies 292
    McDonald's Testimony 299
    Mulloy's Testimony 300
Presidential Commission Findings 315
Management Decision Ambiguities 315

Contents

Additional Commission Findings 320
    Post-Investigation Developments 321
Space Shuttle *Columbia* Disaster 322
Technical and Organizational Issues 323
Shuttle Flight Risk Management 323
Concluding Remarks 324
Satellite Project Failure: Another STEP Case Example 325
Death of a Spy Satellite 325
    Response to Soviet Threat 327
    A Company Trying to Diversify 329
    Multiple Design Challenges 330
    Winning Bid Is Announced 331
    Signs of a Project in Trouble 333
    Search for Lessons 334

Appendix A 337
Appendix B 343
Appendix C 349
Index 393

Preface

This book introduces the STEP methodology for managing science, technology, and engineering projects. The focus is to provide tools and techniques for executing projects in the domains of science, technology, and engineering (STE). STEP refers to science, technology, and engineering projects and STEP methodology refers to the specialized process of managing such projects. This book presents a step-by-step application of project management techniques to managing STEPs. It uses *Project Management Body of Knowledge (PMBOK™)* as the platform for the topics covered. Technical project management is the basis for sustainable national advancement. Thus, managing technical projects effectively is essential for economic vitality. Project management is the process of managing, allocating, and timing resources to achieve a given goal in an efficient and expeditious manner. The objectives that constitute the specified goal may be in terms of time, costs, or technical results. A project can range from the very simple to the very complex. Due to its expanding utility and relevance, project management has emerged as a separate body of knowledge that is embraced by various disciplines ranging from engineering and business to social services. Project management techniques are widely used in many endeavors, including construction management, banking, manufacturing, engineering management, marketing, health care delivery systems, transportation, research and development, defense, and public services. The application of project management is particularly of high value in science, technology, and engineering undertakings. In today's fast-changing IT-based and competitive global market, every enterprise must strive to get ahead of the competition through effective project management in all facets of operations.

Project management represents an excellent basis for integrating various management techniques such as statistics, operations research, six sigma, computer simulation, and so on. The purpose of this book is to present an integrated approach to project management for science, technology, and engineering projects. The integrated approach covers the concepts, tools, and techniques (both new and tested) of project management. The elements of the *PMBOK* provide a unifying platform for the topics covered in the book.

This book is intended to serve as a reference for planners of science, technology, and engineering projects; stakeholders; designers; project managers; business managers; consultants; project analysts; senior executives; project team members; members of project management offices; project customers; functional managers; trainers; and researchers. It can also serve as a guidebook for technical consultants and as a textbook resource for students and educators. It is also useful as supplementary reading for practicing engineers and as a handbook for project operators. It will appeal to technical professionals because of its focused treatment of STEPs. STEP project management will be beneficial for a variety of professional groups and specialty areas including the following:

Acquisitions management
Aerospace engineering

Applied research and development
Composite engineering
Engineering infrastructure design
Facilities engineering
Financial management
Industrial engineering
Information systems analysts
Logistics engineering
Maintenance engineering
Materials science and engineering
Mechanical engineering
Operations research analyst
Process engineering
Project engineering
Science and technology consultancy

The following process areas can also benefit from the application of the techniques of STEP methodology:

Contracting support
Cost estimating
Global logistics support
Global supply chain
Integrated logistics support
Interoperability
Life cycle cost
Performance-based logistics
Product design
Supportability analysis
Systems analysis
Total ownership cost

This book uses a mixed-mode tools-and-techniques approach that combines managerial, organizational, and quantitative methodologies into a logical sequence of project implementation steps.

Adedeji Badiru, PhD, PE, PMP

Acknowledgment

The only way I could have embarked upon and completed this latest massive project was to enter into partnership with my wife, Iswat. She allowed me to turn our bedroom, kitchen, family room, and every nook and corner of our home into writing centers. She participated actively by helping to type, design graphics, format equations, and critique textual materials. My highest expression of gratitude goes to her.

Adedeji Badiru

Author

Adedeji Badiru is a professor and department head of the Department of Systems and Engineering Management at the Air Force Institute of Technology, Dayton, Ohio. He was previously professor and department head of Industrial and Information Engineering at the University of Tennessee in Knoxville. Prior to that, he was professor of industrial engineering and dean of University College at the University of Oklahoma. He is a registered professional engineer (PE), a certified project management professional (PMP), a fellow of the Institute of Industrial Engineers, and a fellow of the Nigerian Academy of Engineering. He holds a BS in industrial engineering, an MS in mathematics, and an MS in industrial engineering from Tennessee Technological University, and a PhD in industrial engineering from the University of Central Florida. His areas of interest include mathematical modeling, project modeling and analysis, economic analysis, and productivity analysis and improvement. He is the author of several books and technical journal articles. He is the editor of the *Handbook of Industrial and Systems Engineering*. He is a member of several professional associations including the Institute of Industrial Engineers, Society of Manufacturing Engineers, Institute for Operations Research and Management Science, American Society for Engineering Education, New York Academy of Science, and Project Management Institute.

Professor Badiru has served as a consultant to several organizations around the world including in Russia, Mexico, Taiwan, Nigeria, and Ghana. He has conducted customized training workshops for numerous organizations including Sony, AT&T, Seagate Technology, U.S. Air Force, Oklahoma Gas & Electric, Oklahoma Asphalt Pavement Association, Hitachi, Nigeria National Petroleum Corporation, and ExxonMobil. He has won several awards for his teaching, research, publications, administration, and professional accomplishments. He holds a leadership certificate from the University of Tennessee Leadership Institute. Professor Badiru has served as a technical project reviewer, curriculum reviewer, and proposal reviewer for several organizations including the Third-World Network of Scientific Organizations, Italy; National Science Foundation; National Research Council; and the American Council on Education. He is on the editorial and review boards of several technical journals and book publishers. Professor Badiru has also served as an industrial development consultant to the United Nations Development Program.

1 Science, Technology, and Engineering Project Methodology

The whole of science is nothing more than a refinement of everyday thinking.

–**Albert Einstein**

This book uses a mixed-mode tools-and-techniques approach that combines managerial, organizational, and quantitative methodologies into a logical sequence of project implementation steps. According to the Albert Einstein's quote at the beginning of this chapter, we must constantly refine our everyday thinking along the dimensions of science, technology, and engineering (STE). The consistent theme of the book is to couple technological requirements and managerial principles in every project endeavor every day. Solutions to societal challenges revolve around the applications of science, technology, engineering, and policy. Policy implementations are actualized through strategic project management (PM). Project formulations and development constitute the crossroads of STE endeavors. As Albert Einstein's quote suggests, everyday thinking is influenced by science and vice versa. As recent world developments would indicate, we are entering a golden age of STE. Every organization, both public and private, appears to be focusing on strategic visions that include STE projects (STEPs) in one form or another. Even organizations that have traditionally been conservative or slow in embracing STE are now scrambling to keep up with the emerging wave. Determined not to be left behind, business and industrial establishments are exploring ways to initiate projects that will bring them the benefits of STE and advance their operations ahead of competition. It thus stands to reason that a dedicated book be developed to guide STEPs and forge the development of science and technology priorities within an organization. World organizations foresee a shortage of STE-skilled workers in the coming years. Technical projects that these workers will work with must be managed in a way that matches the intellectual expectations of the workers. Workers of tomorrow will be knowledge workers as compared to the brawn workers of the past. Thus, knowledge-oriented PM strategies must be developed now for ongoing project challenges as well as challenges of the future. Figure 1.1 illustrates the premise of this book in presenting step-by-step PM across the domains of STE using the foundation of people, process, and tools. Martin (2007) presents the theory of "techonomics" as a simple framework to observe, describe, analyze, and predict organizational changes by methodically tracking technological advancement. Such technological tracking is what influenced the writing of this book as a comprehensive STEP methodology.

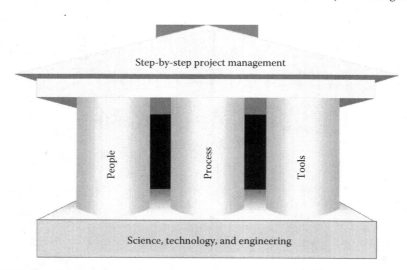

FIGURE 1.1 Step-by-step project management for science, technology, and engineering.

INTRODUCTION TO STEP METHODOLOGY

This book presents tools and techniques for executing projects in the domains of STE. STEP refers to science, technology, and engineering project and STEP methodology refers to the specialized process of managing such projects. The tools and techniques of PM can be utilized to preempt, mitigate, avoid, or prevent unique problems that often develop in a STEP. The key to managing STEPs effectively is to develop and apply specialized practices to improve products, services, and results. Such specialized practices have their foundation in the standard techniques of PM. Application potentials range from agriculture and manufacturing to modern technology. Although many project practitioners usually know what to do, they are often lacking in awareness on the selection of tools and techniques to use. The approach of this book in collating tools and techniques for adopting at the various stages of a project will thus help practitioners to be more effective in learning about what is available for each project requirement.

IMPORTANCE OF GENERAL PROJECT MANAGEMENT KNOWLEDGE

The rapid evolution of science and technology in a dynamic global market creates increasing challenges for those who plan and execute complex projects. STE professionals often focus too narrowly on specific operational requirements of their respective industries without the benefit of exposure to managerial skills for project execution. But it is clear that good PM skills can make a technical professional more effective and more versatile in the overall scheme of a business. A "Quick Quiz" posting on the Project Management Institute (PMI) online *Community Post* points out the importance of general PM skills, which can be applied to different industries such as STE. The quiz is paraphrased below:

The Quick Quiz (adapted from PMI online *Community Post* by Barbee Davis, September 12, 2008)

"To be a good project manager, how much do you have to know about the industry or business that you are serving?"

A. It is more important to have a good PM foundation than to know the business.
B. Each business is so different, in-depth knowledge in the field is key to a successful project.
C. Organizational politics drive project success, so focus on your ability to sway management.
D. Project success is random, so all you can do is work with the skill sets you have.

The answer:

The answer surprisingly is **A**. It is more important to have good general PM knowledge than to know the specific business. Both PM abilities and industry-specific knowledge are advantageous for the project manager. However, it is increasingly common for organizations to appoint someone who can run the project adeptly over someone who has in-depth knowledge of a specific domain of operation. This can be better seen in the examples presented in the online posting for different sizes of projects.

LARGE PROJECTS (LARGE BUDGET: HAS HIGH VISIBILITY AND EXTERNAL CONSEQUENCES)

For a large project that spans many departments, has multiple stakeholders, involves outside suppliers, and provides substantial organizational impact, it is best to have a leader with both PM and industry-specific knowledge. However, if forced to choose between the two areas of expertise, PM skills should weigh more heavily. It may be impossible to find someone with expertise in all of the departments, supplier industries, and stakeholder points of view. But a person with strong training and experience in PM will be able to rely on key experts in each area to manage this sizeable project to a successful conclusion. In order words, subject matter experts (SMEs) are available and could be useful to the project manager.

In very large projects, the project manager should not be doing any of the actual project tasks or activities to produce the project deliverables. He or she should instead be directing, guiding, mentoring, and supervising those actually doing the tasks.

MIDSIZE PROJECTS (MEDIUM BUDGET: HAS AVERAGE VISIBILITY)

For a midsize project, it may be possible to find someone with both the industry knowledge and the PM training to handle the requirements. This project needs fewer inputs from other departments and suppliers. However, the project manager may handle several projects at the same time within a specific area of the organization. So, PM skills remain generically vital. For example, many STEPs may be run simultaneously by the same individual. Ideally, this project manager should not be personally responsible to complete project activities.

SMALL PROJECTS (NO FORMAL BUDGET: FOCUSED ON AN INTERNAL GOAL)

With small projects, it may be typical for the project manager to also create all or most of the work of the project, in addition to managing a few additional resources. This type of internal work may be done using current employees without the cost of their time being calculated. In this situation, the industry and business knowledge of the project leader moves to the forefront. However, these types of projects can certainly be improved to add extra business value by using sound PM practices.

Nowadays, many organizations use a projectized structure in which all work is arranged as projects that are led by trained project managers. This approach highlights the value of a skilled person to guide the work necessary to achieve the organization's business goals.

SYSTEMS ENGINEERING AND PROGRAM MANAGEMENT

More than anything else, science and technology projects require coordinated application of systems engineering and program management. Systems engineering helps to identify an inclusive framework under which all the components of S & T would work together. Program management provides the mechanisms through which specific work and objectives within the framework would be accomplished. Systems engineering can mean different things to different people. This is evident in the variety of definitions available in the literature. Systems engineering can be defined as an interdisciplinary approach linking the entire set of science, technology, and management requirements needed to provide products, service, or results to meet customer needs. Badiru (2006) presents the following definition:

> Systems engineering involves a recognition, appreciation, and integration of all aspects of an organization or a facility. A system is defined as a collection of interrelated elements working together in synergy to produce a composite output that is greater than the sum of the individual outputs of the components. A systems view of a process facilitates a comprehensive inclusion of all the factors involved in the process.

Systems engineering requires an interdisciplinary approach. This involves everyone, not just engineers, scientists, and technologists. It requires a coordinated and cooperative effort from professionals representing program management; engineering; finance; contracting; logistics; test and evaluation; configuration management; data management; environmental assessment, safety, and health; equipment maintenance, clerical personnel; and so on.

Systems engineering encourages an integrated and balanced life cycle view. From the identification of the need to the development and utilization of the product, systems engineering continuously integrates and balances the requirements, cost, and schedule to provide an operationally effective system throughout a project's life cycle (Badiru and Omitaomu, 2007).

Science, Technology, and Engineering Project Methodology

Systems engineering raises questions regarding basic research, applied technology, scientific basis, sustainability, resource requirements, operational logistics, and component interfaces. Without integrated answers to these questions, a project manager will be hard-pressed to manage deadlines and track projects effectively. Far too many silos exist in science and engineering domains, with each specialized area claiming sole ownership of their expertise and resisting "interference" from other areas while fighting off temptation to exercise "incursion" into other areas of expertise.

Technical professionals are a proud lot and will not be bashful in defending an assertion that loudly pronounces "that is not our work." Embracing a systems engineering perspective, everyone can realize that any "work" is everyone's "work" within the organization. Work is so interdependent in any science and technology environment. So, all components must work together toward a common goal. This book emphasizes the importance of applying systems engineering to all phases of PM including acquisition and sustainment programs. In the step-by-step framework, readers can appreciate an overview of systems concepts, process, and interactions between systems engineering and other functional areas.

SYSTEMS ARCHITECTURE FOR PROJECT MANAGEMENT

The military has been a major force in establishing systems engineering platform for executing projects. Through the Department of Defense Architecture Framework (DODAF), the U.S. military executes projects on a consistent platform of an "architecture" perspective borrowed from the conventional physical infrastructure architectural design processes. DODAF is used to standardize the format for architecture descriptions. It seeks to provide a mechanism for operating more efficiently while attending to multiple requirements spread out in multiple and diverse geographical locations. One approach of DODAF adapts traditional architecture to something called capability architecture. The reasoning for this is the widespread belief that scores of defense systems are either redundant or do not meet operational needs. As a result, many recent acquisition reform efforts have been aimed at pursuing interoperable and cost-effective joint military capabilities.

Traditional architects integrate structure and function with the environment. Their end products, the blueprints, merge various stakeholders' visions and requirements into an acceptable product. They provide sheets, or views, that correspond to the homeowner, the plumber, the electrician, the framer, the painter, the residents, and even the neighbors. By contract, the application of systems architecting in the military is not centered on a place of abode (i.e., house), but rather on interoperable weapon systems and diverse spectrum of warfare. This requires a lot of intercomponent coordination. Only a systems view can provide this level of comprehensive appreciation of capability, interdependency, and symbiosis.

Systems architecture supports logical interface of capabilities, operations planning, resource requirements, tool development, portfolio management, goal formulation, acquisition, information management, and project phase out. Some specific requirements for applying systems architecting to program management within the military include the following:

- The Joint Capabilities Integration and Development System (JCIDS) requires that each Capabilities Document contain an annex with a standard DOD-formatted architecture. Users and program offices partner to provide the architecture descriptions.
- The Defense Acquisition System (DAS) requires architecture to develop systems and manage interoperability of components.
- Systems that communicate must have Information Support Plans (ISP), each accompanied by a complete integrated architecture.
- DOD and U.S. Congress require systems architecture to be used for defense business information systems that cost at least $1 million.

Just as the traditional home architect provides specific views to different subcontractors involved in the construction of a house, DODAF prescribes views for various stakeholders involved in a given capability or requirement. There are 26 total views in DODAF organized into three categories:

1. Operational views (OVs)
2. Systems and services views (SVs)
3. Technical standards views (TVs)

The views are a combination of pictures, diagrams, and spreadsheets maintained in an electronic database. The OVs communicate mission-level information and document operational requirements from a user standpoint. The SVs communicate design-level information for use by designers and maintenance personnel. Finally, the TVs document the information technology standards (construction codes) that have been developed for networking compatibility (net-centricity).

DODAF architecture descriptions are the blueprints for linking key inputs and capabilities for planners, designers, and acquirers. For everyone involved in a large and complex project, a consistent architecture framework can guide the systems-of-systems engineering process. DODAF-integrated architectures provide insight into complex operational relationships, interoperability requirements, and systems-related structure.

The step-by-step framework used in this book facilitates a comprehensive project implementation that exhibit the integrative and inclusive processes of systems engineering coupled with the framework provided by the PM body of knowledge.

GUIDE TO USING THIS BOOK

This book introduces the STEP methodology for managing complex projects in STE. The book is based on (and extends) the structural framework of the Project Management Body of Knowledge (PMBOK) as presented by the PMI. The tools and techniques presented are applicable to all process areas of STEPs including contracting support, cost estimating, global logistics support (GLS), global supply chain (GSC), integrated logistics support (ILS), interoperability, life cycle cost (LCC), performance-based logistics (PBL), product design, supportability analysis, systems analysis, and total ownership cost (TOC). The comprehensive contents are presented as general guides for project practitioners. Each user can adapt and customize the guides to the needs of the specific project under consideration. Not all steps will be applicable to all projects; some projects may be subject to extraneous and unique issues not covered in this book. But in any case, the general framework shows what needs to done to effectively manage each complex project. The steps in the framework may also be interchanged to meet in-house organizational practices or specialized needs. In order words, users can omit, add, extend, expand, or modify steps in the framework.

BUILDING BLOCKS FOR STEP METHODOLOGY

The domains of STE form the juncture for the application of PM techniques in this book. The building blocks for STEP are derived from the common concatenation of S, T, E, and M as described below:

Science (S): Science, derived from the Latin word "scientia," meaning "knowledge," is the study of how to increase human understanding of how the physical world works. This is the building block for everything around us.

Technology (T): Technology, derived from the Greek words "technologia" (craft) and "logia" (expression), is broadly defined as the process of using science and engineering to create physical objects of use to humanity, such as machines, hardware, and tools as well as systems of operation, methods of organization, and techniques.

Engineering (E): Engineering is the application of scientific knowledge, natural laws, and physical objects to design and utilize materials, structures, machines, devices, systems, and processes that satisfy a desired objective within specified criteria.

Mathematics (M): Mathematics is the science of numbers, logic, and concepts to form the body of knowledge involving physical quantity, structure, space, and change, and the resulting conclusions drawn from such elements.

Science and Technology (ST): ST refers to the combination of science and technology.

Science, Technology, and Engineering (STE): STE refers to implementations involving the use of science, technology, and engineering to achieve desired products, services, or results.

Science, Technology, Engineering, and Mathematics (STEM): STEM refers to the incorporation of the fundamental principles of mathematics as the foundation for linking science, technology, and engineering.

Science, Technology, and Engineering Project (STEP): STEP refers to the formulation of STE endeavors as project entities to be planned, organized, scheduled, and controlled using the tools and techniques of PM. This approach is based on the belief that we can shape the success of STE applications by employing systematic PM approaches. PM has benefits in the research, development, testing, delivering, modernizing, integrating, and sustaining of technology assets. Through PM, we can evaluate, acquire, maintain, and sustain STE assets. Some key pursuits in managing STEPs include the following:

- Multiorganizational team formation
- Management of technology intellectual capital
- Technology implementation roadmapping
- Technology assessment tools development and deployment

ENGINEERING CHALLENGES FOR TWENTY-FIRST CENTURY

The National Academy of Engineering (NAE), in February 2008, released a list of the 14 grand challenges for engineering in the coming years. Each area of challenges constitutes a complex project that must be planned and executed strategically. The 14 challenges, which can be viewed as STEM areas, are listed below:

1. Make solar energy affordable
2. Provide energy from fusion
3. Develop carbon sequestration methods
4. Manage the nitrogen cycle
5. Provide access to clean water
6. Restore and improve urban infrastructure
7. Advance health informatics
8. Engineer better medicines
9. Reverse-engineer the brain
10. Prevent nuclear terror
11. Secure cyberspace
12. Enhance virtual reality
13. Advance personalized learning
14. Engineer the tools for scientific discovery

The above list of existing and forthcoming engineering challenges indicate an urgent need to apply PM to bring about new products, services, and results efficiently within cost and schedule constraints. PM can effectively be applied to science projects, technology projects, engineering projects, and math projects. As it will become apparent as the STEP PM process unfolds in the following chapters, the challenge areas can be defined and structured as multidimensional projects. The project specifications in each case are presented in STEM initiatives as highlighted in the section that follows.

STEM INITIATIVES AND PROJECT CONCEPTIONS

Although the NAE list focuses on engineering challenges, the fact is that every item on the list has the involvement of general areas of science, technology, and mathematics in one form or another. The STEM elements of each area of engineering challenge are contained in the project definitions below:

Make solar energy economical: Solar energy provides less than 1% of the world's total energy, but it has the potential to provide much, much more.

Provide energy from fusion: Human-engineered fusion has been demonstrated on a small scale. The challenge is to scale up the process to commercial proportions in an efficient, economical, and environmentally benign way.

Develop carbon sequestration methods: Engineers are working on ways to capture and store excess carbon dioxide to prevent global warming.

Manage the nitrogen cycle: Engineers can help restore balance to the nitrogen cycle with better fertilization technologies and by capturing and recycling waste.

Provide access to clean water: The world's water supplies are facing new threats; affordable, advanced technologies could make a difference to millions of people around the world.

Restore and improve urban infrastructure: Good design and advanced materials can improve transportation and energy, water, and waste systems, and also create more sustainable urban environments.

Advance health informatics: Stronger health information systems not only improve everyday medical visits, but they are essential to counter pandemics and biological or chemical attacks.

Engineer better medicines: Engineers are developing new systems to use genetic information, sense small changes in the body, assess new drugs, and deliver vaccines.

Reverse-engineer the brain: The intersection of engineering and neuroscience promises great advances in health care, manufacturing, and communication.

Prevent nuclear terror: The need for technologies to prevent and respond to a nuclear attack is growing.

Secure cyberspace: It is more than preventing identity theft. Critical systems in banking, national security, and physical infrastructure may be at risk.

Enhance virtual reality: True virtual reality creates the illusion of actually being in a different space. It can be used for training, treatment, and communication.

Advance personalized learning: Instruction can be individualized based on learning styles, speeds, and interests to make learning more reliable.

Engineer the tools of scientific discovery: In the century ahead, engineers will continue to be partners with scientists in the great quest for understanding many unanswered questions of nature.

Society will be tackling these grand challenges for the foreseeable decades; PM is one avenue through which we can ensure that the desired products, services, and results can be achieved. With the positive outcomes of these projects achieved, we can improve the quality of life for everyone and our entire world can benefit positively. In the context of tackling the grand challenges as STEPs, some of the critical issues to address are

- Strategic implementation plans
- Strategic communication
- Knowledge management
- Evolution of virtual operating environment
- Structural analysis of projects
- Analysis of integrative functional areas
- Project concept mapping

- Prudent application of technology
- Scientific control
- Engineering research and development

EMERGENCE OF PROJECT MANAGEMENT FOR STEP

PM has quickly evolved into a cohesive body of knowledge dedicated to helping organizations achieve their goals and objectives. In fact, the envisioned goal of the PMI says "Worldwide, organizations will embrace, value and utilize PM and attribute their success to it." This vision is already being realized in many parts of the world. The world has become very interconnected and PM represents the common language of operation for creating products, generating services, and achieving results. The unique characteristics of STEP require specialized care throughout the sequence of planning, organizing, scheduling, tracking, control, and phase out.

There is a growing need to apply better PM to major projects. Press headlines in April 2008 highlight "Defense needs better management of projects." This is in the wake of government audit that reveals gross inefficiencies in managing large defense projects. In a front-page story of the *Washington Post* on April 1, 2008, it was reported that auditors at the Government Accountability Office (GAO) issued a scathing review of dozens of the Pentagon's biggest weapons systems, citing that ships, aircraft, and satellites are billions of dollars over budget and years behind schedule. According to the review, "95 major systems have exceeded their original budgets by a total of $295 billion and are delivered almost 2 years late on average." Further, "none of the systems that the GAO looked at had met all of the standards for best management practices during their development stages." Among programs noted for increased development costs were the "Joint Strike Fighter and Future Combat Systems." The costs of these programs has risen "36% and 40%, respectively," while C-130 avionics modernization costs have risen 323%. And, while defense department officials have tried to improve the procurement process, the GAO added that "significant policy changes have not yet translated into best practices on individual programs." A summary of the report of the accounting office reads

> Every dollar spent inefficiently in developing and procuring weapon systems is less money available for many other internal and external budget priorities, such as the global war on terror and growing entitlement programs. These inefficiencies also often result in the delivery of less capability than initially planned, either in the form of fewer quantities or delayed delivery to the warfighter.

This book covers PM with specific focus on STEPs and how to mitigate or preempt the types of problems noted in the examples above. The book presents a step-by-step guide for managing STEP using the basic elements of the PMBOK as presented by the PMI. STEPs are unique and require special attention in managing the technical and human resources associated with them. Of particular importance is the need for strategic S,T&E manpower development. The STEP orientation of the book is derived both from the acronym as well as the step-by-step presentation of the PM knowledge areas. The transition from engineering to science and technology and vice versa makes it essential to use an integrated approach to managing STEPs.

Science, Technology, and Engineering Project Methodology

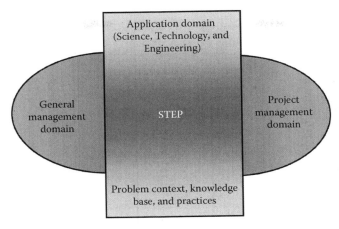

FIGURE 1.2 Intersection of knowledge, practices, and application context.

The bodies of knowledge driving engineering, science, and technology now overlap and should be managed with an overall systems view. Figure 1.2 illustrates the intersecting themes of general management, PM, and specific application context. The general management skills cover the following areas:

- Leading the organization
- Communicating
- Negotiating
- Problem solving
- Influencing the organization

These skills directly affect traditional project requirements for planning, organizing, scheduling, and control. The framework for cross-functional application of PM is illustrated in Figure 1.3. People, process, and technology assets (science and engineering) form the basis for implementing organizational goals. Human resources constitute crucial capital that must be recruited, developed, and preserved. Organizational work process must take advantage of the latest tools and techniques such as business process reengineering (BPR), continuous process improvement (CPI), Lean, Six Sigma, and systems thinking. The coordinated infrastructure represents the envelope of operations and includes physical structures, energy, leadership, operating culture, and movement of materials. The ability of an organization to leverage science and technology to move up the global value chain requires the softer side of PM in addition to the technical techniques. Another key benefit of applying integrative PM to STEPs centers around systems safety. STE undertakings can be volatile and subject to safety violations through one of the following actions:

1. Systems or individuals who deliberately, knowingly, willfully, or negligently violate embedded safety requirements in STEP
2. Systems or individuals who inadvertently, accidentally, or carelessly compromise safety requirements in STEP

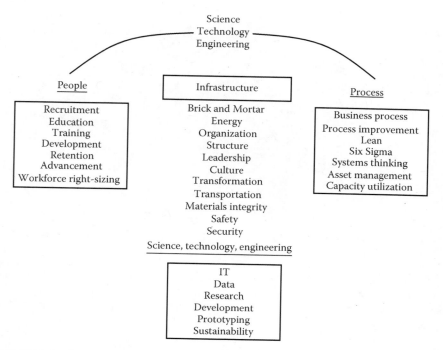

FIGURE 1.3 Framework for cross-functional application of project management.

The above potential avenues for safety violation make safety training, education, practice, safety monitoring, and ethics very essential. An integrative approach to PM helps to cover all the possible ways for safety compromise.

PROJECT DEFINITIONS

Project: A project is traditionally defined as a unique one-of-kind endeavor with a specific goal that has a definite beginning and a definite end. PMBOK defines a project as a temporary endeavor undertaken to create a unique product, service, or result. Temporary means having a defined beginning and a definite end. The term "unique" implies that the project is different from other projects in terms of characteristics.

Project management: This author defines project management (PM) as the process of managing, allocating, and timing resources to achieve a given goal in an efficient and expeditious manner.

PMBOK defines PM as the application of knowledge, skills, tools, and techniques to project activities to achieve project objectives.

Other sources define PM as the collection of skills, tools, and management processes essential for executing a project successfully.

Project management methodology: A project management methodology (PMM) defines a process that a project team uses in executing a project from planning through phase out.

Project management information system: A project management information system (PMIS) refers to an automated system or computer software used by the PM team as a tool for the execution of the activities contained in the PM plan.

Project management system: A project management system (PMS) is the set of interrelated project elements whose collective output, through synergy, exceeds the sum of the individual outputs of the elements.

Composition of a program: A program is defined as a recurring group of interrelated projects managed in a coordinated and synergistic manner to obtain integrated results that are better than what is possible by managing the projects individually. Programs often include elements of collateral work outside the scope of the individual projects. Thus, a program is akin to having a system of systems of projects, whereby an entire enterprise might be affected. While projects have definite endpoints, programs often have unbounded life spans. Figure 1.4 shows the hierarchy of project systems from organizational enterprise to work breakdown structure (WBS) elements.

Identification of stakeholders: Stakeholders are individuals or organizations whose interests may be positively or negatively impacted by a project. Stakeholders must be identified by the project team for every project. A common deficiency in this requirement is that the organization's employees are often ignored, neglected, or taken for granted as stakeholders in projects going on in the organization. As the definition of stakeholders clearly suggests, if the interests of the employees can be positively or negatively affected by a project, then the employees must be viewed as stakeholders. All those who have a vested interest in the project are stakeholders and this might include the following:

- Customers
- Project sponsor
- Users
- Associated companies
- Community

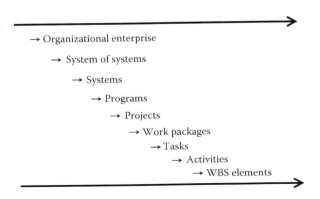

FIGURE 1.4 Hierarchy of project systems.

- Project manager
- Owner
- Project team members
- Shareholders

PROJECT MANAGEMENT KNOWLEDGE AREAS

This chapter covers general principles of PM and applicability to STEP. The subsequent chapters present a lattice of PM topics under the umbrella of PM knowledge areas as defined by the PMI. The knowledge areas are compiled in PMI's PMBOK and are organized into the following broad areas, encoded as IST-CQH-CRP:

1. Project integration management
2. Project scope management
3. Project time management
4. Project cost management
5. Project quality management
6. Project human resource management
7. Project communications management
8. Project risk management
9. Project procurement management

The above segments of the body of knowledge of PM cover the range of functions associated with any project, particularly complex ones. Multinational projects particularly pose unique challenges pertaining to reliable power supply, efficient communication systems, credible government support, dependable procurement processes, consistent availability of technology, progressive industrial climate, trustworthy risk mitigation infrastructure, regular supply of skilled labor, uniform focus on quality of work, global consciousness, hassle-free bureaucratic processes, coherent safety and security system, steady law and order, unflinching focus on customer satisfaction, and fair labor relations. Assessing and resolving concerns about these issues in a step-by-step fashion will create a foundation of success for a large project. While no system can be perfect and satisfactory in all aspects, a tolerable trade-off on the factors is essential for project success.

COMPONENTS OF THE KNOWLEDGE AREAS

The major components of PMBOK are summarized below. This book introduces characteristic or mnemonic symbols to represent each knowledge area as shown below:

Integration	Scope	Time
♪	☼	∞

Science, Technology, and Engineering Project Methodology

Cost	Quality	Human Resource
$	♣	☺

Communication	Scope	Time
♀	♂	♪

The symbols help to connect requirements to actual practice. The symbol for integration signifies harmony of the elements of the project. The symbol for scope signifies globular boundary of the project. The symbol for time connotes movement in the direction of infinity, indicating that time that passes is never recovered. The symbol for cost is intuitively represented by the dollar sign. The symbol for quality signifies uniformity of product attributes, as in the lobes of Club playing card. The symbol for human resource management is intuitively the dingbat character of human face. The symbol for communication is based on a telecommunication-type character of ball and cross. The symbol for risk signifies a projectile strike. The symbol for procurement is based on a resource-hook concept. The key components of each element of the body of knowledge are summarized below:

- Integration
    - Integrative project charter
    - Project scope statement
    - Project management plan
    - Project execution management
    - Change control
- Scope management
    - Focused scope statements
    - Cost/benefit analysis
    - Project constraints
    - Work breakdown structure
    - Responsibility breakdown structure
    - Change control
- Time management
    - Schedule planning and control
    - PERT and Gantt charts
    - Critical path method
    - Network models
    - Resource loading
    - Reporting

- Cost management
  - Financial analysis
  - Cost estimating
  - Forecasting
  - Cost control
  - Cost reporting
- Quality management
  - Total quality management
  - Quality assurance
  - Quality control
  - Cost of quality
  - Quality conformance
- Human resources management
  - Leadership skill development
  - Team building
  - Motivation
  - Conflict management
  - Compensation
  - Organizational structures
- Communications
  - Communication matrix
  - Communication vehicles
  - Listening and presenting skills
  - Communication barriers and facilitators
- Risk management
  - Risk identification
  - Risk analysis
  - Risk mitigation
  - Contingency planning
- Procurement and subcontracts
  - Material selection
  - Vendor prequalification
  - Contract types
  - Contract risk assessment
  - Contract negotiation
  - Contract change orders

PROJECT MANAGEMENT PROCESSES

The major knowledge areas of PM are administered in a structured outline covering six basic clusters as depicted in Figure 1.5. The implementation clusters represent five process groups that are followed throughout the project life cycle. Each cluster itself consists of several functions and operational steps. When the clusters are overlaid on the nine knowledge areas, we obtain a two-dimensional matrix that spans 44 major process steps. Table 1.1 shows an overlay of the PM knowledge areas and the implementation clusters. The monitoring and controlling clusters are usually

Science, Technology, and Engineering Project Methodology

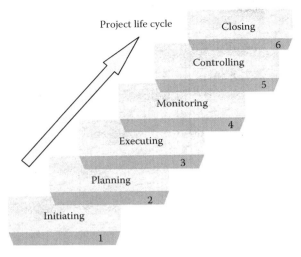

FIGURE 1.5 Implementation clusters for project life cycle.

administered as one lumped process group (monitoring and controlling). In some cases, it may be helpful to separate them to highlight the essential attributes of each cluster of functions over the project life cycle. In practice, the processes and clusters do overlap. Thus, there is no crisp demarcation of when and where one process ends and where another one begins over the project life cycle. In general, project life cycle defines the following:

1. Resources that will be needed in each phase of the project life cycle
2. Specific work to be accomplished in each phase of the project life cycle

Figure 1.6 shows the major phases of project life cycle going from the conceptual phase through the close-out phase. It should be noted that project life cycle is distinguished from product life cycle. Project life cycle does not explicitly address operational issues whereas product life cycle is mostly about operational issues starting from the product's delivery to the end of its useful life. Note that for STEPs, the shape of the life cycle curve may be expedited due to the rapid developments that often occur in STE activities. For example, for a high-technology project, the entire life cycle may be shortened, with a very rapid initial phase, even though the conceptualization stage may be very long. Typical characteristics of project life cycle include the following:

1. Cost and staffing requirements are lowest at the beginning of the project and ramp up during the initial and development stages.
2. The probability of successfully completing the project is lowest at the beginning and highest at the end. This is because many unknowns (risks and uncertainties) exist at the beginning of the project. As the project nears its end, there are fewer opportunities for risks and uncertainties.

TABLE 1.1
Overlay of Project Management Areas and Implementation Clusters

Knowledge Areas	Project Management Process Clusters				
	Initiating	Planning	Executing	Monitoring and Controlling	Closing
Project integration	Develop project charter Develop preliminary project scope	Develop project management plan	Direct and manage project execution	Monitor and control project work Integrated change control	
Scope		Scope planning Scope definition Create WBS		Scope verification Scope control	
Time		Activity definition Activity sequencing Activity resource estimating Activity duration estimating Schedule development		Schedule control	
Cost		Cost estimating Cost budgeting		Cost control	
Quality		Quality planning	Perform quality assurance	Perform quality control	
Human resources		Human resource planning	Acquire project team Develop project team	Manage project team	
Communication		Communication planning	Information distribution	Performance reporting Manage stakeholders	
Risk		Risk management planning Risk identification Qualitative risk analysis Quantitative risk analysis Risk response planning		Risk monitoring and control	
Procurement		Plan purchases and acquisitions Plan contracting	Request seller responses Select sellers	Contract administration	Contract closure

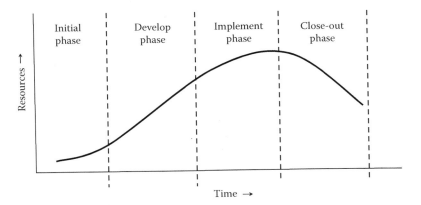

FIGURE 1.6 Phases of project life cycle.

3. The risks to the project organization (project owner) are lowest at the beginning and highest at the end. This is because not much investment has gone into the project at the beginning, whereas much has been committed by the end of the project. There is a higher sunk cost manifested at the end of the project.
4. The ability of the stakeholders to influence the final project outcome (cost, quality, and schedule) is highest at the beginning and gets progressively lower toward the end of the project. This is intuitive because influence is best exerted at the beginning of an endeavor.
5. Value of scope changes decreases over time during the project life cycle while the cost of scope changes increases over time. The suggestion is to decide and finalize scope as early as possible. If there are to be scope changes, do them as early as possible.

The specific application context will determine the essential elements contained in the life cycle of the endeavor. Life cycles of business entities, products, and projects have their own nuances that must be understood and managed within the prevailing organizational strategic plan. The components of corporate, product, and project life cycles are summarized as follows:

Corporate (business) life cycle:
Policy planning → Needs identification → Business conceptualization → Realization → Portfolio management

Product life cycle:
Feasibility studies → Development → Operations → Product obsolescence

Project life cycle:
Initiation → Planning → Execution → Monitoring and Control → Close-out

This book covers the knowledge areas sequentially in Chapters 2 through 10 in the order listed above. There is no strict sequence for the application of the knowledge

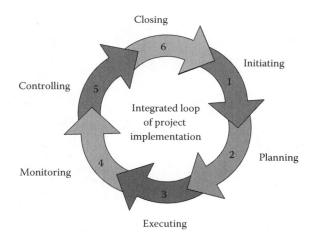

FIGURE 1.7 Integrated loop of project implementation.

areas to a specific project. The areas represent a mixed bag of processes that must be followed in order to achieve a successful project. Thus, some aspects of integration may be found under the knowledge area for communications. In a similar vein, a project may start with the risk management process before proceeding into the integration process. The knowledge areas provide general guidelines. Each project must adapt and tailor the recommended techniques to the specific need and unique circumstances of the project. PMBOK seeks to standardize PM terms and definitions by presenting a common lexicon for PM activities. It is important to implement the steps of PM in an integrated loop as shown in Figure 1.7.

Specific strategic, operational, and tactical goals and objectives are embedded within each step in the loop. For example, "initiating" may consist of project conceptualization and description. Part of "executing" may include resource allocation and scheduling. "Monitoring" may involve project tracking, data collection, and parameter measurement. "Controlling" implies taking corrective action based on the items that are monitored and evaluated. "Closing" involves phasing out or terminating a project. Closing does not necessarily mean a death sentence for a project, as the end of one project may be used as the stepping stone to the next series of endeavors.

PROJECTS AND OPERATIONS

There is sometimes confusion between project management and operations management. Operations are ongoing and repetitive components of normal business functions. By contrast, projects are temporary and unique. Thus, managing day-to-day business operations differs from managing single and distinct projects. However, operations and projects share the common elements of people, process, and tools (Technology) as shown in Figure 1.8. It is particularly important to clarify the distinction between projects and operations for STEPs where the

Science, Technology, and Engineering Project Methodology

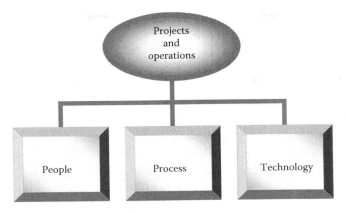

FIGURE 1.8 Common elements of projects and operations.

TABLE 1.2
Characteristics of Projects and Operations

Characteristics of Projects	Characteristics of Operations
1. Projects have unique goals and charters	1. Operations usually have unstructured charter
2. Projects often have discernible organizational structure	2. Operations depend on prevailing organizational structures rather than having individual organizational structure
3. Projects generate unique product, service, or result	3. Operations follow existing policies, procedures, and practices within the organization
4. Projects have clear start and end points	4. Operations produce standardized products and services
5. Projects often involve mixed and diverse teams	5. Operations are continuous and recurring

---

boundaries of science and technology may be fuzzy. Table 1.2 summarizes the major differences.

FACTORS OF STEP SUCCESS OR FAILURE

There are several factors that impinge on the success or failure of a project. In STEPs, factors that enhance project success include the following:

- Well-defined scope
- Communication among project team members
- Cooperation of project teams
- Coordination of project efforts
- Proactive management support
- Measurable metrics of project performance
- Identifiable points of accountability
- Realistic time, cost, and requirements

The facts and realities of STEPs are that

- Scopes are getting larger due to emerging technological opportunities
- Scopes are becoming more complex due to new and uncharted territories of project endeavors
- Scopes require shorter timescales due to rapid changes in science and technology
- Number of project requirements continues to grow due to expanding market demand

When projects fail, it is often due to a combination of the following factors related to project requirements:

- Requirements are incomplete
- Poor definition of project objectives
- Poor definition of scope and premature acceptance
- Requirements are unrealistic
- Requirements are ambiguous
- Requirements are inconsistent
- Changes in requirements are unbudgeted
- Poor management support
- Lack of alignment of project objectives with organizational objectives
- Poor communication
- Lack of cooperation
- Deficient coordination of project efforts

WORK BREAKDOWN STRUCTURE

WBS represents the foundation over which a project is developed and managed. WBS refers to the itemization of a project for planning, scheduling, and control purposes. WBS defines the scope of the project. In the project implementation template, WBS is developed within the scope knowledge area under the planning cluster. The WBS diagram presents the inherent components of a project in a structured block diagram or interrelationship flowchart. WBS shows the relative hierarchies of parts (phases, segments, milestone, etc.) of the project. The purpose of constructing a WBS is to analyze the elemental components of the project in detail. If a project is properly designed through the application of WBS at the project planning stage, it becomes easier to estimate cost and time requirements of the project. Project control is also enhanced by the ability to identify how components of the project link together. Tasks that are contained in the WBS collectively describe the overall project goal. Overall project planning and control can be improved by using a WBS approach. A large project may be broken down into smaller subprojects that may, in turn, be systematically broken down into task groups. Thus, WBS permits the implementation of a "divide and conquer" concept for project control.

Individual components in a WBS are referred to as WBS elements, and the hierarchy of each is designated by a level identifier. Elements at the same level of subdivision are said to be of the same WBS level. Descending levels provide increasingly detailed definition of project tasks. The complexity of a project and the degree of control desired determine the number of levels in the WBS. Each component is successively broken down into smaller details at lower levels. The process may continue until specific project activities (WBS elements) are reached. In effect, the structure of the WBS looks very much like an organizational chart. But it should be emphasized that WBS is not an organization chart. The basic approach for preparing a WBS is as follows:

Level 1 WBS
This contains only the final goal of the project. This item should be identifiable directly as an organizational budget item.

Level 2 WBS
This level contains the major subsections of the project. These subsections are usually identified by their contiguous location or by their related purposes.

Level 3 WBS
Level 3 of the WBS structure contains definable components of the level 2 subsections. In technical terms, this may be referred to as the finite element level of the project.

Subsequent levels of WBS are constructed in more specific details depending on the span of control desired. If a complete WBS becomes too crowded, separate WBS layouts may be drawn for the level 2 components. A statement of work (SOW) or WBS summary should accompany the WBS. The SOW is a narrative of the work to be done. It should include the objectives of the work, its scope, resource requirements, tentative due date, feasibility statements, and so on. A good analysis of the WBS structure will make it easier to perform scope monitoring, scope verification, and control project work later on in the project. Figure 1.9 shows an example of a WBS structure for a hypothetical design project.

PROJECT ORGANIZATION STRUCTURES

Project organization structure provides the framework for implementing a project across functional units of an organization. Project organization structure facilitates integration of functions through cooperation and synergy. Project organizational structures are used to achieve coordinated and cross-functional efforts to accomplish organizational tasks. There are three basic types of organizational structures for projects:

1. Functional organization structure
2. Projectized organization structure
3. Matrix organization structure

However, some specialized or customized adaptations of the three basic structures are used in practice to meet unique project situations. Before selecting an organizational structure, the project team should assess the nature of the job to be performed and its requirements as contained in the WBS. The structure may be defined in terms of functional specializations, departmental proximity, standard management boundaries, operational relationships, or product requirements.

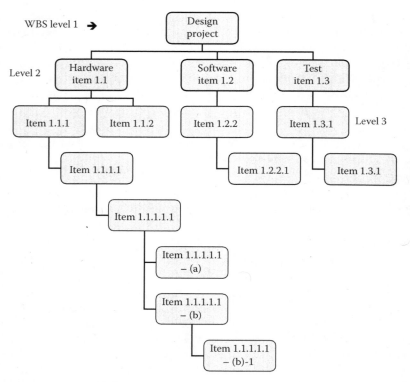

FIGURE 1.9 Example of WBS structure for a design project.

TRADITIONAL FORMAL ORGANIZATION STRUCTURES

Many organizations use the traditional formal or classical organization structures, which show hierarchical relationships between individuals or teams of individuals. Traditional formal organizational structures are effective in service enterprises because groups with similar functional responsibilities are clustered at the same level of the structure. A formal organizational structure represents the officially sanctioned structure of a functional area. An informal organizational structure, on the other hand, develops when people organize themselves in an unofficial way to accomplish a project objective. The informal organization is often very subtle in that not everyone in the organization is aware of its existence. Both formal and informal organizations exist within every project. Positive characteristics of the traditional formal organizational structure include the following:

- Availability of broad manpower base
- Identifiable technical line of control
- Grouping of specialists to share technical knowledge
- Collective line of responsibility
- Possibility of assigning personnel to several different projects
- Clear hierarchy for supervision

- Continuity and consistency of functional disciplines
- Possibility for the establishment of departmental policies, procedures, and missions

However, the traditional formal structure does have some shortcomings as summarized below:

- No one individual is directly responsible for the total project.
- Project-oriented planning may be impeded.
- There may not be a clear line of reporting up from the lower levels.
- Coordination is complex.
- A higher level of cooperation is required between adjacent levels.
- The strongest functional group may wrongfully claim project authority.

FUNCTIONAL ORGANIZATION

The most common type of formal organization is known as the functional organization, whereby people are organized into groups dedicated to particular functions. This structure highlights the need for specialized areas of responsibilities, such as marketing, finance, accounting, engineering, production, design, and administration. In a functional organization, personnel are grouped by job function. While organizational integration is usually desired in an enterprise, there still exists a need to have service differentiation. This helps to distinguish between business units and functional responsibilities. Depending on the size and the type of auxiliary activities involved, several minor, but supporting, functional units can be developed for a project. Projects that are organized along functional lines normally reside in a specific department or area of specialization. The project home office or headquarters is located in the specific functional department. Figure 1.10 shows examples of projects that are organized under the functional structure. The advantages of a functional organization structure are presented below:

- Improved accountability
- Personnel within the structure have one clear chain of command (supervision)
- Discernible lines of control
- Individuals perform projects only within the boundaries of their respective functions
- Flexibility in manpower utilization
- Enhanced comradeship of technical staff
- Improved productivity of specially skilled personnel
- Potential for staff advancement along functional path
- Ability of the home office to serve as a refuge for project problems

The disadvantages of a functional organization structure include

- Potential division of attention between project goals and regular functions
- Conflict between project objectives and regular functions

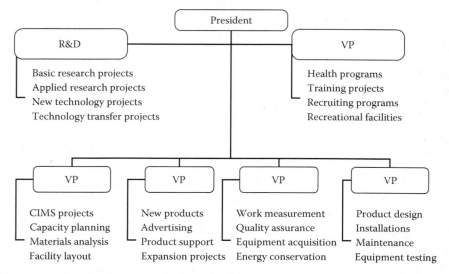

FIGURE 1.10 Functional organization structure.

- Poor coordination similar project responsibilities
- Unreceptive attitudes on the part of the surrogate department
- Multiple layers of management
- Lack of concentrated effort

PROJECTIZED ORGANIZATION

Another approach to organizing a project is to use the end product or goal of the project as the determining factor for personnel structure. This is known as the projectized structure, but often referred to as pure project organization or product organization, whereby the project is organized around a particular product (e.g., project deliverable, goal). The project is set up as a unique entity within the parent organization. It has its own dedicated technical staff and administration. It is linked to the rest of the system through progress reports, organizational policies, procedures, and funding. The interface between product-organized projects and other elements of the organization may be strict or liberal depending on the organization. An example of a pure project organization is shown in Figure 1.11. Projects A, B, C, and D in the figure may directly represent Product Types A, B, C, and D. Projectized organization structure is suitable for two categories of companies:

1. Companies that use management-by-projects as a philosophy of their operations
2. Companies that derive most of their revenues from performing projects for a fee

Such organizations normally have performance systems in place to monitor, track, and control projects. For these companies, the personnel are often colocated.

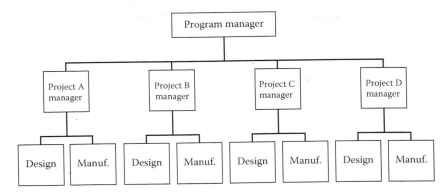

FIGURE 1.11 Projectized organization structure.

The project organization is common in industries that have multiple product lines. Unlike the functional organization, the project organization decentralizes functions. It creates a unit consisting of specialized skills around a given project or product. Sometimes referred to as a team, task force, or product group, the project organization is common in public, research, and manufacturing organizations where specially organized and designated groups are assigned specific functions. A major advantage of the product organization is that it gives the project members a feeling of dedication to and identification with a particular goal.

A possible shortcoming of the project organization is the requirement that the product group be sufficiently funded to be able to stand alone. The product group may be viewed as an ad hoc unit that is formed for the purpose of a specific product. The personnel involved in the project are dedicated to the particular mission at hand. At the conclusion of the mission (e.g., product phase out), the personnel may be reassigned to other projects. Product organization can facilitate the most diverse and flexible grouping of project participants. It has the following advantages:

- Simplicity of structure
- Unity of project purpose
- Localization of project failures
- Condensed and focused communication lines
- Full authority of the project manager
- Quicker decisions due to centralized authority
- Skill development due to project specialization
- Improved motivation, commitment, and concentration
- Flexibility in determining time, cost, performance trade-offs
- Project team's reporting directly to one project manager or boss
- Ability of individuals to acquire and maintain expertise on a given project

The disadvantages of product organization are

- Narrow view on the part of project personnel (as opposed to a global organizational view)

- The same functional expertise is replicated (or duplicated) in multiple projects
- Mutually exclusive allocation of resources (one worker to one project)
- Duplication of efforts on different but similar projects
- Monopoly of organizational resources
- Project team members may have concerns about life-after-the-project
- Reduced skill diversification

One other disadvantage of the product organization is the difficulty supervisors have in assessing the technical competence of individual team members. Since managers may supervise functional personnel in fields foreign to them, it is difficult for them to assess technical capability. For example, a project manager in a projectized structure may supervise personnel from accounting, engineering, design, manufacturing, sales, marketing, and so on. Many major organizations face this problem.

Matrix Organization Structure

The matrix organization structure is a blend of functional and projectized structures. It is a frequently used organization structure in business and industry. It is used where there is multiple managerial accountability and responsibility for a project. It combines the advantages of the traditional structure and the product organization structure. The hybrid configuration of the matrix structure facilitates maximum resource utilization and increased performance within time, cost, and performance constraints. There are usually two chains of command involving both horizontal and vertical reporting lines. The horizontal line deals with the functional line of responsibility while the vertical line deals with the project line of responsibility. An example of a matrix structure is shown in Figure 1.12. The personnel along each vertical line of reporting cross over horizontally to work on the "matrixed" project. The matrix structure is said to be *strong* if it is more closely aligned with projectized organization structure and it is said to be a weak matrix structure if it is more closely aligned to a functional structure. A balanced matrix structure blends projectized and functional structures equally. Figure 1.13 shows the strength relationships of the three structures.

Advantages of matrix organization include the following:

- Good team interaction
- Consolidation of objectives
- Multilateral flow of information
- Lateral mobility for job advancement
- Individuals have an opportunity to work on a variety of projects
- Efficient sharing and utilization of resources
- Reduced project cost due to sharing of personnel
- Continuity of functions after project completion
- Stimulating interactions with other functional teams
- Functional lines rally to support the project efforts
- Each person has a "home" office after project completion
- Company knowledge base is equally available to all projects

Science, Technology, and Engineering Project Methodology

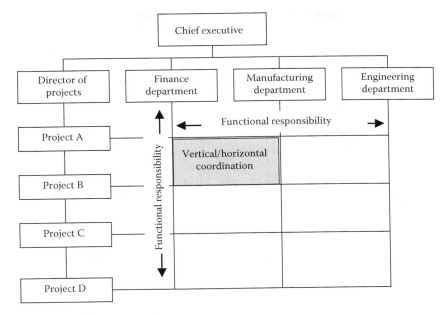

FIGURE 1.12 Matrix organization structure.

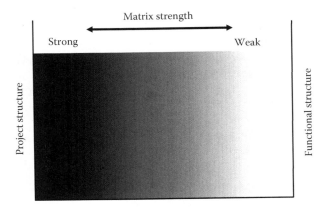

FIGURE 1.13 Matrix blend of project and functional structures.

Some of the disadvantages of matrix organization are summarized below:

- Matrix response time may be slow for fast-paced projects
- Each project organization operates independently
- Overhead cost due to additional lines of command
- Potential conflict of project priorities
- Problems inherent in having multiple bosses
- Complexity of the structure

TABLE 1.3
Levels of Project Characteristics under Different Organizational Structures

Project Characteristics	Organizational Structures				
	Functional	Weak Matrix	Balanced Matrix	Strong Matrix	Projectized
Project manager's authority	Low	Limited	Low to moderate	Moderate to high	High
Resource availability	Low	Limited	Low to moderate	Moderate to high	High
Control of project budget	Functional manager	Functional manager	Mixed	Project manager	Project manager
Role of project manager	Part-time	Part-time	Part-time	Full time	Full time
Project management staff	Part-time	Part-time	Part-time	Full time	Full time

Traditionally, industrial projects are conducted in serial functional implementations such as R&D, engineering, manufacturing, and marketing. At each stage, unique specifications and work patterns may be used without consulting the preceding and succeeding phases. The consequence is that the end product may not possess the original intended characteristics. For example, the first project in the series might involve the production of one component while the subsequent projects might involve the production of other components. The composite product may not achieve the desired performance because the components were not designed and produced from a unified point of view. The major appeal of matrix organization is that it attempts to provide synergy within groups in an organization. Table 1.3 summarizes the levels of responsibilities and project characteristics under different organizational structures. In a projectized structure, the project manager enjoys high to almost total power for project authority and resource availability, whereas, he or she will have little power on project authority and resource availability under a functional structure.

ELEMENTS OF A PROJECT PLAN

A project plan represents the roadmap for executing a project. It contains the outline of the series actions needed to accomplish the project goal. Project planning determines how to initiate a project and execute its objectives. It may be a simple statement of a project goal or it may be a detailed account of procedures to be followed during the project life cycle. In a project plan, all roles and responsibilities must be clearly defined. A project plan is not a bar chart or Gantt chart. The project manager must be versatile enough to have knowledge of most of the components of a project plan. The usual components of a detailed project plan include the following:

- Scope planning
- Scope definition

Science, Technology, and Engineering Project Methodology

- WBS
- Activity definition
- Activity sequencing
- Activity resource estimating
- Activity duration estimating
- Schedule development
- Cost estimating
- Cost budgeting
- Quality plan
- Human resource plan
- Communications plan
- Risk management plan
- Risk identification
- Qualitative and quantitative risk analysis
- Risk response planning
- Contingencies
- Purchase plan
- Acquisition plan
- Contracting plan

GENERAL APPLICABILITY OF PROJECT MANAGEMENT

PM has general applicability to every human endeavor and its use continues to grow rapidly. Ancient and contemporary projects have benefited from PM practices. Records indicate that even the technology of the ancient world practiced PM. The need to develop effective management tools increases with increasing complexity of new technologies and processes. The life cycle of a new product to be introduced into a competitive market is a good example of a complex process that must be managed with integrative PM approaches. The product will encounter management functions as it goes from one stage to another. PM will be needed throughout the design and production stages of the product. PM will be needed in developing marketing, transportation, and supply chain strategies for the product. When the product finally gets to the customer, PM will be needed to integrate its use with those of other products within the customer's organization. The need for a PM approach is established by the fact that a project will always tend to increase in size even if its scope is narrowing. An integrated PM approach can help diminish the adverse impacts of project complexity through good project planning, organizing, scheduling, and control.

PM represents an excellent basis for integrating various management techniques such as finance, economics, operations research, operations management, forecasting, quality control, queuing analysis, and simulation. Traditional approaches to PM use these techniques in a disjointed fashion, thus ignoring the potential interplay between the techniques. The need for integrated PM worldwide is evidenced by repeated reports from the World Bank, which acknowledges that there is an increasing trend of failed projects around the world. The bank has loaned billions of dollars to developing countries over the last half century only to face one failed project after another. The lack of an integrated approach to managing the projects has been cited as one of the major causes of project failures. This is particularly crucial for STEPs. STEPs require a systematic integration of technical, human, and financial resources to achieve organizational goals and objectives.

DOCUMENTING PROJECT LESSONS LEARNED

Mistakes are an essential part of learning and learning is essential for future project success. Plan the project. Execute the project as planned. Learn from the project and document lessons learned as well as best practices. It is essential to close out a project forthrightly. Not closing out a project promptly often leads to project failure. Use the project close-out to plan and initiate the next project. This process is summarized in the PELC (Plan-Execute-Learn-Close) quadrants of project success presented in Figure 1.14. A companion process is the PDCA (Plan, Do, Check, Act) loop presented in Figure 1.15. This is used frequently for quality planning and process control. Another frequently used tool for project implementation is the DMAIC (Define, Measure, Analyze, Improve, Control) process, which is shown in Figure 1.16.

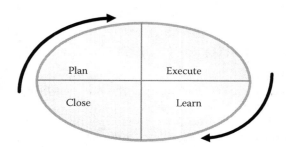

FIGURE 1.14 PELC quadrants of project success.

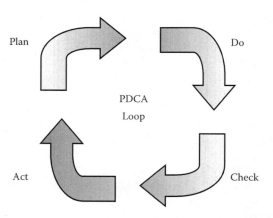

FIGURE 1.15 PDCA loop for project management process interactions.

Science, Technology, and Engineering Project Methodology

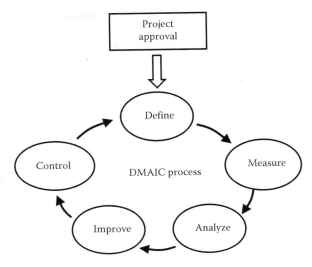

FIGURE 1.16 Application of DMAIC process to project control.

MARVELS OF STEPs

Engineering projects have had far-reaching impacts on human society for centuries. The management of engineering projects continues to require careful attention and specialized supervision. Engineering requires the application of science and technology. Thus, much analytical work is required for executing engineering projects successfully. STEP PM takes care of this through integrated application of managerial and analytical tools and techniques. Contemporary application of proven tools and techniques of PM is essential to address the complexity and intertwined nature of STEPs. Specific examples of engineering projects include the following, some ancient and some modern:

- Roads
- Canals
- Bridges
- Railroads
- Pipelines
- Windmills
- Hydroelectric plants
- Dams
- Solar systems
- Towers
- Oil platforms
- Ocean liners
- Cruise ships
- Large aircraft
- Carriers

- Tunnels
- Sky-crappers
- Mega housing complex
- Malls
- Shopping centers
- Schools
- Correctional facilities
- Bunkers
- Silos
- Storage containers
- Factories
- Sports arenas
- Convention halls
- Pyramids
- Temples
- Domes
- Cathedrals
- Aqueducts
- Viaducts

Some common elements for managing these projects include managing or overcoming constraints of distance, height, depth, security, protection, environment, harsh weather, difficult terrains, and cultural diversity. Case examples available in the popular press and archival literature point to the need for better project coordination and across-the-board product integration. A case in point is the 2007 announcement by aerospace giant Boeing that the delivery of its much-touted 787 Dreamliner airplane would be pushed back by more than 6 months. In the highly competitive and volatile aerospace market, such an announcement could spell death sentence for specific programs. One would think that Boeing had all its ducks in row and project coordination finalized before launching a worldwide marketing blitz for an imminent delivery of the flagship 787 product. Prospective buyers who enjoyed the bliss of the impending product and salivated on what the market would bring were suddenly plunged into the abyss of product anticipation by the unexpected announcement in late Fall 2007. Boeing cited lack of timely and quality delivery from outsourced subcontractors. The company had outsourced about 70% of the production work, with the expectation that that would speed up the delivery of the final product. But that level of contracting out required unprecedented amount of coordination internally and externally. Somehow the coordination did not happen as envisioned. The loss of market goodwill that might result from this type of missed promise can be economically devastating for any organization.

In many instances, jobs that were earlier outsourced are now being reversed-sourced to the original points of operation, particularly in the United States. Outsourcing should not be based on labor savings alone. There are several other factors of importance to be considered from a PM perspective. Some of these include quality issues, lack of coordination, communication gaps, noncompliance with regulatory requirements, lack of conformance to operating standards, trade

Science, Technology, and Engineering Project Methodology

barriers, and political incongruence. The Boeing example is cited to show that even large corporations do not always have their PM right. Thus, more PM is needed at all levels. The techniques of Triple C approach discussed in chapter 8 can help improve communication, cooperation, and coordination requirements throughout any PM endeavor.

ADVANCEMENT OF SOCIETY ON THE BACK OF STE

Engineering evolved from the application of mathematics, science, and technology to solve societal challenges, as society has moved onward from the various developmental stages. The various historical ages of development are summarized below:

1. Stone Age
2. Ice Age
3. Ceramics Age
4. Bronze Age
5. Iron Age
6. Nuclear Age
7. Silicon Age

The Stone Age is the fundamental period of human history whereby the widespread use of "technology" was first embraced. The name originated from the fact that most human tools of the era were made of stone. The expansion of the world from the Savannas of East Africa to new worlds typifies human evolution and advancement. The period led to the emergence of agriculture, the domestication of animals, and the smelting of copper ore to produce metal, and the use of metal to form various tools. The Stone Age is often termed prehistoric because recorded human history had not yet started. Although tools made of wood and animal parts (bone, skeleton, dried skin, tendon, etc.) were also in use during that period, they were rarely preserved and minimally documented. A distinction is often made between the New Stone Age (Neolithic; 7000 BC) and the Old Stone Age (Paleolithic), which spanned about the preceding one million years. The intermediate period between the Old and New Stone Ages is termed the Mesolithic Age. Human evolution from this rudimentary society to the present age shows evidence of progressive advancement toward the use of tools of science and technology. The Silicon Age, which emerged around 1968, appears to be the culmination, but not quite. The development of new materials in ever decreasing size and speed is leading to new ages of human advancement. As these developmental efforts become more intertwined, integrated PM must be put into place in order to maximize operational efficiencies. Figure 1.17 shows the interrelationships of science and technology components. A modern project should leverage all the components maximize overall project output.

LEVELS OF PROJECT EXECUTION

Managing large projects is a multidimensional undertaking that encompasses many facets of the organization. Figure 1.18 presents typical interfaces of domains

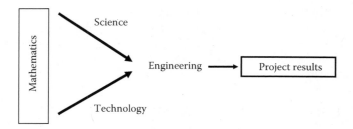

FIGURE 1.17 Interrelationships of math, science, technology, and engineering for project results.

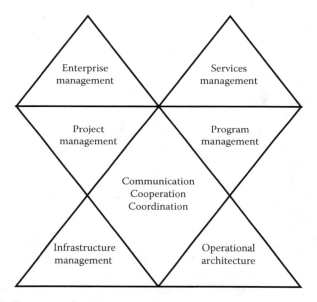

FIGURE 1.18 Framework for project execution.

of operation within an organization. It is seen in the figure that communication, cooperation, and coordination constitute the centerpiece of the domains of operation covering enterprise management, services management, infrastructure management, operational architecture, program management, and project management. Within all of these, the levels of project execution cover the following areas of focus:

- Enterprise-wide impact

This is often management driven.

- Process improvement impact

This is normally driven by changing teams at departmental levels.

- Task-level impact

This is facilitated and effected at workforce level. This is operator driven.

CATEGORIES OF PROJECT OUTPUTS

Recognizing the category of products expected from an enterprise is an essential part of managing projects more effectively. Project outputs are categorized into the classes below. This makes the application of PM process applicable to every undertaking because each effort is expected to generate an output in one or more of the following product categories.

- Product
(Physical products, e.g., new hospital facility)
- Service
(Business process, e.g., new operating procedure)
- Result
(Knowledge creation, e.g., research, education, training)

UNIQUE ASPECTS OF STEPs

The wealth and economic vitality of a nation is ultimately defined by the level of outputs of STEP. STEPs often result in end products that last for decades beyond the life cycle of the project itself. The products may involve long-lasting social, economic, environmental, and political impacts on the society. It is the sustainability of impacts that makes STEP to be of great interest to practitioners and policy makers. In spite of the long history of STEP, reports after reports indicate that most projects are still not completed on time, are over budget, or do not meet performance objectives. Many STEPs are not even completed at all. A step-by-step guide, as proposed in this book, may help alleviate or mitigate adverse impacts of the complexity of STEPs. STEP Systems Logistics is a key element covered throughout this book. For this purpose, the book adopts the following definition:

STEP logistics

- Planning and implementation of a complex science, technology, or engineering task
- Planning and control of the flow of goods and materials through an organization or a STEP process
- Planning and organization of the movement of people, equipment, and supplies across STEP functions

Complex projects represent a hierarchical system of operations. A STEP system is a collection of interrelated technical projects all serving a common end goal. STEP system logistics involves the planning, implementation, movement, scheduling, and control of people, equipment, goods, materials, and supplies across the interfacing boundaries of several related technical projects. Conventional PM must be modified and expanded to address the unique logistics of STEP systems. STEPs can originate from a variety of sources. Some examples are

1. Market requirement for a new product, service, or result
2. Societal demand
3. Organizational need
4. Customer order
5. Technological capability
6. Legislative action

INTEGRATED SYSTEMS APPROACH TO STEPs

PM tools for STEPs can be classified into three major categories described below:

1. Qualitative tools: These are the managerial tools that aid in the interpersonal and organizational processes required for PM.
2. Quantitative tools: These are analytical techniques that aid in the computational aspects of PM.
3. Computer tools: These are computer software and hardware tools that simplify the process of planning, organizing, scheduling, and controlling a project. Software tools can help in with both the qualitative and quantitative analyses needed for PM.

Figure 1.19 illustrates the integrated application of quantitative, qualitative, and software tools to manage STEPs. The approach considers not only the management of the project itself, but also the management of all the functions that support the project. A systems approach helps to increase the intersection of the three categories of PM tools and hence improve overall management effectiveness. Crisis should not be the motivation for the use of PM techniques. PM approaches should be used up front to preempt problems rather than being used to fight project "fires." Figure 1.20 lists examples of typical tools in each of the tool categories of quantitative, qualitative, and computer software.

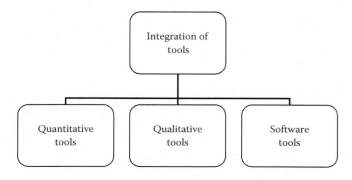

FIGURE 1.19 Integration of STEP tools.

Science, Technology, and Engineering Project Methodology

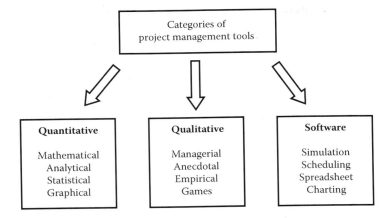

FIGURE 1.20 Quantitative, qualitative, and software tools.

STEP PROJECT MANAGEMENT STEPS

The steps of managing STEPs are presented in Figure 1.21. Figure 1.22 presents a generic flowchart for the execution of STEPs.

MANAGING PROJECT REQUIREMENTS

It is often said that Henry Ford offered his Model T automobile customers only one color option by saying that customers could have "any color they want, as long as it is black." But the fact is that Ford initially offered three colors: Black, bright red,

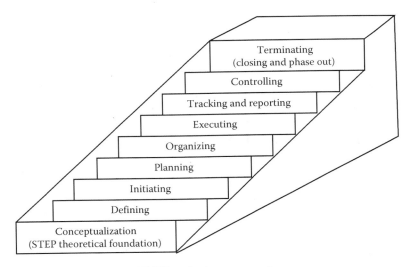

FIGURE 1.21 Components of STEP project management.

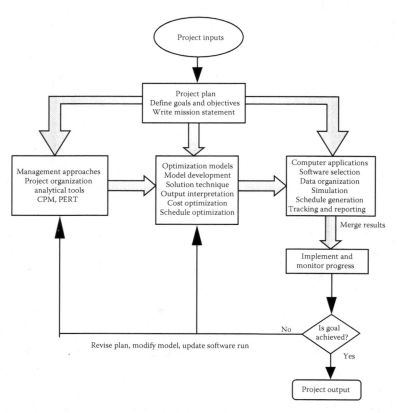

FIGURE 1.22 Flowchart of integrated STEP project management.

and green from 1908 through 1914. But when his production technology advanced to the stage of mass production on moving assembly line, the new process required a fast-drying paint and only one particular black paint pigment met the requirements. Thus, as a result of the emergence of fast-moving mass production lines, Ford was forced to limit color options to black only. This led to the need for the famous quote. The black-only era spanned the period from 1914 through 1925, when further painting advances made it possible to have more color options. This represents a classic example of how technology limitations might dictate the execution of project requirements. Under STEP PM, an organization must remain flexible with operational choices. Figure 1.23 presents process control flowchart for the application of PM to managing production technology.

This chapter has presented a general introduction to the various aspects of STEP PM. These aspects form the foundation for the specific application of the steps contained in the PM body of knowledge. Table 1.4 shows a site map of the step-by-step implementation for STEPs. The steps are outlined in Chapters 2 through 10, with one chapter devoted to each knowledge area. We leave this chapter with

Science, Technology, and Engineering Project Methodology

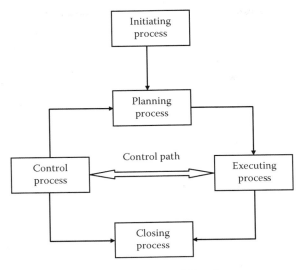

FIGURE 1.23 Project management process control flowchart.

TABLE 1.4
Site Map for Step-by-Step STEP Implementation

Knowledge Areas	Integrative Implementation Steps
Project integration	→ → STEPS 1, 2, 3, 4, 5, 6, 7
Project scope management	→ → STEPS 1, 2, 3, 4, 5
Project time management	→ → STEPS 1, 2, 3, 4, 5, 6
Project cost management	→ → STEPS 1, 2, 3
Project quality management	→ → STEPS 1, 2, 3
Project human resource management	→ → STEPS 1, 2, 3, 4
Project communication management	→ → STEPS 1, 2, 3, 4
Project risk management	→ → STEPS 1, 2, 3, 4, 5, 6
Project procurement management	→ → STEPS 1, 2, 3, 4, 5

---

a conceptual representation of the merging of STE, and contextual systems, as shown in Figure 1.24. Chapter 2 covers project integration management, which demands that STE requirements be integrated across project operations within the spectrum of STEP. The distribution of the adoption and utilization of STE products is summarized in Figure 1.25. The innovators and early adopters create the benchmarks for others to follow. But their success will depend on how well they implement their projects of STE innovation, creation, acquisition, integration, distribution, application, and enhancement.

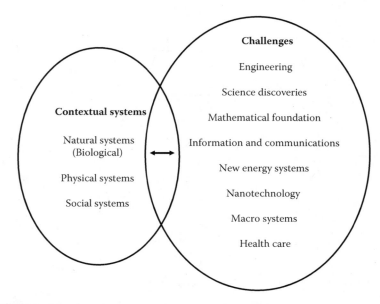

FIGURE 1.24 Merging of science, technology, and engineering systems.

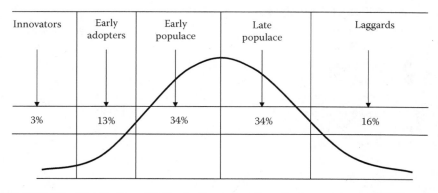

FIGURE 1.25 Distribution of STE adoption categories with the spectrum of STEP.

REFERENCES

Badiru, A. B., Editor, *Handbook of Industrial & Systems Engineering*, Taylor & Francis/CRC Press, Boca Raton, FL, 2006.

Badiru, A. B. and O. A. Omitaomu, *Computational Economic Analysis for Engineering and Industry*, Taylor & Francis/CRC Press, Boca Raton, FL, 2007.

Martin, H. L., *TECHONOMICS: The Theory of Industrial Evolution*, Taylor & Francis/CRC Press, Boca Raton, FL, 2007.

2 STEP Integration

> Coming together is a beginning.
> Keeping together is progress.
> Working together is success.
>
> —Henry Ford

Project integration management specifies how the various parts of a project come together to make up the complete project. This knowledge area recognizes the importance of linking several aspects of a project into an integrated whole. The Henry Ford quote at the beginning of this chapter emphasizes the importance of "togetherness" in any project environment. Project integration management area includes the processes and activities needed to identify, define, combine, unify, and coordinate the various processes and project activities. The traditional concepts of systems analysis are applicable to project processes. The definition of a project system and its components refers to the collection of interrelated elements organized for the purpose of achieving a common goal. The elements are organized to work synergistically together to generate a unified output that is greater than the sum of the individual outputs of the components. The harmony of project integration is evident in the characteristic symbol that this book uses to denote this area of project management knowledge.

While the knowledge areas of project management, as discussed in Chapter 1, overlap and can be implemented in alternate orders, it is still apparent that project integration management is the first step of the project effort. This is particularly based on the fact that the project charter and the project scope statement are developed under the project integration process. In order to achieve a complete and unified execution of a project, both qualitative and quantitative skills must come into play. Figure 2.1 presents the interfacing interactions of qualitative (soft) skills and quantitative (hard) skills. Soft skills (i.e., people skills) can be just as hard as they are easy. Similarly, hard skill can be just as easy as it is difficult. In fact, human resource management issues are often the most difficult to handle; yet they are typically classified as falling under the banner of soft skills. Quantitative modeling skills (hard skills), on the other hand, can be easy once all the input parameters are known and accounted for in the modeling process. Thereafter, the utilization of the model becomes an easy repetition of a proven model. By contrast, the nuances of human emotion, sentiments, and psychological variability make qualitative management not easy at all. A good approach is to embrace the interplay of both qualitative and quantitative management within the project life cycle as illustrated in Figure 2.1.

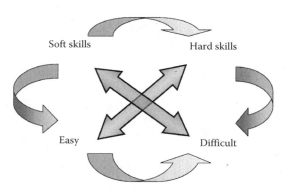

FIGURE 2.1 Interface of soft and hard skills for project integration.

PROJECT INTEGRATION: STEP-BY-STEP IMPLEMENTATION

The integration component of the body of knowledge consists of the elements shown in the block diagram in Figure 2.2. The seven elements in the block diagram are carried out across the process groups presented earlier in Chapter 1. The overview of the elements and the process groups are shown in Table 2.1. Thus, under the knowledge area of integration, the following are the required steps:

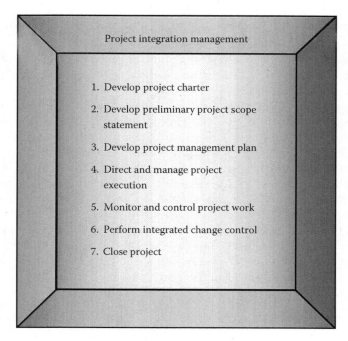

FIGURE 2.2 Block diagram of project integration management.

STEP Integration

TABLE 2.1
Implementation of Project Integration Elements across Process Groups

	Initiating	Planning	Executing	Monitoring and Controlling	Closing
Project integration	Develop project charter Develop preliminary project scope	Develop project management plan	Direct and manage project execution	Monitor and project work Integrated change control	Close project

Step 1: Develop project charter
Step 2: Develop preliminary project scope
Step 3: Develop project management plan
Step 4: Direct and manage project execution
Step 5: Monitor and control project work
Step 6: Perform integrated change control
Step 7: Close project

Each step is carried out in a structure of inputs–tools and techniques–output analysis as shown in Figure 2.3. In addition to the standard *Project Management Body of Knowledge* (*PMBOK*) inputs, tools, techniques, and outputs, the project team will add in-house items of interest to the STEPs presented in this book. Such in-house items are summarized below:

- Inputs: Other in-house (custom) factors of relevance and interest
- Tools and techniques: Other in-house (custom) tools and techniques
- Outputs: Other in-house outputs, reports, and data inferences of interest to the organization

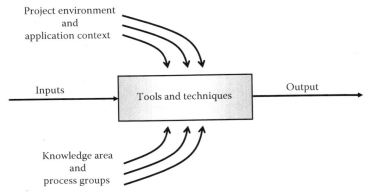

FIGURE 2.3 Input–process–output format for implementing STEP.

Table 2.2 shows the input-to-output items for developing project charter. The tabular format is useful for explicitly identifying what the project analyst needs to do or use for each step of the project management process. Tables 2.3 through 2.8 present the input-to-output entries for the other steps under integration management.

Figure 2.4 illustrates the various mental and physical challenges that must be integrated within the project life cycle. These challenges include project charter, facilities, costs, schedules, organization charts, scope, procurement, and budget.

TABLE 2.2
Tools and Techniques for Developing Project Charter within Integration Management

Step 1: Develop Project Charter

Inputs	Tools and Techniques	Output(s)
Project contract (if applicable)	Project selection methods	Project charter
Project statement of work	AHP (analytic hierarchy process)	Other in-house outputs, reports, and data inferences of interest to the organization
Enterprise environmental factors	Project management methodology	
Organizational process assets	Project management information system	
Other in-house (custom) factors of relevance and interest	Expert judgment	
	Balanced scorecard	
	Process control charts	
	Other in-house (custom) tools and techniques	

TABLE 2.3
Tools and Techniques for Developing Preliminary Project Scope Statement within Integration Management

Step 2: Develop Preliminary Project Scope Statement

Inputs	Tools and Techniques	Output(s)
Project charter	Project management methodology	Preliminary project scope statement
Project SOW	Project management information system	Other in-house outputs, reports, and data inferences of interest to the organization
Enterprise environmental factors	Expert judgment	
Organizational process assets	CMMI (capability maturity model integration)	
Other in-house (custom) factors of relevance and interest	Critical chain	
	Process control charts	
	Other in-house (custom) tools and techniques	

STEP Integration

TABLE 2.4
Tools and Techniques for Developing Project Management Plan within Integration Management

Step 3: Develop Project Management Plan

Inputs	Tools and Techniques	Output(s)
Preliminary project scope statement	Project management methodology	Project management plan
Project management processes	Project management information system	
Enterprise environmental factors	Expert judgment	
Organizational process assets		
Other in-house (custom) factors of relevance and interest		

TABLE 2.5
Tools and Techniques for Managing Project Execution within Integration Management

Step 4: Direct and Manage Project Execution

Inputs	Tools and Techniques	Output(s)
Project management plan	Project management methodology	Project deliverables
Approved corrective actions	Project management information system	Requested changes
Approved preventive actions	Process flow diagram	Implemented change requests
Approved change requests	Other in-house (custom) tools and techniques	Implemented corrective actions
Approved defect repair		Implemented preventive actions
Validated defect repair		Implemented defect repair
Administrative closure procedure		Work performance information
Other in-house (custom) factors of relevance and interest		Other in-house outputs, reports, and data inferences of interest to the organization

Figure 2.5 shows the conventional elements of project management. But it is obvious that each individual element contains additional considerations beyond the broad categories shown in the figure. These additional considerations are captured in the process groups presented by the *PMBOK* approach.

STEP 1: DEVELOPING PROJECT CHARTER

Project charter formally authorizes a project. It is a document that provides authority to the project manager and it is usually issued by a project initiator or sponsor external to the project organization. The purpose of a charter is to define at a high level what the project is about, what the project will deliver, what resources are

TABLE 2.6
Tools and Techniques for Monitoring and Controlling Project Work within Integration Management

Step 5: Monitor and Control Project Work

Inputs	Tools and Techniques	Output(s)
Project management plan	Project management methodology	Recommended corrective actions
Work performance information	Project management information system	Recommended preventive actions
Rejected change requests	Earned value management	Forecasts
Other in-house (custom) factors of relevance and interest	Expert judgment	Recommended defect repair
	Other in-house (custom) tools and techniques	Requested changes
		Other in-house outputs, reports, and data inferences of interest to the organization

TABLE 2.7
Tools and Techniques for Integrated Change Control within Integration Management

Step 6: Perform Integrated Change Control

Inputs	Tools and Techniques	Output(s)
Project management plan	Project management methodology	Approved change requests
Requested changes		Rejected change requests
Work performance information	Project management information system	Update project management plan
Recommended preventive actions		Update project scope statement
Recommended corrective actions	Expert judgment	Approved corrective actions
Deliverables	Other in-house (custom) tools and techniques	Approved preventive actions
Other in-house (custom) factors of relevance and interest		Approved defect repair
		Validated defect repair
		Deliverables
		Other in-house outputs, reports, and data inferences of interest to the organization

needed, what resources are available, and how the project is justified. The charter also represents an organizational commitment to dedicate the time and resources to the project. The charter should be shared with all stakeholders as a part of the communication requirement of Triple C approach. Cooperating stakeholders will not only sign-off on the project but also make personal pledges to support the project. Projects are usually chartered by an enterprise, a government agency, a company, a program organization, or a portfolio organization in response to one or more of the following business opportunities or organizational problems:

STEP Integration

TABLE 2.8
Tools and Techniques for Closing Project within Integration Management

Step 7: Close Project

Inputs	Tools and Techniques	Output(s)
Project management plan	Project management methodology	Administrative closure procedure
Contract documentation	Project management information system	Contract closure procedure
Enterprise environmental factors	Expert judgment	Final product, service or result
Organizational process assets	Other in-house (custom) tools and techniques	Updates on organizational process assets
Work performance information		Other in-house outputs, reports, and data inferences of interest to the organization
Deliverables		
Other in-house (custom) factors of relevance and interest		

FIGURE 2.4 Mental and physical challenges within a project life cycle.

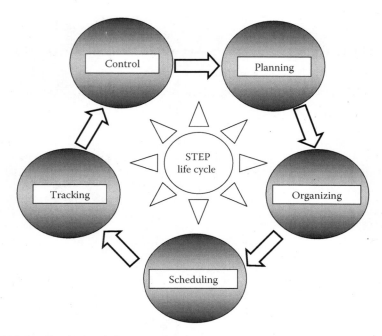

FIGURE 2.5 Conventional elements of project management.

- Market demand
- Response to regulatory development
- Customer request
- Business need
- Exploitation of technological advance
- Legal requirement
- Social need

The driving force for a project charter is the need for an organization to make a decision about which projects to authorize respond to operational threats or opportunities. It is desired that a charter be brief. Depending on the size and complexity of a project, the charter should not be more than two to three pages. Where additional details are warranted, the expatiating details can be provided as addenda to the basic charter document. The longer the basic charter, the less the likelihood that everyone will read and imbibe the contents. So, brevity and conciseness are the desired virtues of good project charters. The charter should succinctly establish the purpose of the project, the participants, and general vision for the project.

The project charter is used as the basis for developing project plans. While it is developed at the outset of a project, a charter should always be fluid. It should be reviewed and updated throughout the life of the project. The components of the project charter are summarized below:

- Project overview
- Assigned project manager and authority level

STEP Integration

- Project requirements
- Business needs
- Project purpose, justification, and goals
- Impact statement
- Constraints (time, cost, performance)
- Assumptions
- Project scope
- Financial implications
- Project approach (policies, procedures)
- Project organization
- Participating organizations and their respective roles and level of participation
- Summary milestone schedule
- Stakeholder influences
- Assumptions and constraints (organizational, environmental, external)
- Business plan and expected return on investment (ROI), if applicable
- Summary budget

The project charter does not include the project plan. Planning documents, which may include project schedule, quality plan, staff plan, communication hierarchy, financial plan, and risk plan, should be prepared and disseminated separately from the charter.

Project overview: The "project overview" provides a brief summary of the entire project charter. It may provide a brief history of the events that led to the project, an explanation of why the project was initiated, a description of project intent, and the identity of the original project owner.

Project goals: "Project goals" identify the most significant reasons for performing a project. Goals should describe improvements the project is expected to accomplish along with who will benefit from these improvements. This section should explain what various benefactors will be able to accomplish due to the project. Note that Triple C approach requires these details as a required step to securing cooperation.

Impact statement: The "impact statement" identifies the influence the project may have on the business, operations, schedule, other projects, current technology, and existing applications. While these topics are beyond the domain of this project, each of these items should be raised for possible action.

Constraints and assumptions: "Constraints and assumptions" identify any deliberate or implied limitations or restrictions placed on the project along with any current or future environment the project must accommodate. These factors will influence many project decisions and strategies. The potential impact of each constraint or assumptions should be identified.

Project scope: "Project scope" defines the operational boundaries for the project. Specific scope components are the areas or functions to be impacted by the project and the work that will be performed. The project scope should identify both what is within the scope of the project and what is outside.

Project objectives: "Project objectives" identify expected deliverables from the project and the criteria that must be satisfied before the project is considered complete.

Financial summary: The "financial summary" provides a recap of expected costs and benefits due to the project. These factors should be more fully defined in the cost–benefit analysis of the project. Project financials must be reforecast during the life of the effort.

Project approach: "Project approach" identifies the general strategy for completing the project and explains any methods or processes, particularly policies and procedures that will be used during the project.

Project organization: The "project organization" identifies the roles and responsibilities needed to create a meaningful and responsive structure that enables the project to be successful. Project organization must identify the people who will play each assigned role. At minimum, this section should identify the people playing the roles of project owner, project manager, and core project team.

A "project owner" is required for each project.

> This role must be filled by one or more individuals who are the fiscal trustee(s) for the project to the larger organization. This person considers the global impact of the project and deems it worthy of the required expenditure of money and time. The project owner communicates the vision for the effort and certifies the initial project charter and project plan. Should changes be required, the project owner confirms these changes and any influence on the project charter and project plan. When project decisions cannot be made at the team level, the project owner must resolve these issues. The project owner must play an active role throughout the project, especially ensuring that the needed resources have been committed to the project and remain available.

A "project manager" is required for each project.

> The project manager is responsible for initiating, planning, executing, and controlling the total project effort. Members of the project team report to the project manager for project assignments and are accountable to the project manager for the completion of their assigned work.

Definition of Inputs to Step 1

Contract: A contract is a contractual agreement between the organization performing the project and the organization requesting the project. It is treated as an input if the project is being done for an external customer.

Project statement of work (SOW): This is a narrative description of products or services to be supplied by the project. For internal projects, it is provided by the project initiator or sponsor. For external projects, it is provided by the customer as part of the bid document. For example, request for proposal, request for information, request for bid, or contract statements may contain specific work to be done. The SOW indicates the following:

- Business need based on required training, market demand, technological advancement, legal requirement, government regulations, industry standards, or trade consensus

STEP Integration

- Product scope description, which documents project requirements and characteristics of the product or service that the project will deliver
- Strategic plan, which ensures that the project supports organization's strategic goals and business tactical actions

Enterprise environmental factors: These are factors that impinge upon the business environment of the organization. They include organizational structure, business culture, governmental standards, industry requirements, quality standards, trade agreements, physical infrastructure, technical assets, proprietary information, existing human resources, personnel administration, internal work authorization system, marketplace profile, competition, stakeholder requirements, stakeholder risk tolerance levels, commercial obligations, access to standardized cost estimating data, industry risk, technology variances, product life cycle, and project management information systems.

Organizational process assets: These refer to the business processes used within an organization. They include standard processes, guidelines, policies, procedures, operational templates, criteria for customizing standards to specific project requirements, organization communication matrix, responsibility matrix, project closure guidelines (e.g., sunset clause), financial controls procedure, defect management procedures, change control procedures, risk control procedures, process for issuing work authorizations, processes for approving work authorizations, management of corporate knowledge base, and so on.

Definition of Tools and Techniques for Step 1

Project selection methods: These methods are used to determine which projects an organization will select for implementation. The methods can range from basic seat-of-the-pants heuristics to highly complex analytical techniques. Some examples are benefit measurement methods, comparative measure of worth analysis, scoring models, benefit contribution, capital rationing approaches, budget allocation methods, and graphical analysis tools. Analytical techniques are mathematical models that use linear programming, nonlinear programming, dynamic programming, integer programming, multiattribute optimization, and other algorithmic tools.

Project management methodology: This defines the set of project management process groups, their collateral processes, and related control functions that are combined for implementation for a particular project. The methodology may or may not follow a project management standard. It may be an adaptation of an existing project implementation template. It can also be a formal mature process or informal technique that aids in effectively developing a project charter.

Project management information system (PMIS): This is a standardized set of automated tools available within the organization and integrated into a system for the purpose of supporting the generation of a project charter, facilitating feedback as the charter is refined, controlling changes to the project charter, or releasing the approved document.

Expert judgment: This is often used to assess the inputs needed to develop the project charter. Expert judgment is available from sources such as experiential database of the organization, knowledge repository, knowledge management practices, knowledge transfer protocol, business units within the organization, consultants, stakeholders, customers, sponsors, professional organizations, technical associations, and industry groups.

Definition of Output of Step 1

Project charter: As defined earlier in this chapter, project charter is a formal document that authorizes a project. It provides authority to the project manager and it is usually issued by a project initiator or sponsor external to the project organization. It empowers the project team to carry out actions needed to accomplish the end goal of the project.

STEP 2: DEVELOP PRELIMINARY PROJECT SCOPE STATEMENT

Project scope presents a definition of what needs to be done. It specifies the characteristics and boundaries of the project and its associated products and services, as well as the methods of acceptance and scope control. Scope is developed based on information provided by the projected initiator or sponsor. Scope statement includes the following:

- Project and product objectives
- Product characteristics
- Service requirements
- Product acceptance criteria
- Project constraints
- Project assumptions
- Initial project organization
- Initial defined risks
- Schedule milestones
- Initial work breakdown structure (WBS)
- Order-of-magnitude cost estimate
- Project configuration management requirements
- Approval requirements

Definition of Inputs to Step 2

Inputs for step 2 are the same as defined for step 1 covering project charter, SOW, environmental factors, and organizational process assets.

Definition of Tools and Techniques for Step 2

The tools and techniques for step 2 are the same as defined for step 1 and cover project management methodology, project management information system, and expert judgment.

Definition of Output of Step 2

The output of step 2 is the preliminary project scope statement, which was defined and described earlier.

STEP 3: DEVELOP PROJECT MANAGEMENT PLAN

A project management plan includes all actions necessary to define, integrate, and coordinate all subsidiary and complementing plans into a cohesive project management plan. It defines how the project is executed, monitored and controlled, and closed. The project management plan is updated and revised through the integrated change control process. In addition, the process of developing project management plan documents the collection of outputs of planning processes and includes the following:

- Project management processes selected by the project management team
- Level of implementation of each selected process
- Descriptions of tools and techniques to be used for accomplishing those processes
- How selected processes will be used to manage the specific project
- How work will be executed to accomplish the project objectives
- How changes will be monitored and controlled
- How configuration management will be performed
- How integrity of the performance measurement baselines will be maintained and used
- The requirements and techniques for communication among stakeholders
- The selected project life cycle and, for multiphase projects, the associated project phases
- Key management reviews for content, extent, and timing

The project management plan can be a summary or integration of relevant subsidiary, auxiliary, and ancillary project plans. All efforts that are expected to contribute to the project goal can be linked into the overall project plan, each with the appropriate level of detail. Examples of subsidiary plans are the following:

- Project scope management plan
- Schedule management plan
- Cost management plan
- Quality management plan
- Process improvement plan
- Staffing management plan
- Communication management plan
- Risk management plan
- Procurement management plan
- Milestone list
- Resource calendar

- Cost baseline
- Quality baseline
- Risk register

Definition of Inputs to Step 3

Inputs to step 3 are the same as defined previously and include preliminary project scope statement, project management processes, enterprise environmental factors, and organizational process assets.

Definition of Tools and Techniques for Step 3

The tools and techniques for step 3 are project management methodology, project information system, and expert judgment. Project management methodology defines a process that aids a project management team in developing and controlling changes to the project plan. Project management information system at this step covers the following segments:

- Automated system, which is used by the project team to do the following:
  - Support generation of the project management plan
  - Facilitate feedback as the document is developed
  - Control changes to the project management plan
  - Release the approved document
- Configuration management system, which is a subsystem that includes subprocesses for accomplishing the following:
  - Submitting proposed changes
  - Tracking systems for reviewing and authorizing changes
  - Providing a method to validate approved changes
  - Implementing change management system
- Configuration control system, which forms a collection of formal procedures used to apply technical and administrative oversight to do the following:
  - Identify and document functional and physical characteristics of a product or component
  - Control any changes to such characteristics
  - Record and report each change and its implementation status
  - Support audit of the products or components to verify conformance to requirements
- Change control system is the segment of project management information system that provides a collection of formal procedures that define how project deliverables and documentation are controlled.

Expert judgment, the third tool for step 3, is applied to develop technical and management details to be included in the project management plan.

Definition of Output of Step 3

The output of step 3 is the project management plan.

Step 4: Direct and Manage Project Execution

Step 4 requires the project manager and project team to perform multiple actions to execute the project plan successfully. Some of the required activities for project execution are summarized below:

- Perform activities to accomplish project objectives
- Expend effort and spend funds
- Staff, train, and manage project team members
- Obtain quotation, bids, offers, or proposals as appropriate
- Implement planned methods and standards
- Create, control, verify, and validate project deliverables
- Manage risks and implement risk response activities
- Manage sellers
- Adapt approved changes into scope, plans, and environment
- Establish and manage external and internal communication channels
- Collect project data and report cost, schedule, technical and quality progress, and status information to facilitate forecasting
- Collect and document lessons learned and implement approved process improvement activities

The process of directing and managing project execution also requires implementation of the following:

- Approved corrective actions that will bring anticipated project performance into compliance with the plan
- Approved preventive actions to reduce the probability of potential negative consequences
- Approved defect repair requests to correct product defects during quality process

Definition of Inputs to Step 4

Inputs to step 4 are summarized as follows:

- Project management plan.
- Approved corrective actions: These are documented, authorized directions required to bring expected future project performance into conformance with the project management plan.
- Approved change requests: These include documented, authorized changes to expand or contract project scope. Can also modify policies, project management plans, procedures, costs, budgets, or revise schedules. Change requests are implemented by the project team.
- Approved defect repair: This is a documented, authorized request for product defect correction found during the quality inspection or the audit process.

- Validated defect repair: This is notification that reinspected repaired items have either been accepted or rejected.
- Administrative closure procedure: This documents all the activities, interactions, and related roles and responsibilities needed in executing the administrative closure procedure for the project.

Definition of Tools and Techniques for Step 4

The tools and techniques for step 4 are project management methodology and project management information system and they were previously defined.

Definition of Outputs of Step 4

- Deliverables
- Requested changes
- Implemented change requests
- Implemented corrective actions
- Implemented preventive actions
- Implemented defect repair
- Work performance information

STEP 5: MONITOR AND CONTROL PROJECT WORK

No organization can be strategic without being quantitative. It is through quantitative measures that a project can be tracked, measured, assessed, and controlled. The need for monitoring and control can be evident in the request for quantification (RFQ) that some project funding agencies use. Some quantifiable performance measures are schedule outcome, cost effectiveness, response time, number of reworks, and lines of computer codes developed. Monitoring and controlling are performed to monitor project processes associated with initiating, planning, executing, and closing and is concerned with the following:

- Comparing actual performance against plan
- Assessing performance to determine whether corrective or preventive actions are required and then recommending those actions as necessary
- Analyzing, tracking, and monitoring project risks to make sure risks are identified, status is reported, response plans are being executed
- Maintaining an accurate timely information base concerning the project's products and associated documentation
- Providing information to support status reporting, progress measurement, and forecasting
- Providing forecasts to update current cost and schedule information
- Monitoring implementation of approved changes

Definition of Inputs to Step 5

Inputs to step 5 include the following:

- Project management plan
- Work performance plan

STEP Integration

- Rejected change requests
  - Change requests
  - Supporting documentation
  - Change review status showing disposition of rejected change requests

Definition of Tools and Techniques for Step 5

- Project management methodology
- Project management information system
- Earned value technique: This measures performance as project moves from initiation through closure. It provides means to forecast future performance based on the past performance
- Expert judgment

Definition of Outputs of Step 5

- Recommended corrective actions: Documented recommendations required to bring expected future project performance into conformance with the project management plan
- Recommended preventive actions: Documented recommendations that reduce the probability of negative consequences associated with project risks
- Forecasts: Estimates or predictions of conditions and events in the project's future based on information available at the time of the forecast
- Recommended defect repair: Some defects found during quality inspection and audit process recommended for correction
- Requested changes

STEP 6: INTEGRATED CHANGE CONTROL

Integrated change control is performed from project inception through completion. It is required because projects rarely run according to plan. Major components of integrated change control include the following:

- Identifying when a change needs to occur or when a change has occurred
- Amending factors that circumvent change control procedures
- Reviewing and approving requested changes
- Managing and regulating flow of approved changes
- Maintaining and approving recommended corrective and preventive actions
- Controlling and updating scope, cost, budget, schedule, and quality requirements based upon approved changes
- Documenting the complete impact of requested changes
- Validating defect repair
- Controlling project quality to standards based on quality reports

Combining configuration management system with integrated change control includes identifying, documenting, and controlling changes to the baseline. Project-wide application of the configuration management system, including change control processes, accomplishes three major objectives:

- Establishes evolutionary method to consistently identify and request changes to established baselines and to assess the value and effectiveness of those changes
- Provides opportunities to continuously validate and improve the project by considering the impact of each change
- Provides the mechanism for the project management team to consistently communicate all changes to the stakeholders

Integrated change control process includes some specific activities of the configuration management as summarized below:

- Configuration identification: This provides the basis from which the configuration of products is defined and verified, products and documents are labeled, changes are managed, and accountability is maintained.
- Configuration status accounting: This involves capturing, storing, and accessing configuration information needed to manage products and product information effectively.
- Configuration verification and auditing: This involves confirming that performance and functional requirements defined in the configuration documentation have been satisfied.

Under integrated change control, every documented requested change must be either accepted or rejected by some authority within the project management team or an external organization representing the initiator, sponsor, or customer. Integrated change control can possibly be controlled by a change control board.

Definition of Inputs to Step 6

The inputs to step 6 include the following items, which were described earlier:

- Project management plan
- Requested changes
- Work performance information
- Recommended preventive actions
- Deliverables

Definition of Tools and Techniques for Step 6

- Project management methodology: This defines a process that helps a project management team in implementing integrated change control for the project.
- Project management information system: This is an automated system used by the team as an aid for the implementation of an integrated change control process for the project. It also facilitates feedback for the project and controls changes across the project.
- Expert judgment: This refers to the process whereby the project team uses stakeholders with expert judgment on the change control board to control and approve all requested changes to any aspect of the project.

STEP Integration

Definition of Outputs of Step 6

The outputs of step 6 include the following:

- Approved change requested
- Rejected change requests
- Project management plan (updates)
- Project scope statement (updates)
- Approved corrective actions
- Approved preventive actions
- Approved defect repair
- Validated defect repair
- Deliverables

STEP 7: CLOSE PROJECT

At its completion, a project must be formally closed. This involves performing the project closure portion of the project management plan or closure of a phase of a multiphase project. There are two main procedures developed to establish interactions necessary to perform the closure function:

- Administrative closure procedure: This provides details of all activities, interactions, and related roles and responsibilities involved in executing the administrative closure of the project. It also covers activities needed to collect project records, analyze project success or failure, gather lessons learned, and archive project information.
- Contract closure procedure: This involves both product verification and administrative closure for any existing contract agreements. Contract closure procedure is an input to the close contract process.

Definition of Inputs to Step 7

The inputs to step 7 are the following:

- Project management plan.
- Contract documentation: This is an input used to perform the contract closure process and includes the contract itself as well as changes to the contract and other documentation, such as technical approach, product description, or deliverable acceptance criteria and procedures.
- Enterprise environmental factors.
- Organizational process assets.
- Work performance information.
- Deliverables, as previously described, and also as approved by the integrated change control process.

Definition of Tools and Techniques of Step 7

- Project management methodology
- Project management information system
- Expert judgment

Definition of Outputs of Step 7

- Administrative closure procedure
    - Procedures to transfer the project products or services to production and/or operations are developed and established at this stage
    - This stage covers a step-by-step methodology for administrative closure that addresses the following:
        - Actions and activities to define the stakeholder approval requirements for changes and all levels of deliverables
        - Actions and activities confirm that the project has met all sponsor, customer, and other stakeholders' requirements
        - Actions and activities to verify that all deliverables have been provided and accepted
        - Actions and activities to validate completion and exit criteria for the project
- Contract closure procedure
    - This stage provides a step-by-step methodology that addresses the terms and conditions of the contracts and any required completion or exit criteria for contract closure.
    - Actions performed at this stage formally close all contracts associated with the completed project.
- Final product, service, or result
    - Formal acceptance and handover of the final product, service, or result that the project was authorized to provide
    - Formal statement confirming that the terms of the contract have been met
- Organizational process assets (updates)
    - Development of the index and location of project documentation using the configuration management system
    - Formal acceptance documentation, which formally indicates that the customer or sponsor has officially accepted the deliverables
    - Project files, which contain all documentation resulting from the project activities
    - Project closure documents, which consist of a formal documentation indicating the completion of the project and transfer of deliverables
    - Historical information, which is transferred to knowledge base of lessons learned for use by future projects
    - Traceability of process steps

APPLICATION OF CMMI AND CARPÉ FUTURUM

Project integration is about linking processes for now and for the future. This can be done through the application of techniques such as capability maturity model integration (CMMI), which is a process improvement approach that provides organizations with the essential elements of effective processes. It can be used to guide process improvement across a project, a division, or an entire organization. CMMI helps integrate traditionally separate organizational functions, set process improvement goals and priorities, provide guidance for quality processes, and provide a point of reference for evaluating current and future processes.

The concept of *Carpé Futurum* (Seize the Future) is very essential for implementing CMMI effectively across planning horizons. Harper and Glew (2008) present *Carpé Futurum* as an anticipatory management approach that should not be compromised. Preemptive management strategy helps to foresee project operational problems so that proactive steps to forestall the problems can be taken. Breakthroughs in science, technology, and engineering (STE) have created new opportunities and challenges in the way we run organizational operations and execute projects. The tools and techniques for managing projects under this contemporary scenario must be elevated to meet the current needs. With anticipatory management, we can be prepared for operational shocks of the future and put in place control actions to preempt adverse impacts of such shocks. Harper and Glew suggest the below-listed questions that are designed to inquire about operational decisions that affect the present and the future:

1. What business should we be in?
2. What will it take to make our vision a reality?
3. What competencies, resources, assets, skills, processes, technology, and facilities will it take for our business to succeed in the next 5 years?
4. How can we preempt our current and potential competitor's actions?
5. How can we anticipate our present and future customer's needs?
6. Do our assumptions about the future reflect emerging realities?
7. Are our predictions and assumptions based on objective analysis or just intuition?
8. How often do we need to update our predictions?
9. Are we monitoring the right trends?
10. Are we looking at the right indicators for what the future may hold?
11. Do we have an effective early-warning system in place to detect variances from expectations?
12. Do we have contingency plans in place to address mission-critical opportunities and threats?
13. What can we do that will change the way the game is played so that we make our competition irrelevant?

There is no doubt that many progressive organizations already have their own internal processes of raising questions similar to the above. The problem often centers around poor execution of responses to the questions. The STEP approach presented in this book can help organize questions and responses within the context of where they really belong within the scope of project management steps. The fast pace of developments in STE makes it imperative to get information fast and disseminate the information to the critical points of needs throughout an organization. Harper and Glew (2008) caution that "It's not what you know that matters. It's how soon you know it, what you plan to do about it and how quickly you deal with it that matters." Project managers who adopt this type of philosophy spend more time preempting problems and advancing their operations rather than fighting fires that erupt due to nonanticipatory inattention.

PROJECT OPR

In integrating the various aspects of a project, the performing organization must clarify the attributes that relate to overall expectations. Expectations are shaped by three key factors of optimism, pessimism, and realism (OPR) summarized as follows:

- Optimistic expectations that demonstrate aggressive pursuit of what the project can deliver
- Pessimistic expectations that maintain a conservative viewpoint of what the project can accomplish
- Realistic expectations that map prevailing project scenarios and work plan to the available data, resources, and personnel

For each case of project assessment, the performing team must be able to distinguish between the left-side story, the right-side story, and the true story behind the project.

PROJECT SUSTAINABILITY

Project efforts must be sustained in other for a project to achieve the intended end results in the long run. Project sustainability is not often addressed in project management but it is very essential particularly for STE type of projects.

Sustainability, in ordinary usage, refers to the capacity to maintain a certain process or state indefinitely. In day-to-day parlance, the concept of sustainability is applied more specifically to living organisms and systems, particularly environmental systems. As applied to the human community, sustainability has been expressed as meeting the needs of the present without compromising the ability of future generations to meet their own needs. The term has its roots in ecology as the ability of an ecosystem to maintain ecological processes, functions, biodiversity and productivity into the future. When applied to systems, sustainability brings out the conventional attributes of a system in terms of having the following capabilities:

- Self-regulation
- Self-adjustment
- Self-correction
- Self-recreation

To be sustainable, nature's resources must only be used at a rate at which they can be replenished naturally. Within the environmental science community, there is a strong belief that the world is progressing on an unsustainable path because the Earth's limited natural resources are being consumed more rapidly than they are being replaced by nature. Consequently, a collective human effort to keep human use of natural resources within the sustainable development aspect of the Earth's finite resource limits has become an issue of urgent importance. Unsustainable management of natural resources puts the Earth's future in jeopardy.

Sustainability has become a widespread, controversial, and complex issue that is applied in many different ways, including the following:

- Sustainability of ecological systems or biological organization (e.g., wetlands, prairies, forests)
- Sustainability of human organization (e.g., ecovillages, ecomunicipalities, sustainable cities)
- Sustainability of human activities and disciplines (e.g., sustainable agriculture, sustainable architecture, sustainable energy)
- Sustainability of projects (e.g., operations, resource allocation, cost control)

For project integration, the concept of sustainability can be applied to facilitate collaboration across project entities. The process of achieving continued improvement in operations, in a sustainable way, requires that engineers create new technologies that facilitate interdisciplinary thought exchanges. Under the STEP methodology of this book, sustainability means asking questions that relate to the consistency and long-term execution of the project plan. Essential questions that should be addressed include the following:

- Is the project plan supportable under current operating conditions?
- Will the estimated cost remain stable within some tolerance bounds?
- Are human resources skills able to keep up with the ever changing requirements of a complex project?
- Will the project team persevere toward the project goal through both rough and smooth times?
- Will interest and enthusiasm for the project be sustained beyond the initial euphoria?

REFERENCES

PMI, *A Guide to the Project Management Body of Knowledge (PMBOK Guide)*, 3rd ed., Project Management Institute, Newtown Square, PA, 2004.

Harper, S. C. and David J. G. Carpé Futurum! *Industrial Engineer*, August 2008, 34–38.

3 STEP Scope Management

Give me six hours to chop down a tree and I will spend the first four sharpening the axe.

–Abraham Lincoln

The Abraham Lincoln quote that opens this chapter illustrates the importance of putting a project in the proper context of what needs to be accomplished. That is, proper scoping of a project with the appropriate tools and resources constitute the fundamental requirement for project success. Scope management refers to the process of directing and controlling the entire scope of the project with respect to a specific goal. Establishment and clear definition of project goals and objectives form the foundation of scope management. The scope and plans form the baseline against which changes or deviations can be monitored and controlled. Proper scoping of a project right at the outset can help prevent problems later on. A project that is out of scope may not have the potential for a successful completion. Project scope management includes the processes required to ensure that the project includes all the work required, and only the work required, to complete the project successfully (PMI, 2004). This implies that scoping prunes a project so that extraneous work is not undertaken. Scope management is primarily concerned with defining and controlling what is and what is not included in the project. A Pareto analysis or ABC analysis can be used to determine and rank the essential elements that should be included in the project scope. Figure 3.1 shows a Pareto distribution of important and unimportant contents of a project scope.

SCOPE DEFINITIONS

It should be emphasized that project scope differs from product scope. Product scope describes the product to be delivered while project scope describes the work required to deliver the product. Product scope addresses the question of how a product works (How does the product work?), while project scope addresses the question of how a project performs (How did the project do?). Product scope is measured against product definition and specification. Project scope is measured against the project plan. Project scope is related to the project charter by the fact that the project charter specifies what is in the project scope. The span of project scope management is evident in the characteristic symbol that this book uses to denote this area of project management knowledge. The budget and schedule of a project are directly impacted by the scope of the project. Figure 3.2 shows the requirement to do a trade-off analysis with respect to the available budget and the desired project schedule.

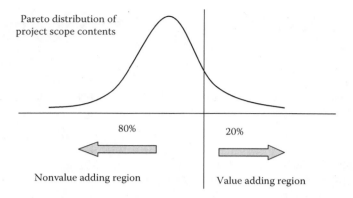

FIGURE 3.1 Pareto distribution of project scope contents.

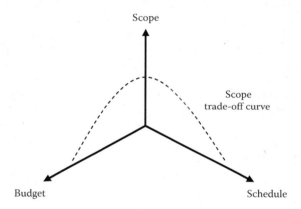

FIGURE 3.2 Scope trade-off curve for budget and schedule.

SCOPE MANAGEMENT: STEP-BY-STEP IMPLEMENTATION

The project scope component of the body of knowledge consists of the elements shown in the block diagram in Figure 3.3. The five elements in the block diagram are executed across the process groups introduced in Chapter 1. The overlay of the elements and the process groups are shown in Table 3.1. Thus, under the knowledge area of scope management, the required steps are

Step 1: Scope planning
Step 2: Scope definition
Step 3: Create work breakdown structure (WBS)
Step 4: Scope verification
Step 5: Scope control

STEP Scope Management

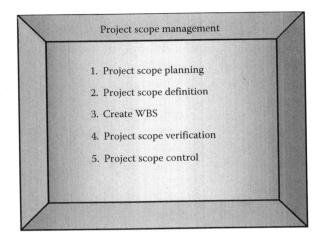

FIGURE 3.3 Block diagram of project scope management.

TABLE 3.1
Implementation of Project Scope Management across Process Groups

	Initiating	Planning	Executing	Monitoring and Controlling	Closing
Project Scope Management		Scope planning Scope definition Create WBS		Scope verification Scope control	

Each step is carried out in a structure of inputs–tools and techniques–output analysis. Table 3.2 shows the input-to-output items for project scope planning. The tabular format is useful for explicitly identifying what the project analyst needs to do or use for each step of the project management process. Tables 3.3 through 3.6 present the input-to-output entries for the other steps under project scope management.

Project scope management plan provides guidance on how project scope will be defined, documented, verified, managed, and controlled. The components of the scope management plan include the following:

- Process to prepare a detailed project scope statement
- Process to create, maintain, and approve WBS
- Process for specifying formal verification and acceptance of deliverables
- Process to control change requests to detailed project scope statement

The primary purpose of project scope definition is to explore deeper details of stakeholder needs, wants, desires, and expectations for the purpose of creating project requirements. It requires an outline of additional constraints and assumptions for

TABLE 3.2
Tools and Techniques for Project Scope Planning within Scope Management

STEP 1: Project Scope Planning

Inputs	Tools and Techniques	Output(s)
Enterprise environmental factors	Expert judgment	Scope management plan
Organizational process assets	Templates, forms, standards	Other in-house outputs, reports,
Project charter	Other in-house (custom) tools	and data inferences of interest
Preliminary project scope statement	and techniques	to the organization
Project management plan		
Other in-house (custom) factors of relevance and interest		

TABLE 3.3
Tools and Techniques for Project Scope Definition within Scope Management

STEP 2: Project Scope Definition

Inputs	Tools and Techniques	Output(s)
Organizational process assets	Product analysis	Project scope statement
Project charter	Identification of alternatives	Requested change
Preliminary project scope statement	Expert judgment	Scope management plan (updated)
Project scope management plan	Stakeholder analysis	Other in-house outputs, reports,
Approved change requests	SIPOC	and data inferences of interest to
Other in-house (custom) factors of relevance and interest	DMAIC	the organization
	Other in-house (custom) tools and techniques	

TABLE 3.4
Tools and Techniques for Creating WBS within Scope Management

STEP 3: Create WBS

Inputs	Tools and Techniques	Output(s)
Organizational process assets	WBS templates	Scope statement (updated)
Scope statement	Decomposition	Work breakdown structure
Scope management plan	Work partitioning	WBS dictionary
Approved change requests	Other in-house (custom) tools and techniques	Scope baseline
Other in-house (custom) factors of relevance and interest		Scope management plan (updated)
		Requested changes
		Other in-house outputs, reports, and data inferences of interest to the organization

TABLE 3.5
Tools and Techniques for Project Scope Verification within Scope Management

STEP 4: Project Scope Verification

Inputs	Tools and Techniques	Output(s)
Scope statement	Inspection	Accepted deliverables
WBS dictionary	Other in-house (custom) tools and techniques	Requested changes
Scope management plan		Recommended corrective actions
Deliverables		Other in-house outputs, reports, and data inferences of interest to the organization
Other in-house (custom) factors of relevance and interest		

TABLE 3.6
Tools and Techniques for Project Scope Control within Scope Management

STEP 5: Project Scope Control

Inputs	Tools and Techniques	Output(s)
Scope statement	Change control system	Project scope statement (updates)
WBS	Variance analysis	WBS updates
WBS dictionary	Replanning	WBS dictionary updates
Scope management plan	Configuration management system	Scope baseline updates
Performance reports	Other in-house (custom) tools and techniques	Requested changes
Approved change requests		Recommended corrective action
Work performance information		Organizational process assets
Other in-house (custom) factors of relevance and interest		Project management plan updates
		Other in-house outputs, reports, and data inferences of interest to the organization

implementing the project. Product analysis develops a better understanding of the product of the project. Stakeholder analysis identifies influence and interests of stakeholders and documents their needs in order to create project requirements. In addition, project scope definition provides a documented basis for making future project decisions and it covers the following:

- Justification—Why is the project needed?
- Product description—What is the expected product of the project?
- Boundaries—What is included and not included in the project?
- Constraints and assumptions

- Deliverables—What are the project deliverables?
- Objectives—How will the success of the project be assessed?

WBS contains deliverables specified as nouns (not verbs), which are identifiable and tangible products of work. Work package in WBS shows the lowest level of WBS component that can be scheduled, cost estimated, monitored, and controlled. The descriptions of work components are compiled in the WBS dictionary. WBS subdivides major project deliverables or subdeliverables into smaller, more manageable components. Project deliverables should be defined in sufficient level of detail to facilitate the development of project activities with cost and duration estimates. If an item is not contained in the WBS, it is not going to be done. That is, it is outside the scope of the project. Every item in the WBS has a verifiable deliverable and it is assigned a unique identifier (code), has a statement of work, responsible organization, and a list of schedule milestones. The scope baseline for a project is composed of the approved detailed project scope statement, associated WBS, and WBS dictionary. Note that WBS is not a quality statement. It is a statement of what needs to be done, not how well it is done.

Scope verification consists of a formal acceptance of project scope. This is typically performed at the end of a phase or project. Stakeholder (i.e., sponsor, client, etc.) formally accepts the scope. The verification ties objectives to WBS. Formal acceptance of scope is documented even though acceptance may be conditional.

For scope control, we should identify changes using the WBS elements. We will evaluate the impact on cost, schedule, resources, and product quality. Stakeholder and project management authorize change. Successful scope management requires documentation. All aspects of the project must be documented in writing. Several industry-based tools and techniques can be used for the purpose of scope control (Badiru, 1993).

USE OF DMAIC FOR SCOPE MANAGEMENT

Figure 3.4 shows an application of Define, Measure, Analyze, Improve, Control (DMAIC) concept to scope management process. DMAIC is a basic component of the Six Sigma methodology for improving work processes by eliminating defects and it complements the Lean Approach, which focuses on eliminating waste in work processes. The Six Sigma methodology is widely used in industry and it is quickly finding a place in service and project enterprises. Six Sigma represents a set of practices that improve efficiency and eliminate defects in products and services. Applying DMAIC to project scoping can ensure that a project covers all the elements defined in the scope statement and only the elements defined in the scope statement.

The Define stage of DMAIC puts the project in the context of a specific business case. It is the first stage in the DMAIC process. In this stage, it is important to define specific goals in achieving outcomes that are consistent with both customer demands as well as the project organization's own business strategy. Define lays down a road map for project accomplishment. Definition may cover several items, but particular

STEP Scope Management

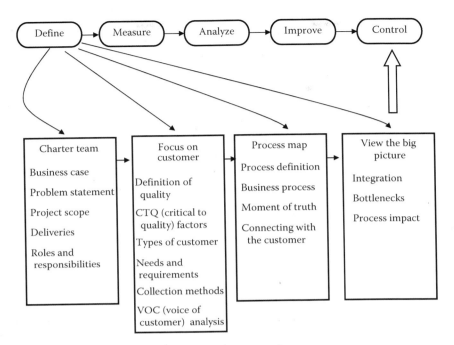

FIGURE 3.4 Application of DMAIC to process control.

elements include team charter, focus on the customer, process map, and systems view of the big picture. Team charter covers business case, problem statement, project scope, deliveries, roles and responsibilities. Also in the define stage we focus on the customer or external constituents of the project. Items addressed in this respect include the definition of quality, critical to quality (CTQ) factors, types of customer, needs and requirements, data gathering methods, and voice of customer (VOC) analysis. The process map portion of Define maps out the steps and elements to be covered during a particular process. The items covered in this stage include process definition, business process outline, confronting the facts, and connecting with the customer. The major benefit of developing a process map is that it highlights the important tasks and functions to be undertaken during the process so that nothing is inadvertently ignored. Viewing the big picture during the Define stage of DMAIC implies using a systems view to analyze how each effort fits into the overall scheme of things.

The Measure stage of DMAIC lays the ground work for measurement of the metrics of project performance. In order to determine whether or not defects have been reduced in a project's output, we need a base measurement. In this stage, accurate measurements must be made and relevant data must be collected and analyzed so that future comparisons can be measured to determine whether or not defects have been reduced. Procedures for measurement are particularly essential for control further down in a project. We must be able to measure an item before we can control or improve it.

The Analyze stage of DMAIC is very important to determine the relationships and the factors of causality in a project process. If the focus of a project is to generate products, services, or results, then we must understand what causes what and how the relationships can be enhanced.

The Improve stage of DMAIC outlines how to plan, pursue, and achieve improvement in the project process. Making improvements or optimizing processes inherent in a project, based on measurements and analysis, will ensure that defects are lowered and work processes are streamlined.

The Control stage is the last step in DMAIC methodology. Control ensures that any variances stand out and are corrected before they can adversely influence a process, thereby causing defects. Controls can be in the form of pilot runs to determine if the processes are capable and then once data are collected, a process can transition into standard work process. Continued measurement and analysis must be undertaken to keep project work processes on track and free of defects below the Six Sigma quality limit.

USE OF SIPOC DIAGRAM FOR SCOPE MANAGEMENT

Suppliers, Inputs, Process, Outputs, Customers (SIPOC) is a diagram that is used to identify all project elements relevant for improvement before the project starts. The process improvement team may also add requirements at the end of the SIPOC diagram to identify the specific customer requirements that are to be satisfied. This helps to obtain clarifications of what, who, where, when, why, and how of improvement efforts. Figure 3.5 shows a flow diagram for SIPOC. The steps for building a SIPOC diagram are summarized below:

1. Select an area that is accessible or visible to the improvement team so that team members can post additions to the SIPOC diagram iteratively. This could be a presentation template projected onto a screen, flip charts with headings (S-I-P-O-C), or Post-it notes mounted onto a wall. Iteratively add items under each heading and proceed through four to five high-level iterations.
2. Identify the suppliers for the project.

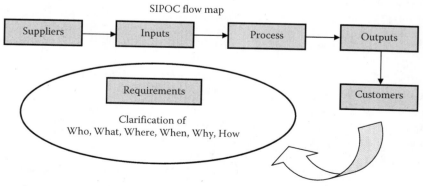

FIGURE 3.5 SIPOC flow diagram for process improvement.

STEP Scope Management

3. Identify and annotate the inputs required from the suppliers.
4. Identify and map the process into which the inputs go. The inputs from the preceding step are required for the process to operate properly.
5. Identify and document the outputs of the process.
6. Identify the customers to whom the outputs are directed.
7. If desired, include the preliminary requirements of the customers for the purpose of clarifying who, what, when, where, why, and how aspects of the project. These clarifications will be verified during a later step of the Six Sigma measurement phase.
8. Discuss the SIPOC contents with the project sponsor, project champion, project management, and stakeholders of the project for verification and validation.

SIPOC helps to define a complex project to ensure that the project is in alignment with the scope statement. SIPOC is often applied at the Measure stage of the DMAIC methodology within the overall Six Sigma effort. SIPOC complements and provides additional details for the usual process mapping and input–output scoping processes of project management. Table 3.7 shows examples of entries for a SIPOC diagram example for technology dealership problem. Such a table can be expanded or customized for specific problems of interest. The SIPOC diagram is particularly useful for project scope verification by addressing the following questions:

- Who supplies inputs to the process?
- What specifications are placed on the inputs?
- Where will functions and operations be performed?
- Who are the true customers of the process?

TABLE 3.7
Entries for SIPOC Diagram Example for Technology Vendor

Suppliers	Inputs	Process	Outputs	Customers	Requirements
Manufacturer	Deliveries	Client consultation	New client account	Technology users	Build to order
Suppliers	Option packages	Assessment of client needs	Purchase order	Technology showroom owners	Operating options
Repair service		Present options to clients	Paperwork to dealer	Service departments	Service contract
Outsource sites		Demo	Paperwork to manufacturer		Installation arrangement
		Client agreement	Payment process		
			Service contract		
			Service notifications		

- What are the requirements of the customers?
- When are the requirements going to be satisfied?
- How will quality performance be ensured?

In a real-life project scoping environment, the scope can be quite volatile particularly for STEPs where dynamic changes can be frequent and profound. Scopes are often enacted, repealed, reenacted, modified, extended, altered, and adjusted for various reasons during the project life cycle. This is where the structured approach of scope planning, scope definition, WBS creation, scope verification, and scope control is very useful for science, technology, and engineering projects.

SCOPE FEASIBILITY ANALYSIS

The scope feasibility of a project should be ascertained in terms of technical factors, managerial potential, economic factors, environmental issues, political expediency, financial analysis, economic realities, and community impact. Scope feasibility can be documented with a report showing all the ramifications of the project. This is particularly essential for multifaceted science, technology, and engineering projects. Technical feasibility refers to the ability of the process to take advantage of the current state of the technology in pursuing further improvement. The technical capability of the personnel as well as the capability of the available technology should be considered. Managerial feasibility involves the capability of the management infrastructure of an organization to achieve and sustain the desired end result. Management support, employee involvement, and commitment are key elements required to ascertain managerial feasibility.

Economic feasibility involves the ability of the proposed project to generate economic benefits. A benefit–cost analysis and a breakeven analysis are important aspects of evaluating the economic feasibility of a STEP from a scoping perspective. The tangible and intangible aspects of the project should be translated into economic terms to facilitate a consistent basis for assessment.

Financial feasibility should be distinguished from economic feasibility. Financial feasibility involves the capability of the project organization to raise the appropriate funds needed to implement the proposed project and maintain it throughout its life cycle. Project financing can be a major obstacle in STEPs because of the level of capital required and the volatility of science and technology. Loan availability, credit worthiness, equity, and loan schedule are important aspects of financial feasibility analysis for project scoping purposes.

Cultural feasibility deals with the compatibility of the proposed project with the cultural atmosphere both from social culture as well as the work environment. In working class communities, STEP functions must be integrated with the local cultural practices and beliefs. For example, an industry that requires the services of females must take into consideration the cultural norms affecting the position of women in the work place in some countries. In rural areas, technology development efforts must not violate culturally sacred grounds that have religious or historical implications.

Social feasibility addresses the influences that a proposed project may have on the social system in the project environment. The ambient social structure may be such that certain categories of workers may be in short supply or nonexistent. The effect of the project on the social status of the project participants must be assessed to ensure compatibility. It should be recognized that workers in certain industries may have certain status symbols within the society.

Community feasibility refers to the general acceptance of the proposed project. Even in cases where a STEP is expected to lead to economic development of a community, there may still be discontent and opposition from local residents, particularly where the "eminent domain" doctrine is exercised by the government. Recent examples include the "bridge to no where" project in Alaska. Apparently, some aspects of the scope of the bridge project were not adequately vetted during the decision process. Another good example is the "Pipeline of Discontent" also called "Line of Conflict" in the Ohio Press. This involves the construction of a $5.6 billion, 1679-mile Rock Express Pipeline (or REX) from Colorado's rocky mountain region through the midwest on to the eastern part of the United States. The pipeline is presented by the Federal Government as a crucial new component of America's energy infrastructure and a boon for consumers. But communities along the path of the pipeline vigorously oppose it. Contentious issues include eminent domain, safety, and environmental impact, in spite of the economic potential that the project would bring.

Safety feasibility is another important aspect that should be considered in STEP planning. Safety feasibility refers to an analysis of whether the project is capable of being implemented, operated, and sustained safely with minimal adverse effects on the environment and safety. Unfortunately, environmental impact assessment is often not adequately and deeply addressed in STEP development projects, often because of the pressure to get a project done before the technology changes.

Politically feasible project may be referred to as a "politically correct project." Political considerations often dictate the direction for a proposed project. This is particularly true for STEP development projects that may have significant government inputs and political implications. For example, political necessity may be a source of support for a project regardless of the project's merits. On the other hand, worthy projects may face insurmountable opposition simply because of political factors. Political feasibility analysis requires an evaluation of the compatibility of project goals with the prevailing goals of the political system.

DIMENSIONS OF SCOPE FEASIBILITY

In general terms, the elements of a feasibility analysis for a STEP should cover the following items:

Need analysis: This indicates the recognition of a need for the project. The need may affect the organization itself, another organization, the public, or the government. A preliminary study should be conducted to confirm and evaluate the need. A proposal of how the need may be satisfied is then developed. Pertinent questions that should be asked include

Is the need significant enough to justify the proposed project?
Will the need still exist by the time the project is completed?
What are the alternate means of satisfying the need?
What is the economic impact of the need?

Process work: This is the preliminary analysis done to determine what will be required to satisfy the need. The work may be performed by a consultant who is a subject matter expert in the project field. The preliminary study often involves system models or prototypes. For STEPs, artist's conception and scaled down models may be used for illustrating the general characteristics of a process.

Engineering and design: This involves a detailed technical study of the proposed project. Written quotations are obtained from suppliers and subcontractors as needed. Technology capabilities are evaluated as needed. Product design, if needed, should be done at this stage.

Cost estimate: This involves estimating project cost to an acceptable level of accuracy. Levels of around -5% to $+15\%$ are common at this level of a project plan. Both the initial and operating costs are included in the cost estimation. Estimates of capital investment, recurring, and nonrecurring costs should also be contained in the cost estimate document.

Financial analysis: This involves an analysis of the cash flow profile of the project. The analysis should consider recapitalization requirements, return on investment, inflation, sources of capital, payback periods, breakeven point, residual values, market volatility, and sensitivity. This is a critical analysis since it determines whether or not and when funds will be available to the project. The project cash flow profile helps to support the economic and financial feasibility of the project.

Project impacts: This portion of scope feasibility analysis provides an assessment of the impact of the proposed project. Environmental, social, cultural, and economic impacts may be some of the factors that will determine how a STEP is perceived by stakeholders. The value-added potential of the project should also be assessed. A value-added tax may be assessed based on the price of a product and the cost of the raw material used in making the product. The tax so collected may be viewed as a contribution to government coffers for reinvestment in the science, technology, and engineering infrastructure of the nation.

Conclusions and recommendations: Scope feasibility analysis should end with the overall outcome of the project analysis. This may indicate an endorsement or disapproval of the project. If disapproved, potential remedies to make it right should be presented. Recommendations on what should be done should be included in this section of the scope feasibility report.

INDUSTRY CONVERSION FOR PROJECT SCOPING

The conversion of an existing industry to a new industry may be a possible approach to satisfy STEP development plans. Industries that are no longer meeting the needs of the society due to economic, social, cultural, or military (defense) requirements may have suitable alternate roles to play in new STEP efforts. For example, economic and military reforms can be leveraged to develop opportunities for private industry in what otherwise would be military-oriented production facilities. Recent changes in the international security environment have prompted several nations to start to investigate how military technology may be converted for industrial purposes. New STEPs should consider the possibility of industry conversion to achieve project goals. Recapitalization of old industry can be channeled toward the development of contemporary science, technology, and engineering industries.

ASSESSMENT OF LOCAL RESOURCES AND WORK FORCE

The feasibility of a STEP should consider an assessment of the resources available locally to support the proposed production operations. Most nations and communities are blessed with abundant natural resources. The sad fact is that these resources are often underdeveloped and underutilized. When the resources are fully developed, it is often through exploitation by external organizations. For example, many countries that enjoyed oil discovery and boom in the 1970s now face economic uncertainty due to problems in that sector of the economy. Lack of diversification to utilize local resources and local work force can spell doom when large-scale STEPs are undertaken.

Blame is often placed on the lack of technical expertise when expressing the chagrin of failed STEPs. But when the expertise is available, it is frequently underused, misapplied, or misappropriated. It is true that local experts working individually accomplish nothing in the overwhelming bureaucracy that engulfs their expertise. In order for local experts to have an impact, there must be a coalition. A strong professional coalition is the only means of bringing about a meaningful change. Task forces should be set up to document the availability of local resources and their respective potentials for generate derivative products. The availability of skilled local workforce should also be factored into project scope feasibility analysis. The products should be prioritized based on pressing needs of the society and global market realities.

DEVELOPING SCOPE-BASED PROJECT PROPOSAL

Once a project is shown to be feasible along most of the appropriate dimensions of its operation, the next step is to issue a request for proposal (RFP) depending on the funding sources involved. Proposals are classified as either solicited or unsolicited. Solicited proposals are those written in response to a request for a proposal while unsolicited ones are those written without a formal invitation from the funding source. Many companies prepare proposals in response to inquiries received from potential clients. Many proposals are written under competitive bids. If an RFP is issued, it should include statements about project scope, funding level, deliverables, performance criteria, and deadlines.

The purpose of the RFP is to identify companies that are qualified to successfully conduct the project in a cost-effective manner. But cost should not be the only basis for selecting the winning bid. Formal RFPs are sometimes issued to only a selected list of bidders who have been preliminarily evaluated as being qualified. These may be referred to as targeted RFPs. In some cases, general or open RFPs are issued and whoever is interested may bid for the project. This, however, has been found to be inefficient in some respect. Ambitious, but unqualified, organizations waste valuable time preparing losing proposals. The proposal recipient, on the other hand, spends much time reviewing and rejecting unqualified proposals. Open proposals do have proponents who praise their "equal opportunity" approach.

In practice, each organization has its own RFP format, content, and procedures. The request is called by different names including procurement invitation (PI), procurement request (PR), request for bid (RFB), or invitation for bids (IFB). In some countries, it is sometimes referred to as request for tender (RFT). Irrespective of the format used, an RFP should request information on bidder's costs, technical capability, management, and other characteristics. It should, in turn, furnish sufficient scope information on the expected work. A typical detailed RFP should include the following:

- Project background: Need, scope, preliminary studies, and results.
- Project deliverables and deadlines: Product, service, or results that are expected from the project, when the products are expected, and how the products will be delivered should be contained in this document.
- Project performance specifications: Sometimes, it may be more advisable to specify system requirements rather than rigid specifications. This gives the systems or project analysts the flexibility to utilize the most updated and most cost-effective technology in meeting the requirements. If rigid specifications are given, what is specified is what will be provided regardless of cost and level of efficiency.
- Funding level: This is sometimes not specified because of nondisclosure policies or because of budget uncertainties. However, whenever possible, the funding level should be indicated in the requirements. This will help responders to map their input of resources to the expected pay out for the effort.
- Reporting requirements: Project reviews, format, number and frequency of written reports, oral communication, financial disclosure, and other requirements should be specified.
- Contract administration: Guidelines for data management, proprietary work, intellectual property rights, progress monitoring, proposal evaluation procedure, requirements for inventions, trade secrets, copyrights, and so on should be included in the RFP.
- Special requirements (as applicable): Facility access restrictions, equal opportunity/affirmative actions, small business support, access facilities for the handicap, false statement penalties, cost sharing, compliance with government regulations, and so on should be included if applicable.
- Boilerplates (as applicable): These are special requirements that specify the specific ways certain project items are handled. Boilerplates are usually written based on organizational policy and are not normally subject to conditional changes. For example, an organization may have a policy that requires that no more than 50% of a contract award will be paid prior to the completion of the contract. Boilerplates are quite common in

government-related projects. Thus, STEPs may need boilerplates dealing with environmental impacts, social contribution, and financial requirements. These are issues that may not normally be considered within the science and technology hustle and bustle to get the job done.

PROPOSAL PREPARATION SCOPE

Whether responding to an RFP or preparing an unsolicited proposal, care must be taken to provide enough detail to permit an accurate assessment of a project proposal. The proposing organization will need to find out the following:

- Project time frame
- Level of competition
- Agency's available budget
- Structure of the funding agency
- Point of contact (POC) within the agency
- Previous contracts awarded by the agency
- Exact procedures used in awarding contracts
- Nature of the work done by the funding agency

The project proposal should present a detailed plan for executing the proposed project. The proposal may be directed to a management team within the same organization or to an external organization. The proposal contents may be written in two parts: a technical section and a management section.

TECHNICAL SECTION OF PROJECT PROPOSAL

Project background
- Organization's expertise in the project area
- Project scope
- Primary objectives
- Secondary objectives

Technical approach
- Required technology
- Available technology
- Problems and their resolutions
- WBS

Work statement
- Task definitions and list
- Expectations

Schedule
- Gantt charts
- Milestones
- Deadlines

Project deliverables

The value of the project
- Significance
- Benefit
- Impact

MANAGEMENT SECTION OF PROJECT PROPOSAL

Project staff and experience
- Personnel credentials

Organization
- Task assignment
- Project manager, liaison, assistants, consultants, etc.

Cost analysis
- Personnel cost
- Equipment and materials
- Computing cost
- Travel
- Documentation preparation
- Cost sharing
- Facilities cost

Delivery dates
- Specified deliverables

Quality control measures
- Rework policy

Progress and performance monitoring
- Productivity measurement

Cost control measures

SCOPE BUDGET PLANNING

Scoping can be an expression of budgeting. Scope determines the extent of budgeting just as budgeting determines the extent of project scope. The budgeting approach employed for a project can be used to express the overall organizational policy and commitment. Budget often specifies the following:

- Performance measures
- Incentives for efficiency
- Project selection criteria
- Expressions of organizational policy
- Plans for how resources are to be expended

STEP Scope Management

- Catalyst for productivity improvement
- Control basis for managers and administrators
- Standardization of operations within a given horizon

The preliminary effort in the preparation of a budget is the collection and proper organization of relevant data. The preparation of a budget for a project is more difficult than the preparation of budgets for regular and permanent organizational endeavors, however. While recurring endeavors usually generate historical data that serve as inputs to subsequent estimating functions, projects, on the other hand, are often one-time undertakings without the benefit of prior data. The input data for the budgeting process may include inflationary trends, cost of capital, standard cost guides, past records, and forecast projections. Budgeting may be done as top-down or bottom-up.

SCOPING TOP-DOWN

This involves collecting data from upper-level sources such as top and middle managers. The cost estimates supplied by the managers may come from their judgments, past experiences, or past data on similar project activities. The cost estimates are passed to lower-level managers, who then break the estimates down into specific work components within the project. These estimates may, in turn, be given to line managers, supervisors, and so on to continue the process. At the end, individual activity costs are developed. The top management presents the overall budget while the line worker generates specific activity budget requirements. One advantage of the top-down budgeting approach is that individual work elements need not be identified prior to approving the overall project budget. Another advantage of the approach is that the aggregate or overall project budget can be reasonably accurate even though specific activity costs may contain substantial errors.

SCOPING BOTTOM-UP

In bottom-up budgeting, elemental activities, their schedules, descriptions, and labor skill requirements are used to construct detailed budget requests. The line workers who are actually performing the activities are asked to supply cost estimates. Estimates are made for each activity in terms of labor time, materials, and machine time. The estimates are then converted to monetary values. The estimates are combined into composite budgets at each successive level up the budgeting hierarchy. If estimate discrepancies develop, they can be resolved through the intervention to senior management, junior management, functional managers, project managers, accountants, or financial consultants. Analytical tools such as learning-curve analysis, work sampling, and statistical estimation may be used in the budgeting process as appropriate to improve the quality of cost estimates. All component costs and departmental budgets are combined into an overall budget and sent to top management for approval. A common problem with bottom-up budgeting is that individuals tend to overstate their needs with the notion that top management may cut the budget by some percentage. It should be noted, however, that sending erroneous and misleading estimates will only lead to a loss of credibility.

Zero-Base Scoping

This is another budgeting approach that bases the level of project funding on previous performance. It is normally applicable to recurring programs, especially those in the public sector. Accomplishments in past funding cycles are weighed against the level of resource expenditure. Programs that are stagnant in terms of their accomplishments relative to budget size do not receive additional budgets. Programs that have suffered decreasing yields are subjected to budget cuts or even elimination. By contrast, programs that have a record of accomplishments are rewarded with larger budgets. A major problem with zero-base budgeting is that it puts participants under tremendous pressure to perform data collection, organization, and program justification. So much time may be spent documenting program accomplishments that productivity improvements on current projects may be compromised. Proponents of zero-base budgeting see it as a good approach of encouraging managers and administrators to be more conscious of their management responsibilities. From a project control perspective, the zero-base budgeting approach may be useful in identifying and eliminating specific activities that have not contributed to project goals in the past.

PROJECT SCOPING WITH WBS

As presented in Chapter 1, project WBS refers to the itemization of a project for planning, scheduling, and control purposes. It essentially communicates the scope of the project by presenting the inherent components of a project in a structured block diagram or interrelationship flowchart. WBS shows the hierarchies of parts (phases, segments, milestone, etc.) of the project. The purpose of constructing a WBS is to analyze the elemental components of the project in detail. If a project is properly designed through the application of WBS at the project planning stage, it becomes easier to estimate cost and time requirements of the project. Project control is also enhanced by the ability to identify how components of the project link together within the scope of the project.

Project Scope Selection Criteria

Project selection is an essential first step in scoping the efforts of an organization. Figure 3.6 presents a simple graphical evaluation of project selection. The vertical axis represents the value-added basis of the project under consideration while the horizontal axis represents the level of complexity associated with the project. In this example, value can range from low to high while complexity can range from easy to difficult. The figure shows four quadrants containing regions of high value with high complexity, low value with high complexity, high value with low complexity, and low value with low complexity. A fuzzy region is identified with an overlay circle. The organization must evaluate each project on the basis of overall organization value streams. The figure can be modified to represent other factors of interest to an organization instead of value-added and project complexity.

STEP Scope Management

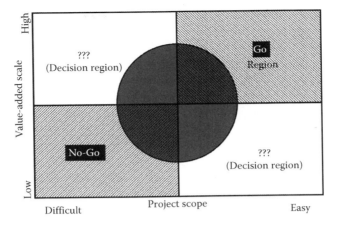

FIGURE 3.6 Project scope selection quadrant: Go or No-Go regions.

CRITERIA FOR PROJECT REVIEW

Some of the specific criteria that may be included in project review and selection are presented below:

- Cost reduction
- Customer satisfaction
- Process improvement
- Revenue growth
- Operational responsiveness
- Resource utilization
- Project duration
- Execution complexity
- Cross-functional efficiency
- Partnering potential

HIERARCHY OF SELECTION

In addition to evaluating an overall project, elements making up the project may need to be evaluated on the basis of the hierarchy presented below. This will facilitate achieving an integrated project management view of the organization's operations.

- System
- Program
- Task
- Work packages
- Activity

Sizing of Projects

Associating a size measure to an industrial project provides a means of determining level of relevance and efforts required. A simple guideline is presented below:

- Major (over 60 man-months of effort)
- Intermediate (6–60 man-months)
- Minor (Less than 6 man-months)

Planning Levels

When selecting projects and its associated work packages, planning should be done in an integrative and hierarchical manner following the levels of planning presented below:

- Supra level
- Macro level
- Micro level

Hammersmith's Project Alert Scale: Red, Yellow, Green Convention

Hammersmith (2006) presented a guideline for alert scale for project tracking and evaluation. He suggested putting projects into categories of RED, YELLOW, or GREEN with the definitions below:

> RED (if not corrected, project will be late and/or over budget)
>
> YELLOW (project is at risk of turning RED)
>
> GREEN (project is on time and on budget)

Product Assurance Concept for Industrial Projects

Product assurance activities will provide the product deliverables throughout a program development period. These specific activities for continuous effort are to

1. Track and incorporate specific technologies: The technology management task will track pertinent technologies through various means (e.g., vendor surveys and literature search). More importantly, the task will determine strategies to incorporate specific technologies.
2. Analyze technology trend and conduct long-range planning: The output of technology assessment should be used to formulate long-range policies, directions, and research activities so as to promote the longevity and evolution.
3. Encourage government and industry leaders' participation: In order to determine long-term strategy, the technical evaluation task needs to work closely with government and industry leaders so as to understand their

STEP Scope Management

long-range plans. Technology panels may be formed to encourage participation from these leaders.
4. Influence industry directions: Like other developing programs, the program management will have the opportunity to influence industry direction and spawn new technologies. Since the effort can be treated as a model, many technologies and products developed can be applied to other similar systems.
5. Conduct prototyping work: Prototyping will be used to evaluate the suitability, feasibility, and cost of incorporating a particular technology. In essence, it provides a less costly mechanism to test a technology before significant investment is spent in the product development process. Technologies that have high risk with high payoffs should be chosen as the primary subjects for prototyping.

All the foregoing discussions can be summarized into the response surface presented in Figure 3.7, whereby the multidimensional factors influencing a STEP are incorporated into the overall scope management process. The success of a project scope is dependent on time, resource availability, and all the other inherent issues pointed out in this chapter. Notice that the response surface is inclined. This indicates that there is a directional development given sufficient time and resources

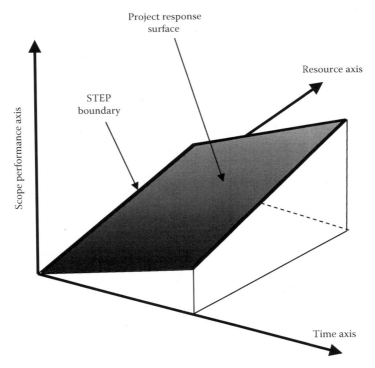

FIGURE 3.7 Multidimensional project scope response surface.

to execute the project. In actual practice, the scope response surface will not be smooth. There will be peaks and valleys that represent angles of compromise as the project moves along within its specified science, technology, and engineering boundaries.

REFERENCES

Badiru, A. B. *Managing Industrial Development Projects: A Project Management Approach*, Van Nostrand Reinhold, New York, 1993.

PMI, *A Guide to the Project Management Body of Knowledge (PMBOK Guide)*, 3rd ed., Project Management Institute, Newtown Square, PA, 2004.

Hammersmith, A. G. Implementing a PMO—The diplomatic pit bull, Workshop Presentation at East Tennessee PMI Chapter Meeting, January 10, 2006.

4 STEP Time Management

Intellectuals solve problems; geniuses prevent them.

–Albert Einstein

Chapters 5 through 7 address Time, cost, and quality. These are often treated as an integrated entity because of the need to perform time–cost–quality trade-offs. As mentioned by Albert Einstein's quote at the beginning of this chapter, problem prevention is a better strategy than the solution approach. This is applicable to prevent and control strategies for time, cost, and quality challenges faced in STEPs. Figure 4.1 shows an integrated view of time, cost, and quality considerations while Figure 4.2 shows the trade-off axes of the three factors, often referred to as the triple constraints. Integrated view is a theme that is seen frequently in many project management related literatures (Collin, 2001; Martin, 2007; Niven, 2002; PMI, 2004). Time management involves the effective and efficient use of time to facilitate the execution of a project expeditiously. Time, in terms of project schedule, is often the most noticeable aspect of a project. Consequently, time management is of utmost importance in project management. This is even more critical for STEPs, which are subject to rapid changes in technology. The first step of good time management is to develop a project plan that represents the process and techniques needed to execute the project satisfactorily. The effectiveness of time management is reflected in schedule performance analysis. Hence, scheduling is a major focus in project management. Many people erroneously view schedule management as project management. But, in fact, schedule management is just one aspect of project management.

TIME MANAGEMENT: STEP-BY-STEP IMPLEMENTATION

The time management component of the Project Management Body of Knowledge consists of the elements shown in the block diagram in Figure 4.3. The six elements in the block diagram are carried out across the process groups presented in Chapter 1. The overlay of the elements and the process groups are shown in Table 4.1. Thus, under the knowledge area of time management, the required steps are

Step 1: Activity definition

Step 2: Activity sequencing

Step 3: Activity resource estimating

Step 4: Activity time estimating

Step 5: Project schedule development

Step 6: Project schedule control

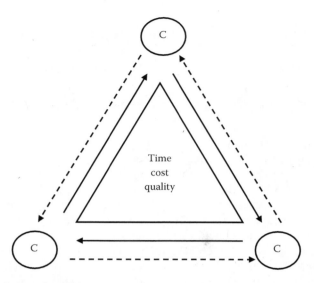

FIGURE 4.1 Integration of time, cost, and quality considerations.

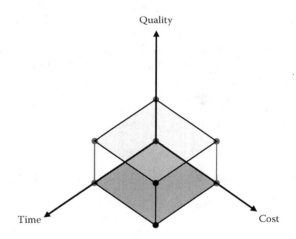

FIGURE 4.2 Time–cost–quality trade-off axes.

Project time management is defined as the set of processes required to accomplish timely completion of the project. As with other steps of project management, time management processes overlap and interact with other cross-functional processes within the knowledge areas. Time management is preceded by the development of project management plan, which is an output of project integration management.

Each step of project time management is carried out in a structure of inputs–tools and techniques–output analysis. Table 4.2 shows the input-to-output items for activity definition. The tabular format is useful for explicitly identifying what the project analyst needs to do or use for each step of the project management process.

STEP Time Management

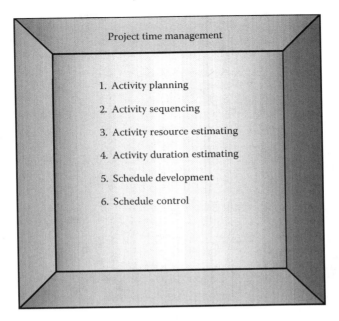

FIGURE 4.3 Block diagram of project time management.

TABLE 4.1
Implementation of Project Time Management across Process Groups

	Initiating	Planning	Executing	Monitoring and Controlling	Closing
Project Time Management		Activity definition		Project schedule control	
		Activity sequencing			
		Activity resource estimating			
		Activity duration estimating			
		Project schedule development			

Tables 4.3 through 4.7 present the input-to-output entries for the other steps under project time management. Activity involves identifying the specific activities that need to be performed to produce the various project deliverables. Under tools and techniques, decomposition defines the final outputs as schedule activities versus deliverables found in the work breakdown structure (WBS). WBS elements are nouns that identify deliverables while schedule activities are verbs indicating actions to be performed to accomplish work elements. Activity definition process identifies deliverables at the lowest level in the WBS. These are called work packages. Activity definition takes the work packages and subdivides or decomposes them into smaller

TABLE 4.2
Tools and Techniques for Activity Definition within Time Management

STEP 1: Activity Definition

Inputs	Tools and Techniques	Output(s)
Enterprise environmental factors	Project decomposition	Activity list
Organizational process assets	Templates, forms, standards	Activity attributes
Project scope statement	Expert judgment	Milestone list
Work breakdown structure (WBS)	Planning component	Requested changes
WBS dictionary	Rolling wave planning	Other in-house outputs,
Project management plan	Other in-house (custom) tools	reports, and data inferences
Other in-house (custom) factors of relevance and interest	and techniques	of interest to the organization

TABLE 4.3
Tools and Techniques for Activity Sequencing within Time Management

STEP 2: Activity Sequencing

Inputs	Tools and Techniques	Output(s)
Project scope statement	Precedence diagramming method (PDM)	Project schedule network diagram
Activity list	Arrow diagramming method (ADM)	Activity list updates
Activity attributes	Schedule network templates	Activity attributes updates
Milestone list	Dependency determination	Requested changes
Approved change requests	Applying leads and lags	Other in-house outputs,
Other in-house (custom) factors of relevance and interest	Process control charts	reports, and data
	Critical chain	inferences of interest to
	Other in-house (custom) tools and techniques	the organization

TABLE 4.4
Tools and Techniques for Activity Resource Estimating within Time Management

STEP 3: Activity Resource Estimating

Inputs	Tools and Techniques	Output(s)
Enterprise environmental factors	Expert judgment	Activity resource requirements
Organizational process assets	Analysis of alternatives	Activity attributes updates
Activity list	Project management software	Resource breakdown structure
Activity attributes	Bottom-up estimating	Resource calendar updates
Resource availability	Goal programming	Required changes
Project management plan	Portfolio management	Other in-house outputs, reports,
Other in-house (custom) factors of relevance and interest	Balanced scorecard	and data inferences of interest
	Other in-house (custom) tools and techniques	to the organization

TABLE 4.5
Tools and Techniques for Activity Duration Estimating within Time Management

STEP 4: Activity Duration Estimating

Inputs	Tools and Techniques	Output(s)
Enterprise environmental factors	Expert judgment	Activity duration estimates
Organizational process assets	Analogous estimating	Activity attribute updates
Project scope statement	Parametric estimating	Other in-house outputs, reports, and data inferences of interest to the organization
Activity list	Three-point estimates	
Activity attributes	Reserve analysis	
Activity resource requirements	Process control charts	
Resource calendar	Goal programming	
Project management plan	Other in-house (custom) tools and techniques	
Other in-house (custom) factors of relevance and interest		

TABLE 4.6
Tools and Techniques for Project Schedule Development within Time Management

STEP 5: Project Schedule Development

Inputs	Tools and Techniques	Output(s)
Organizational process assets	Schedule network analysis	Project schedule
Project scope statement	Critical path method (CPM)	Schedule model data
Activity list	Schedule compression	Schedule baseline
Activity attributes	What-if scenario analysis	Resource requirement updates
Project schedule network diagrams	Resource leveling	Activity attributes updates
Activity resource requirements	Critical chain method	Project calendar updates
Resource calendar	Project management software	Requested changes
Activity duration estimates	Calendar coordination	Project management plan updates
Project management plan risk register	Adjusting leads and lags	Other in-house outputs, reports, and data inferences of interest to the organization
Other in-house (custom) factors of relevance and interest	Schedule model	
	Critical chain	
	Process control charts	
	Other in-house (custom) tools and techniques	

components called schedule activities, which provide the basis for scheduling, executing, monitoring, and controlling during the project life cycle.

As presented in earlier chapters, enterprise environmental factors include existing organizational culture, systems, database repository, infrastructure, standards, and organization structure. Organizational process assets include standard processes, policies, guidelines, communication requirements, financial controls, existing change controls, and risk control. Rolling wave planning is a form of progressive

TABLE 4.7
Tools and Techniques for Project Schedule Control within Time Management

STEP 6: Project Schedule Control

Inputs	Tools and Techniques	Output(s)
Schedule management plan	Progress reporting	Schedule model data updates
Schedule baseline	Schedule change control system	Schedule baseline updates
Performance reports	Performance measurement	Performance measurements
Approved change requests	Project management software	Requested changes
Other in-house (custom) factors of relevance and interest	Variance analysis	Recommended corrective actions
	Schedule comparison bar chart	Organizational process assets updates
	Critical chain	Activity list updates
	Other in-house (custom) tools and techniques	Activity attributes updates
		Project management plan updates
		Other in-house outputs, reports, and data inferences of interest to the organization

elaboration of work. In this case, near-term work is planned in detail while far-term work in the future is planned for at a relatively high (or broad) level. Milestone lists in the project network can be mandatory or optional. Activity sequencing involves identifying and documenting logical relationships among schedule activities. Logical sequencing should highlight precedence relationship and appropriate leads and lags. The three basic types of precedence relationship are

1. Technical precedence requirement
2. Procedural precedence requirement
3. Imposed precedence requirement

Of the three types of precedence constraints, technical precedence is the most difficult to circumvent. The procedural precedence requirement can, in many cases, be relaxed due to prevailing workflow flexibility. The imposed relationship is often due to resource-shortage impositions. Thus, if we can change our workflow concepts and exercise resource allocation options, we may be able to achieve project schedule improvements.

A lead is the amount of time by which the start of an activity leads (or overlaps with) the activity's predecessor. Lag is the amount of time by which an activity waits (or lags behind) after the finish time of the activity's predecessor. Project scope statement includes product characteristics that can affect sequencing. Approved changes are authorized changes to project schedule, budget, or scope. Referring to the three-dimensional (3-D) cube relationship, the scope axis is often used to represent project performance, project quality, or project expectations.

CPM NETWORK SCHEDULING

Project scheduling is often the most visible step in the sequence of steps of project management. The two most common techniques of basic project scheduling are the critical path method (CPM) and program evaluation and review technique (PERT). The network of activities contained in a project provides the basis for scheduling the project and can be represented graphically to show both the contents and objectives of the project. Extensions to CPM and PERT include precedence diagramming method (PDM) and critical resource diagramming (CRD). These extensions were developed to take care of unique project scenarios and requirements. PDM technique permits the relaxation of strict precedence structures in a project so that the project duration can be compressed. CRD handles the project scheduling process by using activity-resource assignments as the primary focus for the scheduling process. This approach facilitates resource-based scheduling rather than activity-based scheduling so that resources can be more effectively assigned and utilized.

CPM network analysis procedures originated from the traditional Gantt chart or bar chart developed during World War I. There have been several mathematical techniques for scheduling activities, especially where resource constraints are a major factor. Unfortunately, the mathematical formulations are not generally practical due to the complexity involved in implementing them for realistically large projects. Even computer implementations of the complex mathematical techniques often become too cumbersome for real-time managerial decisions. Project network diagram is any schematic representation of the logical relationships among project schedule activities. The diagram is typically drawn from left to right. The two major types of network diagrams are

- Arrow diagramming method (ADM) or activity-on-arrow (AOA)
- Precedence diagramming method (PDM) or activity-on-node (AON)

In the AOA approach, arrows are used to represent activities, while nodes represent starting and ending points of activities. In the AON approach, nodes represent activities while arrows represent precedence relationships. For PDM, nodes are normally represented as rectangles. Examples or AOA, AON, and PDM are shown in Figure 4.4. Time, cost, and resource requirement estimates are developed for each activity during the network planning phase and are usually based on historical records, time standards, forecasting, regression functions, or other quantitative models. In AOA networks, dummy activities are denoted by dashed arrows. Dummy activities have zero time durations and zero resource requirements. Only start-to-finish dependency relationships are possible in AOA.

A basic CPM project network analysis is typically implemented in three phases:

- Network planning phase
- Network scheduling phase
- Network control phase

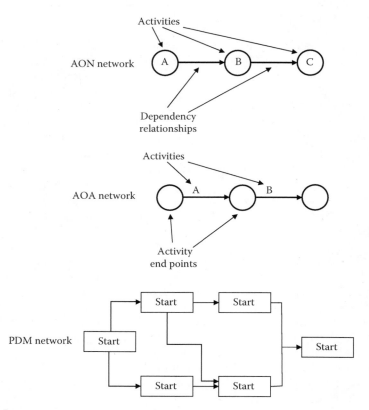

FIGURE 4.4 Types of project network diagrams.

Network planning: In network planning phase, the required activities and their precedence relationships are determined. Precedence requirements may be determined on the basis of the following:

- Physical constraints, which represent mandatory activity dependencies
- Procedural requirements, which represent discretionary activity order or dependencies
- Imposed limitations, which represent externally imposed activity dependencies

An example of a physical constraint is the requirement to erect walls before installing a roof. This is a technical limitation grounded in fixed sequence and can hardly be overcome. Such constraints are inherent in the nature of the project and will be found in any project of the same type. Thus, there is a hard logic associated with physical activity dependencies.

An example of a procedural constraint is a project team preference to have morning meetings prior to starting work. This is defined based on a preferred logic of the team. It may be based on proven process or best-practice process. This often creates arbitrary float or slack times. Thus, there is a soft logic associated with discretionary activity dependencies.

An example of an external constraint is a relationship imposed between project-based and nonproject-based activities, such as the requirement to obtain building permit before starting construction work. Such dependencies are not within the control of the project team because they are externally imposed. Regulatory requirements, trade agreements, and contractual boilerplates are other sources of external dependencies. If we can remove regulatory impediments, we can accomplish relaxation of imposed precedence relationships.

"Network scheduling" is performed by using forward-pass and backward-pass computations. These computations give the earliest and latest starting and finishing times for each activity. The amount of "slack" or "float" associated with each activity is determined during these computations. The activity path that includes the least slack in the network is used to determine the critical activities. This path, being the longest path in the network, also determines the duration of the project. Resource allocation and time–cost trade-offs are sometimes performed during network scheduling.

"Network control" involves tracking the progress of a project on the basis of the network schedule and taking corrective actions when needed. An evaluation of actual performance versus expected performance determines deficiencies in the project progress. The advantages of project network analysis are presented below.

Advantages for communication

- Clarifies project objectives
- Establishes the specifications for project performance
- Provides a starting point for more detailed task analysis
- Presents a documentation of the project plan
- Serves as a visual communication tool

Advantages for control

- Presents a measure for evaluating project performance
- Helps determine what corrective actions are needed
- Gives a clear message of what is expected
- Encourages team interaction

Advantages for team interaction

- Offers a mechanism for a quick introduction to the project
- Specifies functional interfaces on the project
- Facilitates ease of task coordination

Figure 4.5 shows the graphical representation for AON network while Figure 4.6 shows the AOA (ADM) version of the same project network. The usual network components are

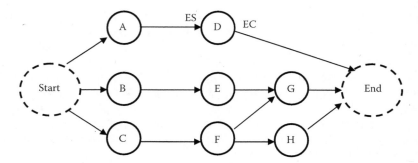

FIGURE 4.5 Graphical representation of AON network.

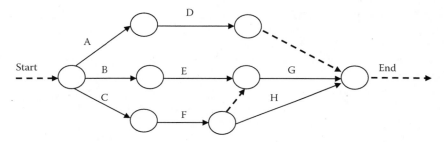

FIGURE 4.6 Graphical representation of AOA (ADM) network.

- *Node*: A node is a circular representation of an activity.
- *Arrow*: An arrow is a line connecting two nodes and having an arrowhead at one end. The arrow implies that the activity at the tail of the arrow precedes the one at the head of the arrow.
- *Activity*: An activity is a time-consuming effort required to perform a part of the overall project. An activity is represented by a node in the AON system or by an arrow in the AOA system. The job the activity represents may be indicated by a short phrase or symbol inside the node or along the arrow.
- *Restriction*: A restriction is a precedence (dependency) relationship that establishes the sequence of activities. When one activity must be completed before another activity can begin, the first is said to be a predecessor of the second.
- *Dummy*: A dummy is used to indicate one event of a significant nature (e.g., milestone). It is denoted by a dashed circle and treated as an activity with zero time duration. A dummy is not required in the AON method. However, it may be included for convenience, network clarification, or to represent a milestone in the progress of the project.
- *Predecessor activity*: A predecessor activity is one which immediately precedes the one being considered.
- *Successor activity*: A successor activity is one that immediately follows the one being considered.

- *Descendent activity*: A descendent activity is any activity restricted by the one under consideration.
- *Antecedent activity*: An antecedent activity is any activity that must precede the one being considered. Activities A and B are antecedents of D. Activity A is antecedent of B and A has no antecedent.
- *Merge point*: A merge point exists when two or more activities are predecessors to a single activity. All activities preceding the merge point must be completed before the merge activity can commence.
- *Burst point*: A burst point exists when two or more activities have a common predecessor. None of the activities emanating from the same predecessor activity can be started until the burst-point activity is completed.
- *Precedence diagram*: A precedence diagram is a graphical representation of the activities making up a project and the precedence requirements needed to complete the project. Time is conventionally shown to be from left to right, but no attempt is made to make the size of the nodes or arrows proportional to the duration of time.

Figure 4.7 shows lead–lag diagrams for the PDM. In the figure, the start-to-start (SS) relationship is referred to as "Lead" and it is specified as a negative (−) quantity. It is the amount of time by which the start of Activity A leads the start of Activity B. The finish-to-finish (FF) relationship is referred to as "Lag" and it is specified as a positive (+) quantity. It is the amount of time by which the completion of Activity B lags behind the completion of Activity A. The start-to-finish (SF) relationship shows time space between the starting time of Activity B and the finishing time of Activity A. The finish-to-start (FS) relationship indicates the time separating the completion of Activity A and the start of Activity B. FS is the most common PDM dependency. It is normally zero, which means that B starts immediately after A finishes. The SS relationship relates the start time of Activity A to the start time of Activity B. A careful study of the PDM constraints can reveal where relaxation of the restrictions are possible so that schedule comprehension can be achieved to reduce overall project duration.

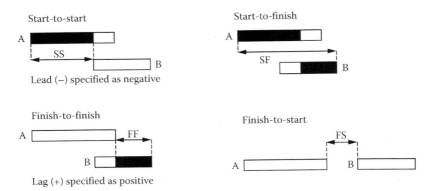

FIGURE 4.7 Lead–lag diagrams for precedence diagramming method.

WORKING WITH ACTIVITY PRECEDENCE RELATIONSHIPS

As mentioned earlier, the precedence relationships in a CPM network fall into three major categories of technical precedence (mandatory), procedural precedence (discretionary), and imposed precedence (external). Technical precedence requirements reflect the technical relationships among activities. For example, in conventional construction, walls must be erected before the roof can be installed. Procedural precedence requirements, however, are determined by policies and procedures that may be arbitrary or subjective and may have no concrete justification. Imposed precedence requirements can be classified as resource-imposed, project status-imposed, or environment-imposed. For example, resource shortage across projects may require that one task be completed before another can begin, or the current status of a project (e.g., percent completion) may determine that one activity be performed before another, or the physical environment of a project, such as weather changes or the effects of concurrent projects, may determine the precedence relationships of the activities in a project.

The primary goal of CPM analysis is to identify the "critical path," which is a determination of the minimum completion time of a project. The computational analysis involves both forward-pass and backward-pass procedures. The forward pass determines the earliest start time and the earliest completion time for each activity in the network. The backward pass determines the latest start time and the latest completion time for each activity.

Network Notations:

A: Activity identification

ES: Earliest starting time

EC: Earliest completion time

LS: Latest starting time

LC: Latest completion time

t: Activity duration

T: Project duration

n: Number of activities in the project network

During the forward pass, it is assumed that each activity will begin at its ES. An activity can begin as soon as the last of its predecessors is finished. The completion of the forward pass determines the EC of the project. The backward-pass analysis is the reverse of the forward-pass analysis. The project begins at its LC and ends at the latest starting time of the first activity in the project network. The steps of CPM network analysis are summarized below.

Step 1: Unless otherwise stated, the starting time of a project is set equal to time zero. That is, the first node, *node* 1, in the network diagram has an earliest start time of zero. Thus,

$$ES(1) = 0.$$

If a desired starting time, t_0, is specified, then $ES(1) = t_0$.

Step 2: The ES for any node (Activity j) is equal to the maximum of the EC of the immediate predecessors of the node. That is,

$$ES(i) = \text{Max } \{EC(j)\}$$
$$j \in P(i)$$

where $P(i)$ = {set of immediate predecessors of activity i}.

Step 3: The EC of Activity i is the activity's ES plus its estimated time t_i. That is,

$$EC(i) = ES(i) + t_i.$$

Step 4: The EC of a project is equal to the EC of the last node, n, in the project network. That is,

$$EC(\text{Project}) = EC(n).$$

Step 5: Unless the LC of a project is explicitly specified, it is set equal to the EC of the project. This is called the zero project slack convention. That is,

$$LC(\text{Project}) = EC(\text{Project}).$$

Step 6: If a desired deadline, T_p, is specified for the project, then

$$LC(\text{Project}) = T_p.$$

It should be noted that a LC or deadline may sometimes be specified for a project on the basis of contractual agreements.

Step 7: The LC for Activity j is the smallest of the latest start times of the activity's immediate successors. That is,

$$LC(j) = \text{Min}$$
$$i \in S(j)$$

where $S(j)$ = {immediate successors of activity j}.

Step 8: The LS for activity j is the LC minus the activity time. That is,

$$LS(j) = LC(j) - t_i.$$

EXAMPLE OF CPM ANALYSIS

Table 4.8 presents the data for a simple project network. The AON network for the example is shown in Figure 4.8. Dummy activities are included in the network to designate single starting and ending points for the network.

TABLE 4.8
Data for Sample Project for CPM Analysis

Activity	Predecessor	Duration (Days)
A	–	2
B	–	6
C	–	4
D	A	3
E	C	5
F	A	4
G	B,D,E	2

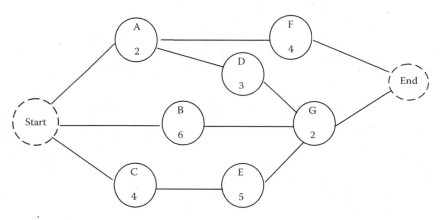

FIGURE 4.8 Example of activity network.

CPM FORWARD PASS

The forward-pass calculations are shown in Figure 4.9. Zero is entered as the ES for the initial node. Since the initial node for the example is a dummy node, its duration is zero. Thus, EC for the starting node is equal to its ES. The ES values for the immediate successors of the starting node are set equal to the EC of the START node and the resulting EC values are computed. Each node is treated as the "start" node for its successor or successors. However, if an activity has more than one predecessor, the maximum of the ECs of the preceding activities is used as the activity's starting time. This happens in the case of activity G, whose ES is determined as Max{6, 5, 9} = 9. The earliest project completion time for the example is 11 days. Note that this is the maximum of the immediately preceding ECs: Max{6, 11} = 11. Since the dummy-ending node has no duration, its EC is set equal to its ES of 11 days.

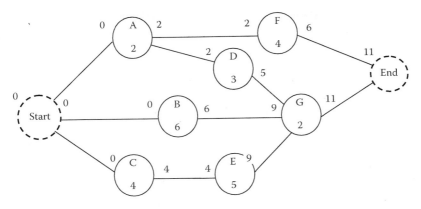

FIGURE 4.9 Forward-pass analysis for CPM example.

CPM BACKWARD PASS

The backward-pass computations establish the LS and LC for each node in the network. The results of the backward-pass computations are shown in Figure 4.10. Since no deadline is specified, the LC of the project is set equal to the EC. By backtracking and using the network analysis rules presented earlier, the latest completion and latest start times are determined for each node. Note that in the case of activity A with two immediate successors, the LC is determined as the minimum of the immediately succeeding LS. That is, Min{6, 7} = 6. A similar situation occurs for the dummy starting node. In that case, the LC of the dummy start node is Min{0, 3, 4} = 0. Since this dummy node has no duration, the LS of the project is set equal to the node's LC. Thus, the project starts at time 0 and is expected to be completed by time 11.

Within a project network, there are usually several possible paths and a number of activities that must be performed sequentially, as well as some activities that may be performed concurrently. If an activity has ES and EC times that are not

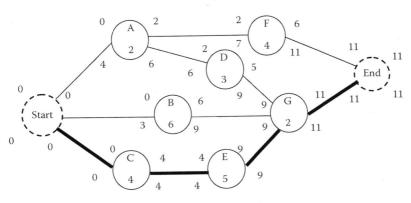

FIGURE 4.10 Backward-pass analysis for CPM example.

equal, then the actual start and completion times of that activity may be flexible. The amount of flexibility an activity possesses is called "slack" time or "float" time. The slack time is used to determine the critical activities in the network, as will be discussed below.

DETERMINATION OF CRITICAL ACTIVITIES

The critical path is defined as the path with the least slack in the network. All activities on the critical path are classified as critical activities. These activities can create bottlenecks in the project if they are delayed. The critical path is also the longest path in the network diagram. In large networks, it is possible to have multiple critical paths. In this case, it may be difficult to visually identify all the critical paths. There are four basic types of activity slack or float. They are described below.

- *Total slack* (TS) is defined as the amount of time an activity may be delayed from its ES without delaying the LC of the project. The total slack of activity j is the difference between the LC and the EC of the activity, or the difference between the LS and the ES of the activity:

$$TS(j) = LC(j) - EC(j) \text{ or } TS(j) = LS(j) - ES(j).$$

- *Free slack* (FS) is the amount of time an activity may be delayed from its ES without delaying the starting time of any of its immediate successors. An activity's free slack is calculated as the difference between the minimum ES of the activity's successors and the EC of the activity.

$$FS(j) = \text{Min}\{ES(i)\} - EC(j)$$
$$j \in S(j)$$

- *Interfering slack* (IS) is the amount of time by which an activity interferes with (or obstructs) its successors when its total slack is fully used. It is computed as the difference between total slack and free slack.

$$IS(j) = TS(j) - FS(j).$$

- *Independent float* (IF) is the amount of float that an activity will always have regardless of the completion times of its predecessors or the starting times of its successors. It is computed as

$$IF = \text{Max}\left\{0, \left(\text{Min ES}_j - \text{Max LC}_i - t_k\right)\right\}$$
$$j \in S(k); \quad i \in P(k)$$

where
   ES_j is the earliest starting time of the succeeding activity
   LC_i is the latest completion time of the preceding activity
   t is the duration of the activity whose independent float is being calculated

Independent float takes a pessimistic view of the situation of an activity. It evaluates the situation assuming that the activity is pressured from both sides—that is, when its predecessors are delayed as late as possible while its successors are to be started as early as possible. Independent float is useful for conservative planning purposes. Activities can be buffered with independent floats as a way to handle contingencies. For Figure 4.10, the total slack and the free slack for activity A are

$$TS = 6 - 2 = 4 \text{ days}$$

$$FS = \text{Min}\{2, 2\} - 2 = 2 - 2 = 0$$

Similarly, the total slack and the free slack for activity F are

$$TS = 11 - 6 = 5 \text{ days}$$

$$FS = \text{Min}\{11\} - 6 = 11 - 6 = 5 \text{ days}$$

Table 4.9 presents a tabulation of the results of the CPM example. The table contains the earliest and latest times for each activity as well as the total and free slacks. The results indicate that the minimum total slack in the network is zero. Thus, activities C, E, and G are identified as the critical activities. The critical path (C-E-G) is highlighted in Figure 4.10 and consists of the following sequence of activities:

START → C → E → G → END.

The total slack for the overall project itself is equal to the total slack observed on the critical path. The minimum slack in most networks will be zero since the ending LC is set equal to the ending EC. If a deadline is specified for a project, then the project's LC should be set to the specified deadline. In that case, the minimum total slack in the network will be given by

$$TS_{\text{Min}} = (\text{Project deadline}) - EC \text{ of the last node.}$$

This minimum total slack will then appear as the total slack for each activity on the critical path. If a specified deadline is lower than the EC at the finish node, then the

TABLE 4.9
Result of CPM Analysis for Sample Project

Activity	Duration (Days)	ES	EC	LS	LC	TS	FS	Critical
A	2	0	2	4	6	4	0	—
B	6	0	6	3	9	3	3	—
C	4	0	4	0	4	0	0	Critical
D	3	2	5	6	9	4	4	—
E	5	4	9	4	9	0	0	Critical
F	4	2	6	7	11	5	5	—
G	2	9	11	9	11	0	0	Critical

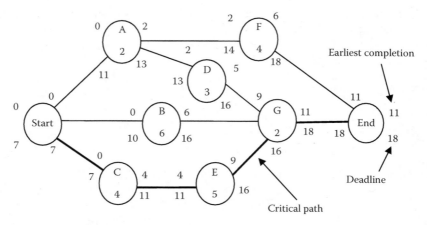

FIGURE 4.11 CPM network with deadline.

project will start out with a negative slack. This means that it will be behind schedule before it even starts. It may then become necessary to expedite some activities (i.e., crashing) in order to overcome the negative slack. Figure 4.11 shows an example with a specified project deadline. In this case, the deadline of 18 days occurs after the EC of the last node in the network.

SUBCRITICAL PATHS

In a large project network, there may be paths that are near critical. Such paths require almost as much attention as the critical path since they have a high risk of becoming critical when changes occur in the network. Analysis of subcritical paths may help in the classification of tasks into ABC categories on the basis of Pareto analysis, which separates the most important activities from the less important ones. This can be used for more targeted allocation of resources. With subcritical analysis, attention can shift from focusing only on the critical path to managing critical and near-critical tasks. Steps for identifying the subcritical paths are

Step 1: Sort activities in increasing order of total slack.

Step 2: Partition the sorted activities into groups based on the magnitudes of total slack.

Step 3: Sort the activities within each group in increasing order of their earliest starting times.

Step 4: Assign the highest level of criticality to the first group of activities (e.g., 100%). This first group represents the usual critical path.

Step 5: Calculate the relative criticality indices for the other groups in decreasing order of criticality.

STEP Time Management

TABLE 4.10
Analysis of Subcritical Paths

Path Number	Activities on Path	Total Slack	λ (%)	λ' (%)
1	A,C,G,H,	0	100	10
2	B,D,E	1	97.56	9.78
3	F,I	5	87.81	8.90
4	J,K,L	9	78.05	8.03
5	O,P,Q,R	10	75.61	7.81
6	M,S,T	25	39.02	4.51
7	N,AA,BB,U	30	26.83	3.42
8	V,W,X	32	21.95	2.98
9	Y,CC,EE	35	17.14	2.54
10	DD,Z,FF	41	0	1.00

Define the following variables:

α_1 the minimum total slack in the network
α_2 the maximum total slack in the network
β total slack for the path whose criticality is to be calculated

Compute the path's criticality level as

$$\lambda = \frac{\alpha_2 - \beta}{\alpha_2 - \alpha_1}(100\%)$$

The above procedure yields relative criticality levels between 0% and 100%. Table 4.10 presents an example of path criticality levels. The criticality level may be converted to a scale between 1 (least critical) and 10 (most critical) by the scaling factor below:

$$\lambda' = 1 + 0.09\lambda$$

SCHEDULE TEMPLATES

Schedule network templates are standard project network diagrams that can be reused. A project analyst can use the entire network or a portion of the network. In fact, the subcritical path elements can be used to compose a subnetwork. Portions of the overall network template are called subnets or "fragnets." Subnets are useful in large projects with repeated tasks. For example, the floors in a high-rise building construction represent a repeated subnetwork of a large project. The schedule developed and executed for each floor is repeated for the other floors.

GANTT CHARTS

A project schedule is developed by mapping the results of CPM analysis to a calendar timeline. The Gantt chart is one of the most widely used tools for presenting project schedules. A Gantt chart can show planned and actual progress of activities. As a project progresses, markers are made on the activity bars to indicate actual work accomplished. Figure 4.12 presents the Gantt chart for the CPM example using the ES from the CPM result table. Figure 4.13 presents the Gantt chart for the example based on the LS. Critical activities are indicated by the shaded bars.

Review of the CPM analysis shows that the starting time of activity F can be delayed from day 2 until day 7 (i.e., TS = 5) without delaying the overall project. Likewise, A, D, or both may be delayed by a combined total of 4 days (TS = 4) without delaying the overall project. If all the 4 days of slack are used

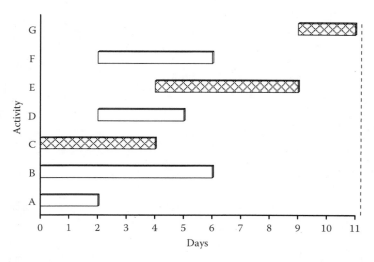

FIGURE 4.12   Gantt chart based on earliest starting times.

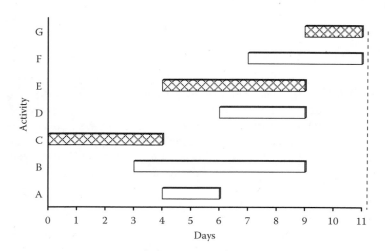

FIGURE 4.13   Gantt chart based on latest starting times.

up by A, then D cannot be delayed. If A is delayed by 1 day, D can be delayed by up to 3 days without causing a delay of G, which determines project completion. The Gantt chart also indicates that activity B may be delayed by up to 3 days without affecting the project's completion time.

In Figure 4.13, the activities are shown scheduled by their LCs. This represents an extreme case where activity slack times are fully used. No activity in this schedule can be delayed without delaying the project. Notice that only one activity is scheduled over the first 3 days. This may be compared to the earliest start schedule, which has three starting activities. The schedule in Figure 4.13 may be useful if there is a situational constraint (e.g., externally imposed restriction) that permits only a few activities to be scheduled in the early stages of the project. Such constraints may involve shortage of project personnel, lack of initial budget, time allocated for project initiation, time allocated for personnel training, an allowance for a learning period, or general resource constraints. Scheduling of activities based on ES times indicates an optimistic view, while scheduling on the basis of LS times represents a pessimistic approach.

PROJECT CRASHING

Crashing is the expediting or compression of activity duration. Crashing is done as a trade-off between shorter task duration and higher task cost. It must be determined whether the total cost savings realized from reducing the project duration is enough to justify the higher costs associated with reducing individual task durations. If there is a delay penalty associated with a project, it may be possible to reduce the total project cost even though crashing increases individual task costs. If the cost savings on the delay penalty is higher than the incremental cost of reducing the project duration, then crashing is justified. Normal task duration refers to the time required to perform a task under normal circumstances. "Crash task duration" refers to the reduced time required to perform a task when additional resources are allocated to it.

If each activity is assigned a range of time and cost estimates, then several combinations of time and cost values will be associated with the overall project. Iterative procedures are used to determine the best time or cost combination for a project. Time–cost trade-off analysis may be conducted, for example, to determine the marginal cost of reducing the duration of the project by one time unit. Table 4.11 presents an extension of the data for the example problem to include normal and crash times as well as normal and crash costs for each activity. The normal duration of the project is 11 days, as seen earlier, and the normal cost is $2775.

If all the activities are reduced to their respective crash durations, the total crash cost of the project will be $3545. In that case, the crash time is found by CPM analysis to be 7 days. The CPM network for the fully crashed project is shown in Figure 4.14. Note that activities C, E, and G remain critical. Sometimes, the crashing of activities may result in additional critical paths. The Gantt chart in Figure 4.15 shows a schedule of the crashed project using the ES times. In practice, one would not crash all activities in a network. Rather, some selection rule would be used to determine which activity should be crashed and by how much. One approach is to crash only the critical activities or those activities with the best ratios of incremental

TABLE 4.11
Normal and Crash Time and Cost Data

Activity	Normal Duration (Days)	Normal Cost ($)	Crash Duration (Days)	Crash Cost ($)	Crashing Ratio
A	2	210	2	210	0
B	6	400	4	600	100
C	4	500	3	750	250
D	3	540	2	600	60
E	5	750	3	950	100
F	4	275	3	310	35
G	2	100	1	125	25
		2775		3545	

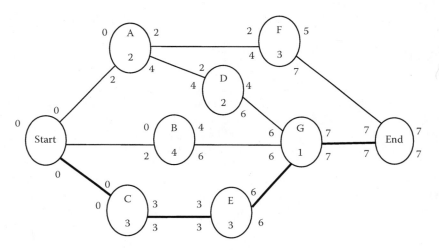

FIGURE 4.14 Example of fully crashed CPM network.

cost versus time reduction. The last column in Table 4.11 presents the respective ratios for the activities in our example. The crashing ratios are computed as

$$r = \frac{\text{Crash cost} - \text{Normal cost}}{\text{Normal duration} - \text{Crash duration}}$$

Activity G offers the lowest cost per unit time reduction of $25. If the preferred approach is to crash only one activity at a time, we may decide to crash activity G first and evaluate the increase in project cost versus the reduction in project duration. The process can then be repeated for the next best candidate for crashing, which is activity F in this case. The project completion time is not reduced any further since activity F is not a critical activity. After F has been crashed, activity D can then be

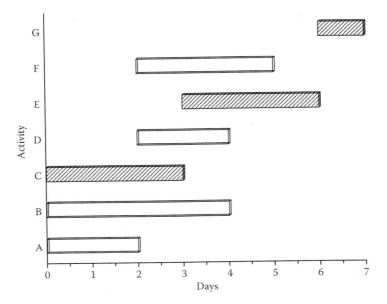

FIGURE 4.15 Gantt chart of fully crashed CPM network.

crashed. This approach is repeated iteratively in order of activity preference until no further reduction in project duration can be achieved or until the total project cost exceeds a specified limit.

A more comprehensive analysis is to evaluate all possible combinations of the activities that can be crashed. However, such a complete enumeration would be prohibitive, since there would be a total of 2^c crashed networks to evaluate, where c is the number of activities that can be crashed out of the n activities in the network ($c \leq n$). For our example, only 6 out of the 7 activities in the network can be crashed. Thus, a complete enumeration will involve $2^6 = 64$ alternate networks. Table 4.12 shows 7 of the 64 crashing options. Activity G, which offers the best crashing ratio,

TABLE 4.12
Selected Crashing Options for CPM Example

Option Number	Activities Crashed	Network Duration (Days)	Time Reduction (Days)	Incremental Cost ($)	Total Cost ($)
1.	None	11	—	—	2775
2.	G	10	1	25	2800
3.	G,F	10	0	35	2835
4.	G,F,D	10	0	60	2895
5.	G,F,D,B	10	0	200	3095
6.	G,F,D,B,E	8	2	200	3295
7.	G,F,D,B,E,C	7	1	250	3545

reduces the project duration by only 1 day. Even though activities F, D, and B are crashed by a total of 4 days at an incremental cost of $295, they do not generate any reduction in project duration. Activity E is crashed by 2 days and generates a reduction of 2 days in project duration. Activity C, which is crashed by 1 day, generates a further reduction of 1 day in the project duration. It should be noted that the activities that generate reductions in project duration are the ones that were earlier identified as the critical activities.

In general, there may be more than one critical path, so the project analyst needs to check for the set of critical activities with the least total crashing ratio in order to minimize the total crashing cost. Also, one needs to update the critical paths every time a set of activities is crashed because new activities may become critical in the meantime. For the network given in Figure 4.15, the path C-E-G is the only critical path. Therefore, we do not need to consider crashing other jobs since the incurred cost will not affect the project completion time. There are 12 possible ways one can crash activities C, G, and E in order to reduce the project time.

Several other approaches exist for determining which activities to crash in a project network. Two alternate approaches are presented below for computing the crashing ratio, r. The first one directly uses the criticality of an activity to determine its crashing ratio while the second one uses the calculation shown below:

$$r = \text{Criticality index}$$

$$r = \frac{\text{Crash cost} - \text{Normal cost}}{(\text{Normal duration} - \text{Crash duration})(\text{Criticality index})}$$

The first approach gives crashing priority to the activity with the highest probability of being on the critical path. In deterministic networks, this refers to the critical activities. In stochastic networks, an activity is expected to fall on the critical path only a percentage of the time. The second approach is a combination of the approach used for the illustrative example and the criticality index approach. It reflects the process of selecting the least-cost expected value. The denominator of the expression represents the expected number of days by which the critical path can be shortened.

CRITICAL CHAIN ANALYSIS

"Critical chain" is the "theory of constraints" (Dettmer 1997; Goldratt 1997; Woeppel 2001) applied to project management specifically for managing and scheduling projects. Constraint management is based on the principle that the performance of a system's constraint will determine the performance of the entire system. If a project's characteristic constraint is effectively managed, the overall project will be effectively managed. This is analogous to the belief that the worst performer of an organization will dictate the performance of the organization. Similarly, the weakest link in a chain determines the strength of the chain. Because overall operation is essentially a series of linkages of activities, one break in the linkage determines a break of the overall operation. That is, it takes only one negative to negate a series of positives: (+)(+)(+)(+)(−)(+)(+) = (−). Looking at this from a production point of

STEP Time Management

view, the bottleneck operation determines the throughput of a production system. From a group operation point of view, the last passenger on a complimentary shuttle bus determines the departure time of the bus. What all these mean in the context of project scheduling is that focus should be on the critical activities in the project network diagram. This means that the critical chain is the most important focus. With respect to applying the theory of constraints, there are three types of constraints:

1. Paradigm constraint (policy-based)
2. Resource constraint (physical limitation)
3. Material constraint (imposition by project environment)

Each constraint type impacts the project differently. For project scheduling purposes, critical chain is used to generate several alterations to the traditional CPM/PERT network. All individual activity slacks (or "buffer") become the project buffer. Each team member, responsible for his or her component of the activity network, creates a duration estimate free from any padding. The typical approach is to estimate based on a 50% probability of success. All activities on the critical chain (path) and feeder chains (noncritical chains in the network) then are linked with minimal time padding. The project buffer now is aggregated and some proportion of the saved time is added to the project. Even adding 50% of the saved time significantly reduces the overall project schedule while requiring team members to be concerned less with activity padding and more with task completion. Even if the project team members miss their delivery date 50% of the time, the overall effect on the project's duration is minimized because of the downstream aggregated buffer. Readers can refer to the References at the end of this chapter for further details on the application of critical chain.

The same approach can also be used for tasks that are not on the critical chain. Accordingly, all feeder path activities are reduced by the same order of magnitude and a feeder buffer is constructed for the overall noncritical chain of activities. It should be noted that critical chain distinguishes between its use of buffer and the traditional project network use of project slack. In CPM/PERT, project slack is a function of the overall completed activity network. In other words, slack is an outcome of the task dependencies, whereas critical chain buffer is used as an a priori (or advance) planning contingency that is based on a logical redesign of each activity and the application of an aggregated project buffer at the end of the project. The following deficiencies have been noted about critical chain vis-à-vis the traditional CPM/PERT network analysis:

1. Lack of project milestones makes coordinated scheduling, particularly with external suppliers, highly problematic. Critics point out that the lack of in-process project milestones adversely affects the ability to coordinate schedule dates with suppliers who provide the external delivery of critical components.
2. Although it may be true that critical chain brings increased discipline to project scheduling, efficient methods for applying this technique to a firm's portfolio of projects are unclear; that is, critical chain offers benefits on a project-by-project basis, but its usefulness at the overall integrated program

level has not been ascertained. Furthermore, because critical chain requires dedicated resources in a multiproject environment where resources are shared, it is impossible to avoid multitasking, which adversely impacts its utility.

3. Evidence of its success is still almost exclusively anecdotal and based on single-case studies. There is no large-scale empirical research to verify its overall effectiveness.

In summary, because of the dynamism of technology and fast-paced scientific evolution, STEPs particularly require new ways of analysis and scheduling activities. The buffering approach offered by critical chain analysis represents another way of looking at the problem. Chapter 5 deals with STEP project cost management.

REFERENCES

Collins, J. *Good to Great*, HarperCollins, New York, 2001.
Dettmer, H. W. *Goldratt's Theory of Constraints: A Systems Approach to Continuous Improvement*, Quality Press, Milwaukee, WI, 1997.
Goldratt, E. M. *Critical Chain*, The North River Press, Great Barrington, MA, 1997.
Martin, H. L. *Techonomics: The Theory of Industrial Evolution*, Taylor & Francis/CRC Press, Boca Raton, FL, 2007.
Niven, P. R. *Balanced Scorecard: Step-by-Step: Maximizing Performance and Maintaining Results*, Wiley, New York, 2002.
PMI, *A Guide to the Project Management Body of Knowledge (PMBOK Guide)*, 3rd ed., Project Management Institute, Newtown Square, PA, 2004.
Woeppel, M. J. *Manufacturer's Guide to Implementing the Theory of Constraints*, St. Lucie Press, Boca Raton, FL, 2001.

5 STEP Cost Management

Follow the money

–All the President's Men

Follow the technology

–H. Lee Martin (Techonomics)

Follow not only the money, but the technology also is a lesson that aptly typifies what STEP cost management epitomizes as suggested by the quotes referenced at the beginning of this chapter. Cost management is a primary function in project management. Cost is a vital criterion for assessing project performance. Cost management involves having an effective control over project costs through the use of reliable techniques of estimation, forecasting, budgeting, and reporting. Cost estimation requires collecting relevant data needed to estimate elemental costs during the life cycle of a project. Cost planning involves developing an adequate budget for the planned work. Cost control involves continual process of monitoring, collecting, analyzing, and reporting cost data. Martin (2007) defines techonomics as the study of how technology affects the economy and a theory of organizational evolution that results from technological advance fueled and selected by economic success. STEP cost management is impacted by the state of technology and the concomitant cost factors. The primary components of cost management within any project undertaking are

- Cost estimating
- Cost budgeting
- Cost control

Cost control must be exercised across the other elements of the project management knowledge areas (PMI, 2004). The technique of earned value management plays a major and direct role in cost management. The technique is covered in detail later in this chapter.

COST MANAGEMENT: STEP-BY-STEP IMPLEMENTATION

The cost management component of the *Project Management Body of Knowledge* (*PMBOK*) consists of the elements shown in the block diagram in Figure 5.1. The three elements in the block diagram are carried out across the process groups

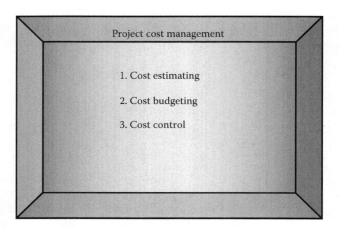

FIGURE 5.1 Block diagram of project cost management.

TABLE 5.1
Implementation of Project Cost-Management across Process Groups

	Initiating	Planning	Executing	Monitoring and Controlling	Closing
Project Cost Management		Cost estimating Cost budgeting		Cost control	

presented earlier in Chapter 1. The overlay of the elements and the process groups are shown in Table 5.1. Thus, under the knowledge area of cost management, the required steps are

Step 1: Cost estimation
Step 2: Cost budgeting
Step 3: Cost control

Tables 5.2 through 5.4 present the inputs, tools, techniques, and outputs of each step.

STEP PORTFOLIO MANAGEMENT

Project portfolio management is the systematic application of the tools and techniques of management to the collection of cost-based element of a project. Examples of STEP portfolios would be planned initiatives, ongoing projects, and ongoing support services, and investment in emerging technology. A formal STEP portfolio management strategy enables measurement and objective evaluation of investment scenarios. Some of the key aspects of an effective STEP portfolio management are

TABLE 5.2
Tools and Techniques for Cost Estimating within Project Cost Management

Step 1: Cost Estimation

Inputs	Tools and Techniques	Output(s)
Enterprise environmental factors	Analogous estimating	Activity cost estimates
Organizational process assets	Resource cost rates	Activity cost supporting detail
Project scope statement	Goal programming	Requested changes
WBS	Return on investment analysis	Cost management plan (updates)
WBS dictionary	Bottom-up estimating	Other in-house outputs, reports, and data inferences of interest to the organization
Project management plan	Parametric estimating	
Other in-house (custom) factors of relevance and interest	Project management cost software	
	Vendor bid analysis	
	Reserve analysis	
	Cost of quality	
	CMMI (Capability Maturity Model Integration)	
	Other in-house (custom) tools and techniques	

TABLE 5.3
Tools and Techniques for Cost Budgeting within Project Cost Management

Step 2: Cost Budgeting

Inputs	Tools and Techniques	Output(s)
Project scope statement	Cost aggregation	Cost baseline
Work breakdown structure (WBS)	Portfolio management	Project funding requirements
WBS dictionary	Reserve analysis	Cost management plan (updates)
Activity cost estimates	Parametric estimating	Requested changes
Activity cost estimate supporting detail	Funding limit reconciliation	Other in-house outputs, reports, and data inferences of interest to the organization
Project schedule	Balanced scorecard	
Resource calendars	Critical chain elements budgeting	
Contract	Other in-house (custom) tools and techniques	
Cost management plan		
Other in-house (custom) factors of relevance and interest		

1. Define the project, supporting program, and enabling system as well as the required portfolio.
2. Define business value and desired return on investment (ROI) and prioritize projects.
3. Define an overall project portfolio management methodology for STEP.

TABLE 5.4
Tools and Techniques for Cost Control within Project Cost Management

Step 3: Cost Control

Inputs	Tools and Techniques	Output(s)
Cost baseline	Process control charts	Cost estimates (updates)
Project funding requirements	Cost change control system	Cost baseline (estimates)
Performance reports	Performance measurement analysis	Performance measurements
Work performance information	Forecasting	Forecasted completion
Approved change requests	Trend analysis	Requested changes
Project management plan	Project performance reviews	Recommended corrective actions
Other in-house (custom) factors of relevance and interest	Project management software	Organizational process assets (updates)
	Variance analysis	Project management plan (updates)
	Variance management	
	Earned value management	
	Other in-house (custom) tools and techniques	Other in-house outputs, reports, and data inferences of interest to the organization

4. Delineate an overall project portfolio in translating strategy into results.
5. Introduce a balanced scorecard that synthesizes and integrates the numerous and complex metrics related to different STEP portfolio management processes into one framework.
6. Clarify projects that will provide effective allocation and management of limited resources.
7. Introduce progressive project assessment approach including initial project assessment, mid-cycle project assessment, and closing project assessment.
8. Employ quantitative techniques to objectively assess a project for its absolute merit and relative merit against other projects.
9. Utilize weighted scoring models to quantify intangible benefits of the project.
10. Evaluate project decision techniques that clarify choices involving both risks and opportunities.
11. Build a business case for each project and rank order projects based on strategic fit, risks, opportunities, and the changing nature of science and technology.
12. Establish criteria for phasing out a project when it is no longer serving the desired purpose.

PROJECT COST ELEMENTS

Cost management in a project environment refers to the functions required to maintain effective financial control of the project throughout its life cycle. There are several cost concepts that influence the economic aspects of managing industrial

STEP Cost Management

projects. Within a given scope of analysis, there will be a combination of different types of cost factors as defined below.

Actual cost of work performed: The cost actually incurred and recorded in accomplishing the work performed within a given time period.

Applied direct cost: The amounts recognized in the time period associated with the consumption of labor, material, and other direct resources without regard to the date of commitment or the date of payment. These amounts are to be charged to work-in-process (WIP) when resources are actually consumed, material resources are withdrawn from inventory for use, or material resources are received and scheduled for use within 60 days.

Budgeted cost for work performed: The sum of the budgets for completed work plus the appropriate portion of the budgets for level of effort and apportioned effort. Apportioned effort is effort that by itself is not readily divisible into short-span work packages but is related in direct proportion to measured effort.

Budgeted cost for work scheduled: The sum of budgets for all work packages and planning packages scheduled to be accomplished (including work in process) plus the amount of level of effort and apportioned effort scheduled to be accomplished within a given period of time.

Burdened costs: Burdened costs are cost components that are fully loaded with overhead charges as well as other pertinent charges. This includes cost of management and other costs associated with running the business.

Cost baseline: The cost baseline is used to measure and monitor project cost and schedule performance. It presents a summation of costs by period. It is used to measure cost and schedule performance and sometimes called performance measurement baseline (PMB).

Diminishing returns: The law of diminishing returns refers to the phenomenon of successively less output for each incremental resource input.

Direct cost: Cost that is directly associated with actual operations of a project. Typical sources of direct costs are direct material costs and direct labor costs. Direct costs are those that can be reasonably measured and allocated to a specific component of a project.

Economies of scale: This is a term referring to the reduction of the relative weight of the fixed cost in total cost, achieved by increasing the quantity of output. Economies of scale help to reduce the final unit cost of a product and are often simply referred to as the savings due to mass production.

Estimated cost at completion: This refers to the sum of actual direct costs, plus indirect costs that can be allocated to a contract plus the estimate of costs (direct and indirect) for authorized work remaining to be done.

First cost: The total initial investment required to initiate a project or the total initial cost of the equipment needed to start the project.

Fixed cost: Costs incurred regardless of the level of operation of a project. Fixed costs do not vary in proportion to the quantity of output. Examples of costs that make up the fixed cost of a project are administrative expenses, certain types of taxes, insurance cost, depreciation cost, and debt servicing cost. These costs usually do not vary in proportion to quantity of output.

Incremental cost: The additional cost of changing the production output from one level to another. Incremental costs are normally variable costs.

Indirect cost: This is a cost that is indirectly associated with project operations. Indirect costs are those that are difficult to assign to specific components of a project. An example of an indirect cost is the cost of computer hardware and software needed to manage project operations. Indirect costs are usually calculated as a percentage of a component of direct costs. For example, the indirect costs in an organization may be computed as 10% of direct labor costs.

Life cycle cost: This is the sum of all costs, recurring and nonrecurring, associated with a project during its entire life cycle.

Maintenance cost: This is a cost that occurs intermittently or periodically for the purpose of keeping project equipment in good operating condition.

Marginal cost: Marginal cost is the additional cost of increasing production output by one additional unit. The marginal cost is equal to the slope of the total cost curve or line at the current operating level.

Operating cost: This is a recurring cost needed to keep a project in operation during its life cycle. Operating costs may consist of such items as labor, material, and energy costs.

Opportunity cost: This refers to the cost of foregoing the opportunity to invest in a venture that, if pursued, would have produced an economic advantage. Opportunity costs are usually incurred due to limited resources that make it impossible to take advantage of all investment opportunities. It is often defined as the cost of the best-rejected opportunity. Opportunity costs can also be incurred due to a missed opportunity rather than due to an intentional rejection. In many cases, opportunity costs are hidden or implied because they typically relate to future events that cannot be accurately predicted.

Overhead cost: These are costs incurred for activities performed in support of the operations of a project. The activities that generate overhead costs support the project efforts rather than contributing directly to the project goal. The handling of overhead costs varies widely from company to company. Typical overhead items are electric power cost, insurance premiums, cost of security, and inventory carrying cost.

Standard cost: This is a cost that represents the normal or expected cost of a unit of the output of an operation. Standard costs are established in advance. They are developed as a composite of several component costs, such as direct labor cost per unit, material cost per unit, and allowable overhead charge per unit.

Sunk cost: Sunk cost is a cost that occurred in the past and cannot be recovered under the present analysis. Sunk costs should have no bearing on the prevailing economic analysis and project decisions. Ignoring sunk costs can be a difficult task for analysts.

STEP Cost Management

For example, if $950,000 was spent 4 years ago to buy a piece of equipment for a technology-based project, a decision on whether or not to replace the equipment now should not consider that initial cost. But uncompromising analysts might find it difficult to ignore that much money. Similarly, an individual making a decision on selling a personal automobile would typically try to relate the asking price to what was paid for the automobile when it was acquired. This is wrong under the strict concept of sunk costs.

Total cost: This is the sum of all the variable and fixed costs associated with a project.

Variable cost: This cost varies in direct proportion to the level of operation or quantity of output. For example, the costs of material and labor required to make an item will be classified as variable costs since they vary with changes in level of output.

BASIC CASH FLOW ANALYSIS

Economic analysis is performed when a choice must be made between mutually exclusive projects that compete for limited resources. The cost performance of each project will depend on the timing and levels of its expenditures. The techniques of computing cash flow equivalence permit us to bring competing project cash flows to a common basis for comparison. The common basis depends on the prevailing interest rate. Two cash flows that are equivalent at a given interest rate will not be equivalent at a different interest rate. The basic techniques for converting cash flows from one point in time to another are presented in the following sections.

Time Value of Money Calculations

Cash flow conversion involves the transfer of project funds from one point in time to another. The following notation is used for the variables involved in the conversion process:

i = interest rate per period
n = number of interest periods
P = a present sum of money
F = a future sum of money
A = a uniform end-of-period cash receipt or disbursement
G = a uniform arithmetic gradient increase in period-by-period payments or disbursements

In many cases, the interest rate used in performing economic analysis is set equal to the minimum attractive rate of return (MARR) of the decision maker. The MARR is also sometimes referred to as hurdle rate, required internal rate of return (IRR), ROI, or discount rate. The value of MARR is chosen for a project based on the objective of maximizing the economic performance of the project.

Calculations with Compound Amount Factor

The procedure for the single payment compound amount factor finds a future amount, F, that is equivalent to a present amount, P, at a specified interest rate, i, after n periods. This is calculated by the following formula:

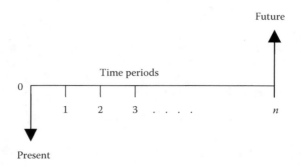

FIGURE 5.2 Single payment compound amount cash flow.

$$F = P(1+i)^n$$

A graphic representation of the relationship between P and F is shown in Figure 5.2.

Example: A sum of $5000 is deposited in a project account and left there to earn interest for 15 years. If the interest rate per year is 12%, the compound amount after 15 years can be calculated as follows:

$$F = \$5000(1+0.12)^{15} = \$27,367.85$$

Calculations with Present Value Factor

Present value (*PV* or *P*), also called present worth, is the present-day at-hand value of a cash flow. The present value factor computes *PV* when *F* is given. The present value factor is obtained by solving for *P* in the equation for the compound amount factor. That is,

$$P = F(1+i)^{-n}$$

Supposing it is estimated that $15,000 would be needed to complete the implementation of a project 5 years from now, how much should be deposited in a special project fund now so that the fund would accrue to the required $15,000 exactly 5 years from now? If the special project fund pays interest at 9.2% per year, the required deposit would be

$$P = \$15,000(1+0.092)^{-5} = \$9,660.03$$

Calculations with Uniform Series Present Worth Factor

The uniform series present worth factor is used to calculate the present worth equivalent, *P*, of a series of equal end-of-period amounts, *A*. Figure 5.3 shows the uniform

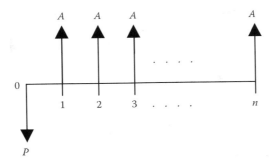

FIGURE 5.3 Uniform series cash flow.

series cash flow. The derivation of the formula uses the finite sum of the present worth values of the individual amounts in the uniform series cash flow as shown below:

$$P = \sum_{t=1}^{n} A(1+i)^{-t}$$
$$= A\left[\frac{(1+i)^n - 1}{i(1+i)^n}\right]$$

Example: Suppose a sum of $12,000 must be withdrawn from an account to meet the annual operating expenses of a multiyear project. The project account pays interest at 7.5% per year compounded on an annual basis. If the project is expected to last 10 years, how much must be deposited in the project account now so that the operating expenses of $12,000 can be withdrawn at the end of every year for 10 years? The project fund is expected to be depleted to zero by the end of the last year of the project. The first withdrawal will be made 1 year after the project account is opened and no additional deposits will be made in the account during the project life cycle. The required deposit is calculated in this way:

$$P = \$12,000\left[\frac{(1+0.075)^{10} - 1}{0.075(1+0.075)^{10}}\right]$$
$$= \$82,368.92$$

Calculations with Uniform Series Capital Recovery Factor

The capital recovery formula is used to calculate the uniform series of equal end-of-period payments, A, that are equivalent to a given present amount, P. This is the converse of the uniform series present amount factor. The equation for the uniform series capital recovery factor is obtained by solving for A in the uniform series present amount factor. That is,

$$A = P\left[\frac{i(1+i)^n}{(1+i)^n - 1}\right]$$

Example: Suppose a piece of equipment needed to launch a project must be purchased at a cost of $50,000. The entire cost is to be financed at 13.5% per year and repaid on a monthly installment schedule over 4 years. It is desired to calculate what the monthly loan payments will be. It is assumed that the first loan payment will be made exactly 1 month after the equipment is financed. If the interest rate of 13.5% per year is compounded monthly, then the interest rate per month will be 13.5%/12 = 1.125% per month. The number of interest periods over which the loan will be repaid is 4(12) = 48 months. Consequently, the monthly loan payments are calculated to be

$$A = \$50,000 \left[\frac{0.01125(1+0.01123)^{48}}{(1+0.01125)^{48} - 1} \right]$$

$$= \$1353.82$$

Calculations with Uniform Series Compound Amount Factor

The series compound amount factor is used to calculate a single future amount that is equivalent to a uniform series of equal end-of-period payments. The cash flow is shown in Figure 5.4. Note that the future amount occurs at the same point in time as the last amount in the uniform series of payments. The factor is derived as shown below:

$$F = \sum_{t=1}^{n} A(1+i)^{n-t}$$

$$= A \left[\frac{(1+i)^n - 1}{i} \right]$$

Example: If equal end-of-year deposits of $5000 are made to a project fund paying 8% per year for 10 years, how much can be expected to be available for withdrawal from the account for capital expenditure immediately after the last deposit is made?

$$F = \$5,000 \left[\frac{(1+0.08)^{10} - 1}{0.08} \right]$$

$$= \$72,432.50$$

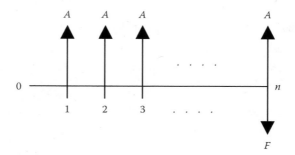

FIGURE 5.4 Uniform series compound amount cash flow.

Calculations with Uniform Series Sinking Fund Factor

The sinking fund factor is used to calculate the uniform series of equal end-of-period amounts, A, that are equivalent to a single future amount, F. This is the reverse of the uniform series compound amount factor. The formula for the sinking fund is obtained by solving for A in the formula for the uniform series compound amount factor. That is,

$$A = F\left[\frac{i}{(1+i)^n - 1}\right]$$

Example: How large are the end-of-year equal amounts that must be deposited into a project account so that a balance of $75,000 will be available for withdrawal immediately after the 12th annual deposit is made? The initial balance in the account is zero at the beginning of the first year. The account pays 10% interest per year. Using the formula for the sinking fund factor, the required annual deposits are

$$A = \$75,000\left[\frac{0.10}{(1+0.10)^{12} - 1}\right]$$
$$= \$3,507.25$$

Calculations with Capitalized Cost Formula

Capitalized cost refers to the present value of a single amount that is equivalent to a perpetual series of equal end-of-period payments. This is an extension of the series present worth factor with an infinitely large number of periods. This is shown graphically in Figure 5.5.

Using the limit theorem from calculus as n approaches infinity, the series present worth factor reduces to the following formula for the capitalized cost:

$$P = \frac{A}{i}$$

Example: How much should be deposited in a general fund to service a recurring public service project to the tune of $6500 per year forever if the fund yields an annual interest rate of 11%? Using the capitalized cost formula, the required one-time deposit to the general fund is

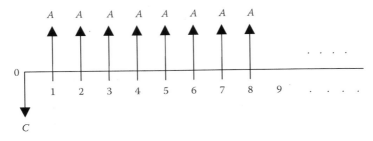

FIGURE 5.5 Capitalized cost cash flow.

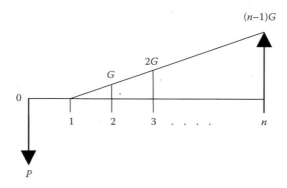

FIGURE 5.6 Arithmetic gradient cash flow with zero-base amount.

$$P = \frac{\$6500}{0.11}$$
$$= \$59{,}090.91$$

Arithmetic Gradient Series

The gradient series cash flow involves an increase of a fixed amount in the cash flow at the end of each period. Thus, the amount at a given point in time is greater than the amount at the preceding period by a constant amount. This constant amount is denoted by G. Figure 5.6 shows the basic gradient series in which the base amount at the end of the first period is zero. The size of the cash flow in the gradient series at the end of period t is calculated as

$$A_t = (t-1)G, \quad t = 1, 2, \ldots, n$$

The total present value of the gradient series is calculated by using the present amount factor to convert each individual amount from time t to time 0 at an interest rate of $i\%$ per period and then summing up the resulting present values. The finite summation reduces to a closed form as shown below:

$$P = \sum_{t=1}^{n} A_t (1+i)^{-t}$$
$$= G\left[\frac{(1+i)^n - (1+ni)}{i^2(1+i)^n}\right]$$

Example: The cost of supplies for a 10-year project increases by $1500 every year starting at the end of year two. There is no cost for supplies at the end of the first year. If interest rate is 8% per year, determine the present amount that must be set aside at time zero to take care of all the future supplies expenditures. We have $G = 1500$, $i = 0.08$, and $n = 10$. Using the arithmetic gradient formula, we obtain

STEP Cost Management

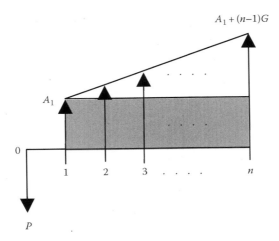

FIGURE 5.7 Arithmetic gradient cash flow with nonzero-base amount.

$$P = 1500 \left\{ \frac{1-[1+10\,(0.08)](1+0.08)^{-10}}{(0.08)^2} \right\}$$

$$= \$1500(25.9768)$$

$$= \$38,965.20$$

In many cases, an arithmetic gradient starts with some base amount at the end of the first period and then increases by a constant amount thereafter. The nonzero base amount is denoted as A_1. Figure 5.7 shows this type of cash flow.

The calculation of the present amount for such cash flows requires breaking the cash flow into a uniform series cash flow of amount A_1 and an arithmetic gradient cash flow with zero base amount. The uniform series present worth formula is used to calculate the present worth of the uniform series portion while the basic gradient series formula is used to calculate the gradient portion. The overall present worth is then calculated:

$$P = P_{\text{uniform series}} + P_{\text{gradient series}}$$

$$= A_1 \left[\frac{(1+i)^n - 1}{i(1+i)^n} \right] + G \left[\frac{(1+i)^n - (1+ni)}{i^2(1+i)^n} \right]$$

Internal Rate of Return

The IRR for a cash flow is defined as the interest rate that equates the future worth at time n or present worth at time 0 of the cash flow to zero. If we let i^* denote the IRR, then we have:

$$FW_{t=n} = \sum_{t=0}^{n} (\pm A_t)(1+i^*)^{n-t} = 0$$

$$PW_{t=0} = \sum_{t=0}^{n} (\pm A_t)(1+i^*)^{-t} = 0$$

where
"+" is used in the summation for positive cash flow amounts or receipts and "−" is used for negative cash flow amounts or disbursements
A_t denotes the cash flow amount at time t, which may be a receipt (+) or a disbursement (−)
The value of i^* is referred to as *discounted cash flow rate of return, IRR,* or *true rate of return*

The procedure above essentially calculates the net future worth or the net present worth of the cash flow. That is,

Net future worth = Future worth of receipts − future worth of disbursements

NFW = FW(receipts) − FW(disbursements)

Net present worth = Present worth of receipts − present worth of disbursements

NPW = PW(receipts) − PW(disbursements)

Setting the NPW or NFW equal to zero and solving for the unknown variable i determines the IRR of the cash flow.

Benefit–Cost Ratio Analysis

The benefit cost ratio of a cash flow is the ratio of the present worth of benefits to the present worth of costs. This is defined as follows:

$$B/C = \frac{\sum_{t=0}^{n} B_t (1+i)^{-t}}{\sum_{t=0}^{n} C_t (1+i)^{-t}} = \frac{PW_{benefits}}{PW_{costs}}$$

where
B_t is the benefit (receipt) at time t
C_t is the cost (disbursement) at time t

If the benefit–cost ratio is greater than one, then the investment is acceptable. If the ratio is less than one, the investment is not acceptable. A ratio of one indicates a breakeven situation for the project.

Simple Payback Period

Payback period refers to the length of time it will take to recover an initial investment. The approach does not consider the impact of the time value of money. Consequently, it is not an accurate method of evaluating the worth of an investment. However, it is a simple technique that is used widely to perform a "quick-and-dirty" assessment of investment performance. Another limitation of the technique is that it considers only the initial cost. Other costs that may occur after time zero are not included in the calculation. The payback period is defined as the smallest value of n (n_{min}) that satisfies the following expression:

$$\sum_{t=1}^{n_{min}} R_t \geq C$$

where
  R_t is the revenue at time t
  C_0 is the initial investment

The procedure calls for a simple addition of the revenues period by period until enough total has been accumulated to offset the initial investment.

Example: An organization is considering installing a new computer system that will generate significant savings in material and labor requirements for order processing. The system has an initial cost of $50,000. It is expected to save the organization $20,000 a year. The system has an anticipated useful life of 5 years with a salvage value of $5000. Determine how long it would take for the system to pay for itself from the savings it is expected to generate. Since the annual savings are uniform, we can calculate the payback period by simply dividing the initial cost by the annual savings. That is,

$$n_{min} = \frac{\$50,000}{\$20,000}$$
$$= 2.5 \text{ years}$$

Note that the salvage value of $5000 is not included in the above calculation since the amount is not realized until the end of the useful life of the asset (i.e., after 5 years). In some cases, it may be desired to consider the salvage value. In that case, the amount to be offset by the annual savings will be the net cost of the asset. In that case, we would have the following:

$$n_{min} = \frac{\$50,000 - \$5000}{\$20,000}$$
$$= 2.25 \text{ years}$$

If there are tax liabilities associated with the annual savings, those liabilities must be deducted from the savings before the payback period is calculated.

Discounted Payback Period

In this book, we introduce the *discounted payback period* approach, in which the revenues are reinvested at a certain interest rate. The payback period is determined when enough money has been accumulated at the given interest rate to offset the initial cost as well as other interim costs. In this case, the calculation is done by the following expression:

$$\sum_{t=1}^{n_{min}} R_t (1+i)^{n_{min}-1} \geq \sum_{t=0}^{n_{min}} C_t$$

Example: A new solar cell unit is to be installed in an office complex at an initial cost of $150,000. It is expected that the system will generate annual cost savings of $22,500 on the electricity bill. The solar cell unit will need to be overhauled every 5 years at a cost of $5000 per overhaul. If the annual interest rate is 10%, find the discounted payback period for the solar cell unit considering the time value of money. The costs of overhaul are to be considered in calculating the discounted payback period.

Solution: Using the single payment compound amount factor for one period iteratively, the following set of solutions is obtained for cumulative savings for each time period:

Period 1: $22,500
Period 2: $22,500 + $22,500 (1.10)^1 = $47,250
Period 3: $22,500 + $47,250 (1.10)^1 = $74,475
Period 4: $22,500 + $74,475 (1.10)^1 = $104,422.50
Period 5: $22,500 + $104,422.50 (1.10)^1 − $5000 = $132,364.75
Period 6: $22,500 + $132,364.75 (1.10)^1 = $168,101.23

The initial investment is $150,000. By the end of period 6, we have accumulated $168,101.23, which is more than the initial cost. Interpolating between period 5 and period 6 results in n_{min} of 5.49 years. That is, it will take 5.5 years to recover the initial investment. The calculation is shown below:

$$n_{min} = 5 + \frac{150,000 - 132,364.75}{168,101.25 - 132,364.75}(6-5)$$
$$= 5.49$$

Time Required to Double Investment

It is sometimes of interest to determine how long it will take a given investment to reach a certain multiple of its initial level. The "Rule of 72" is one simple approach to calculate the time required to for an investment to double in value at a given interest rate per period. The Rule of 72 gives the following formula for estimating the time required:

$$n = \frac{72}{i}$$

where i is the interest rate expressed in percentage. Referring to the single payment compound amount factor, we can set the future amount equal to twice the present amount and then solve for n. That is, $F = 2P$. Thus,

STEP Cost Management

$$2P = P(1+i)^n$$

Solving for n in the above equation yields an expression for calculating the exact number of periods required to double P:

$$n = \frac{\ln(2)}{\ln(1+i)}$$

where i is the interest rate expressed in decimals. In general, the length of time it would take to accumulate m multiples of P is expressed as

$$n = \frac{\ln(m)}{\ln(1+i)}$$

where m is the desired multiple. For example, at an interest rate of 5% per year, the time it would take an amount, P, to double in value ($m = 2$) is 14.21 years. This, of course, assumes that the interest rate will remain constant throughout the planning horizon. Table 5.5 presents a tabulation of the values calculated from both approaches. Figure 5.8 shows a graphical comparison of the Rule of 72 to the exact calculation.

Effects of Inflation on Project Costing

Inflation can be defined as the decline in purchasing power of money, and as such, is a major player in the financial and economic analysis of projects. Multiyear projects are particularly subject to the effects of inflation. Some of the most common causes of inflation include the following:

TABLE 5.5
Evaluation of the Rule of 72

i%	n(Rule of 72)	n(Exact Value)
0.25	288.00	277.61
0.50	144.00	138.98
1.00	72.00	69.66
2.00	36.00	35.00
5.00	14.20	17.67
8.00	9.00	9.01
10.00	7.20	7.27
12.00	6.00	6.12
15.00	4.80	4.96
18.00	4.00	4.19
20.00	3.60	3.80
25.00	2.88	3.12
30.00	2.40	2.64

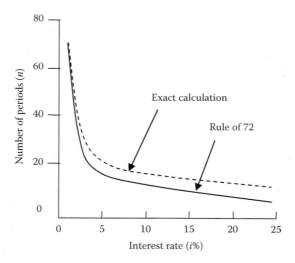

FIGURE 5.8 Evaluation of investment life for double return.

- An increase in the amount of currency in circulation
- A shortage of consumer goods
- An escalation of the cost of production
- An arbitrary increase in prices set by resellers

The general effects of inflation are felt in terms of an increase in the prices of goods and a decrease in the worth of currency. In cash flow analysis, ROI for a project will be affected by time value of money as well as inflation. The real interest rate (d) is defined as the desired rate of return in the absence of inflation. When we talk of "today's dollars" or "constant dollars," we are referring to the use of the real interest rate. The combined interest rate (i) is the rate of return combining the real interest rate and the inflation rate. If we denote the inflation rate as j, then the relationship between the different rates can be expressed as shown below:

$$1 + i = (1 + d)(1 + j)$$

Thus, the combined interest rate can be expressed as follows:

$$i = d + j + dj$$

Note that if $j = 0$ (i.e., no inflation), then $i = d$. We can also define commodity escalation rate (g) as the rate at which individual commodity prices escalate. This may be greater than or less than the overall inflation rate. In practice, several measures are used to convey inflationary effects. Some of these are the consumer price index, the producer price index, and the wholesale price index. A "market basket" rate is defined as the estimate of inflation based on a weighted average of the annual rates of change in the costs of a wide range of representative commodities. A "then-current" cash flow is a cash flow that explicitly incorporates the impact of inflation.

STEP Cost Management

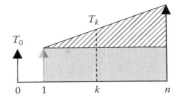

FIGURE 5.9 Cash flows for effects of inflation.

A "constant worth" cash flow is a cash flow that does not incorporate the effect of inflation. The real interest rate, d, is used for analyzing constant worth cash flows. Figure 5.9 shows constant worth and then-current cash flows.

The then-current cash flow in the figure is the equivalent cash flow considering the effect of inflation. C_k is what it would take to buy a certain "basket" of goods after k time periods if there was no inflation. T_k is what it would take to buy the same "basket" in k time period if inflation were taken into account. For the constant worth cash flow, we have

$$C_k = T_0, \quad k = 1, 2, \ldots, n$$

and for the then-current cash flow, we have

$$T_k = T_0(1+j)^k, \quad k = 1, 2, \ldots, n$$

where j is the inflation rate. If $C_k = T_0 = \$100$ under the constant worth cash flow, then we have $\$100$ worth of buying power. If we are using the commodity escalation rate, g, then we will have

$$T_k = T_0(1+g)^k, \quad k = 1, 2, \ldots, n$$

Thus, a then-current cash flow may increase based on both a regular inflation rate (j) and a commodity escalation rate (g). We can convert a then-current cash flow to a constant worth cash flow by using the following relationship:

$$C_k = T_k(1+j)^{-k}, \quad k = 1, 2, \ldots, n$$

If we substitute T_k from the commodity escalation cash flow into the expression for C_k above, we get the following:

$$\begin{aligned}
C_k &= T_k(1+j)^{-k} \\
&= T(1+g)^k(1+j)^{-k} \\
&= T_0\left[(1+g)/(1+j)\right]^k, \quad k = 1, 2, \ldots, n
\end{aligned}$$

Note that if $g = 0$ and $j = 0$, the $C_k = T_0$. That is, there is no inflationary effect. We can now define the effective commodity escalation rate (v):

$$v = [(1 + g)/(1 + j)] - 1$$

The commodity escalation rate (g) can be expressed as follows:

$$g = v + j + vj$$

Inflation can have a significant impact on the financial and economic aspects of an industrial project. Inflation may be defined, in economic terms, as the increase in the amount of currency in circulation. To a producer, inflation means a sudden increase in the cost of items that serve as inputs for the production process (equipment, labor, materials, etc.). To the retailer, inflation implies an imposed higher cost of finished products. To an ordinary citizen, inflation portends a noticeable escalation of prices of consumer goods. All these aspects are intertwined in a project management environment.

The amount of money supply, as a measure of a country's wealth, is controlled by the government. When circumstances dictate such action, governments often feel compelled to create more money or credit to take care of old debts and pay for social programs. When money is generated at a faster rate than the growth of goods and services, it becomes a surplus commodity and its value (i.e., purchasing power) will fall. This means that there will be too much money available to buy only a few goods and services. When the purchasing power of a currency falls, each individual in a product's life cycle (i.e., each person or entity that spends money on a product throughout its life cycle from production through disposal) has to use more of the currency in order to obtain the product. Some of the classic concepts of inflation are discussed below:

1. In *cost-driven* or *cost-push inflation*, increases in producer's costs are passed on to consumers. At each stage of the product's journey from producer to consumer, prices are escalated disproportionately in order to make a good profit. The overall increase, in the product's price is directly proportional to the number of intermediaries it encounters on its way to the consumer.
2. In *demand-driven* or *demand-pull inflation, excessive* spending power of consumers forces an upward trend in prices. This high spending power is usually achieved at the expense of savings. The law of supply and demand dictates that the more the demand, the higher the price. This results in demand-driven or demand-pull inflation.
3. Impact of international economic forces can induce inflation on a local economy. Trade imbalances and fluctuations in currency values are notable examples of international inflationary factors.
4. In wage-driven or wage-push inflation, the increasing base wages of workers generate more disposable income and hence higher demands for goods and services. The high demand consequently creates a pull on prices. Coupled with this, employers pass the additional wage cost on to consumers through higher prices. This type of inflation is very difficult to contain

because wages set by union contracts and prices set by producers almost never fall.
5. Easy availability of credit leads consumers to "buy now and pay later," thereby creating another opportunity for inflation. This is a dangerous type of inflation because the credit not only pushes prices up but also leaves consumers with less money later to pay for the credit. Eventually, many credits become uncollectible debts, which may then drive the economy toward recession.
6. Deficit spending results in an increase in money supply and thereby creates less room for each dollar to get around. The popular saying indicating that "a dollar does not go far anymore" simply refers to inflation in laymen's terms. The different levels of inflation may be categorized as discussed below.

Mild inflation
When inflation is mild (at 2–4%), the economy actually prospers. Producers strive to produce at full capacity in order to take advantage of the high prices to the consumer. Private investments tend to be brisk and more jobs become available. However, the good fortune may only be temporary. Prompted by the prevailing success, employers are tempted to seek larger profits and workers begin to ask for higher wages. They cite their employer's prosperous business as a reason to bargain for bigger shares of the business profit. So, we end up with a vicious cycle where the producer asks for higher prices, the unions ask for higher wages, and inflation starts an upward trend.

Moderate inflation
Moderate inflation occurs when prices increase at 5–9%. Consumers start purchasing more as a hedge against inflation. They would rather spend their money now than watch it decline further in purchasing power. The increased market activity serves to fuel further inflation.

Severe inflation
Severe inflation is indicated by price escalations of 10% or more. Double-digit inflation implies that prices rise much faster than wages do. Debtors tend to be the ones who benefit from this level of inflation because they repay debts with money that is less valuable than when they borrowed.

Hyperinflation
When each price increase signals an increase in wages and costs, which again sends prices further up, the economy has reached a stage of malignant galloping inflation or hyperinflation. Rapid and uncontrollable inflation destroys the economy. The currency becomes economically useless as the government prints it excessively to pay for obligations.

Inflation can affect any industrial project in terms of raw materials procurement, salaries and wages, and/or cost tracking dilemmas. Some effects are immediate and easily observable while others are subtle and pervasive. Whatever form it takes, inflation must be taken into account in long-term project planning and control. Large projects especially may be adversely affected by the effects of inflation in terms of cost overruns and poor resource utilization. Managers should note that the level of inflation will determine the severity of the impact on projects.

BREAKEVEN ANALYSIS

Breakeven analysis refers to the determination of the balanced performance level where project income is equal to project expenditure. The total cost of an operation is expressed as the sum of the fixed and variable costs with respect to output quantity. That is,

$$TC(x) = FC + VC(x)$$

where
  x is the number of units produced
  $TC(x)$ is the total cost of producing x units
  FC is the total fixed cost
  $VC(x)$ is the total variable cost associated with producing x units

The total revenue resulting from the sale of x units is defined as

$$TR(x) = px$$

where p is the price per unit. The profit due to the production and sale of x units of the product is calculated as

$$P(x) = TR(x) - TC(x)$$

The breakeven point of an operation is defined as the value of a given parameter that will result in neither profit nor loss. The parameter of interest may be the number of units produced, the number of hours of operation, the number of units of a resource type allocated, or any other measure of interest. At the breakeven point, we have the following relationship:

$$TR(x) = TC(x) \text{ or } P(x) = 0$$

In some cases, there may be a known mathematical relationship between cost and the parameter of interest. For example, there may be a linear cost relationship between the total cost of a project and the number of units produced. The cost expressions facilitate a straightforward breakeven analysis. Figure 5.10 shows an example of a breakeven point for a single project. Figure 5.11 shows examples of multiple breakeven points that exist when multiple projects are compared. When two project alternatives are compared, the breakeven point refers to the point of indifference between the two alternatives. In Figure 5.11, x_1 represents the point where projects A and B are equally desirable, x_2 represents where A and C are equally desirable, and x_3 represents where B and C are equally desirable. The figure shows that if we are operating below a production level of x_2 units, then project C is the preferred project among the three. If we are operating at a level more than x_2 units, then project A is the best choice.

Example: Three project alternatives are being considered for producing a new product. The required analysis involves determining which alternative should be selected on the basis of how many units of the product are produced per year. Based

STEP Cost Management

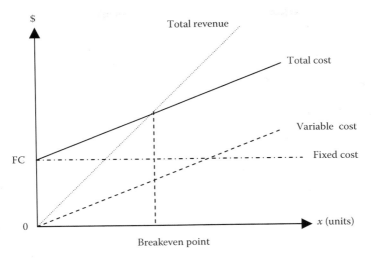

FIGURE 5.10 Breakeven point for a single project.

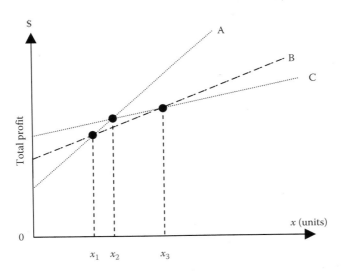

FIGURE 5.11 Breakeven points for multiple projects.

on past records, there is a known relationship between the number of units produced per year, x, and the net annual profit, $P(x)$, from each alternative. The level of production is expected to be between 0 and 250 units per year. The net annual profits (in thousands of dollars) are given below for each alternative:

Project A: $P(x) = 3x - 200$
Project B: $P(x) = x$
Project C: $P(x) = (1/50)x^2 - 300$.

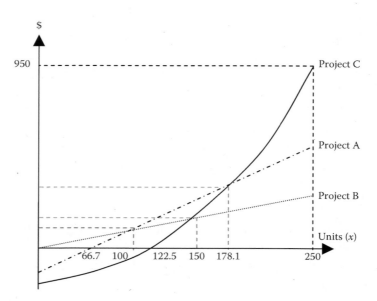

FIGURE 5.12 Plot of profit functions.

This problem can be solved mathematically by finding the intersection points of the profit functions and evaluating the respective profits over the given range of product units. It can also be solved by a graphical approach. Figure 5.12 shows a plot of the profit functions. Such a plot is called a breakeven chart. The plot shows that Project B should be selected if between 0 and 100 units are to be produced, Project A should be selected if between 100 and 178.1 units (178 physical units) are to be produced, and Project C should be selected if more than 178 units are to be produced. It should be noted that if less than 66.7 units (66 physical units) are produced, Project A will generate a net loss rather than a net profit. Similarly, Project C will generate losses if less than 122.5 units (122 physical units) are produced.

Profit Ratio Analysis

Breakeven charts offer opportunities for several different types of analysis. In addition to the breakeven points, other measures of worth or criterion measures may be derived from the charts. A measure called the *profit ratio* is presented here for the purpose of obtaining a further comparative basis for competing projects. A profit ratio is defined as the ratio of the profit area to the sum of the profit and loss areas in a breakeven chart. That is,

$$\text{Profit ratio} = \frac{\text{Area of profit region}}{\text{Area of profit region} + \text{Area of loss region}}$$

For example, suppose that the expected revenue and the expected total cost associated with a project are given, respectively, by the following expressions:

STEP Cost Management

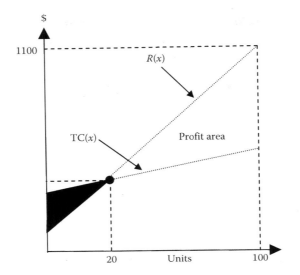

FIGURE 5.13 Area of profit versus area of loss.

$$R(x) = 100 + 10x$$
$$TC(x) = 2.5x + 250$$

where x is the number of units produced and sold from the project. Figure 5.13 shows the breakeven chart for the project. The breakeven point is shown to be 20 units. Net profits are realized from the project if more than 20 units are produced and net losses are realized if less than 20 units are produced. It should be noted that the revenue function in Figure 5.13 represents an unusual case, in which a revenue of $100 is realized when zero units are produced.

Suppose it is desired to calculate the profit ratio for this project if the number of units that can be produced is limited to between 0 and 100 units. From Figure 5.13, the surface area of the profit region and the area of the loss region can be calculated by using the standard formula for finding the area of a triangle: area = (1/2)(base)(height). Using this formula, we have the following:

$$\text{Area of profit region} = \frac{1}{2}(\text{base})(\text{height})$$
$$= \frac{1}{2}(1100 - 500)(100 - 20)$$
$$= 24{,}000 \text{ square units}$$

$$\text{Area of loss region} = \frac{1}{2}(\text{base})(\text{height})$$
$$= \frac{1}{2}(250 - 100)(20)$$
$$= 1500 \text{ square units}$$

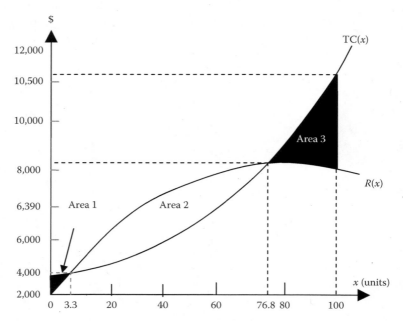

FIGURE 5.14 Breakeven chart for revenue and cost functions.

Thus, the profit ratio is computed as follows:

$$\text{Profit ratio} = 24{,}000/(24{,}000 + 1500) = 0.9411 \equiv 94.11\%$$

The profit ratio may be used as a criterion for selecting among project alternatives. If this is done, the profit ratios for all the alternatives must be calculated over the same values of the independent variable. The project with the highest profit ratio will be selected as the desired project. For example, Figure 5.14 presents the breakeven chart for an alternate project, say Project II. It can be seen that both the revenue and cost functions for the project are nonlinear. The revenue and cost are defined as follows:

$$R(x) = 160x - x^2$$

$$TC(x) = 500 + x^2$$

If the cost and/or revenue functions for a project are not linear, the areas bounded by the functions may not be easily determined. For those cases, it may be necessary to use techniques such as definite integrals to find the areas. Figure 5.14 indicates that the project generates a loss if less than 3.3 units (3 actual units) are produced or if more than 76.8 units (76 actual units) are produced. The respective profit and loss areas on the chart are calculated as shown below:

Area 1 (loss) = 802.80 unit-dollars
Area 2 (profit) = 132,272.08 unit-dollars
Area 3 (loss) = 48,135.98 unit-dollars

Consequently, the profit ratio for Project II is computed as

$$\text{Profit ratio} = \frac{\text{Total area of profit region}}{\text{Total area of profit region} + \text{Total area of loss region}}$$
$$= \frac{132{,}272.08}{802.76 + 132{,}272.08 + 48{,}135.98}$$
$$= 72.99\%$$

The profit ratio approach evaluates the performance of each alternative over a specified range of operating levels. Most of the existing evaluation methods use single-point analysis with the assumption that the operating condition is fixed at a given production level. The profit ratio measure allows an analyst to evaluate the net yield of an alternative, given that the production level may shift from one level to another. An alternative, for example, may operate at a loss for most of its early life, but it may generate large incomes to offset those losses in its later stages. Conventional methods cannot easily capture this type of transition from one performance level to another. In addition to being used to compare alternate projects, the profit ratio may also be used for evaluating the economic feasibility of a single project. In such a case, a decision rule may be developed, such as the following:

If profit ratio is greater than 75%, accept the project.
If profit ratio is less than or equal to 75%, reject the project.

PROJECT COST ESTIMATION

Cost estimation and budgeting help establish a strategy for allocating resources in project planning and control. Based on the desired level of accuracy, there are three major categories of cost estimation for budgeting: order-of-magnitude estimates, preliminary cost estimates, and detailed cost estimates. Order-of-magnitude cost estimates are usually gross estimates based on the experience and judgment of the estimator. They are sometimes called "ballpark" figures. These estimates are typically made without a formal evaluation of the details involved in the project. The level of accuracy associated with order-of-magnitude estimates can range from −50% to +50% of the actual cost. These estimates provide a quick way of getting cost information during the initial stages of a project. The estimation range is summarized as follows:

$50\%(\text{actual cost}) \leq \text{order-of-magnitude estimate} \leq 150\%(\text{actual cost})$

Preliminary cost estimates are also gross estimates, but with a higher level of accuracy. In developing preliminary cost estimates, more attention is paid to some selected details of the project. An example of a preliminary cost estimate is the estimation of expected labor cost. Preliminary estimates are useful for evaluating project alternatives before final commitments are made. The level of accuracy associated with preliminary estimates can range from −20% to +20% of the actual cost, as shown below:

$80\%(\text{actual cost}) \leq \text{preliminary estimate} \leq 120\%(\text{actual cost})$

Detailed cost estimates are developed after careful consideration is given to all the major details of a project. Considerable time is typically needed to obtain detailed cost estimates. Because of the amount of time and effort needed to develop detailed cost estimates, the estimates are usually developed after a firm commitment has been made that the project will take off. Detailed cost estimates are important for evaluating actual cost performance during the project. The level of accuracy associated with detailed estimates normally ranges from −5% to +5% of the actual cost.

$$95\%(\text{actual cost}) \leq \text{detailed cost} \leq 105\%(\text{actual cost})$$

There are two basic approaches to generating cost estimates. The first one is a variant approach, in which cost estimates are based on variations of previous cost records. The other approach is the generative cost estimation, in which cost estimates are developed from scratch without taking previous cost records into consideration.

Optimistic and Pessimistic Cost Estimates

Using an adaptation of the PERT formula, we can combine optimistic and pessimistic cost estimates. If O = optimistic cost estimate, M = most likely cost estimate, and P = pessimistic cost estimate, the estimated cost can be stated as follows:

$$E[C] = \frac{O + 4M + P}{6}$$

and the cost variance can be estimated as follows:

$$V[C] = \left[\frac{P - O}{6}\right]^2$$

Project Budget Allocation

Project budget allocation involves sharing limited resources among competing tasks in a project. The budget allocation process serves the following purposes:

- Plan for resource expenditure
- Project selection criterion
- Projection of project policy
- Basis for project control
- A performance measure
- A standardization of resource allocation
- An incentive for improvement

Top-Down Budgeting

Top-down budgeting involves collecting data from upper-level sources such as top and middle managers. The figures supplied by the managers may come from their personal judgment, past experience, or past data on similar project activities. The cost estimates are passed to lower-level managers, who then break the estimates down into

STEP Cost Management

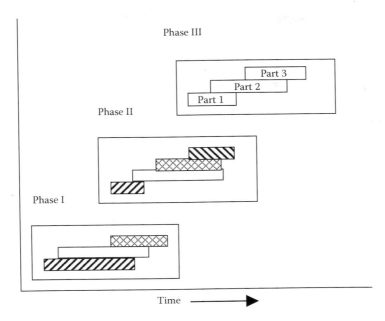

FIGURE 5.15 Budgeting by project phases.

specific work components within the project. These estimates may, in turn, be given to line managers, supervisors, and lead workers to continue the process until individual activity costs are obtained. Thus, top management provides the global budget, while the functional level worker provides specific budget requirements for project items.

BOTTOM-UP BUDGETING

In this method, elemental activities, their schedules, descriptions, and labor skill requirements are used to construct detailed budget requests. Line workers familiar with specific activities are asked to provide cost estimates and then make estimates for each activity in terms of labor time, materials, and machine time. The estimates are then converted to an appropriate cost basis. The dollar estimates are combined into composite budgets at each successive level up the budgeting hierarchy. If estimate discrepancies develop, they can be resolved through the intervention of senior management, middle management, functional managers, project manager, accountants, or standard cost consultants. Figure 5.15 shows the breaking down of a project into phases and parts in order to facilitate bottom-up budgeting and improve both schedule and cost control.

Elemental budgets may be developed on the basis of the timed progress of each part of the project. When all the individual estimates are gathered, we can obtain a composite budget estimate. Figures 5.16 and 5.17 show an example of the various components that may be involved in an overall budget. The bar chart appended to a segment of the pie chart indicates the individual cost components making up that particular segment. To further aid in the process, analytical tools such as learning curve analysis, work sampling, and statistical estimation may be employed in the cost estimation and budgeting processes.

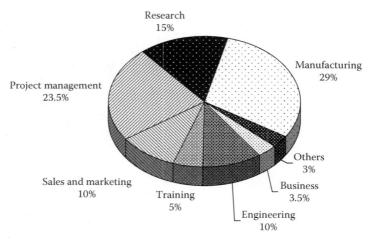

FIGURE 5.16 Pie chart of budget distribution.

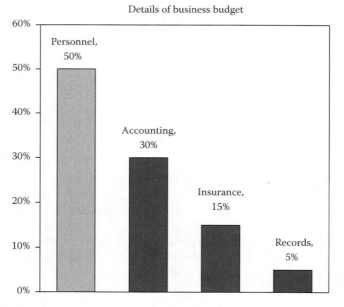

FIGURE 5.17 Bar chart of budget and distribution.

BUDGETING AND RISK ALLOCATION FOR TYPES OF CONTRACT

Budgeting and allocation of risk are handled based on the type of contract involved. The list below carries progressively higher risk to the buyer (customer) while it carries progressively lower risk to the contractor (producer):

    Type 1:   Firm fixed price (FFP)
    Type 2:   FFP with economic adjustment
    Type 3:   Fixed price incentive fee (FPIF)

STEP Cost Management

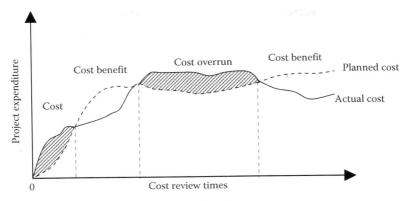

FIGURE 5.18 Evaluation of actual and projected cost.

Type 4:   Cost and cost sharing (CCS)
Type 5:   Cost plus incentive fee (CPIF)
Type 6:   Cost plus award fee (CPFF)
Type 7:   Cost plus fixed fee (CPFF)
Type 8:   Cost plus percentage fee (CPPF)
Type 9:   Indefinite delivery
Type 10: Time and materials
Type 11: Basic agreements (Blanket contract)

Type 1 contract carries the highest risk to the contractor (producer) whereas it carries the lowest risk to the buyer (customer). Type 11 contract carries the lowest risk to the contractor (producer) whereas it carries the highest risk to the buyer (customer). The risk level is progressive in each direction of the list.

Cost Monitoring

As a project progresses, costs can be monitored and evaluated to identify areas of unacceptable cost performance. Figure 5.18 shows a plot of cost versus time for projected cost and actual cost. The plot permits a quick identification of the points at which cost overruns occur in a project.

Plots similar to those presented above may be used to evaluate cost, schedule, and time performance of a project. An approach similar to the profit ratio presented earlier may be used along with the plot to evaluate the overall cost performance of a project over a specified planning horizon. Presented below is a formula for cost performance index (CPI):

$$\text{CPI} = \frac{\text{Area of cost benefit}}{\text{Area of cost benefit} + \text{Area of cost overrun}}$$

As in the case of the profit ratio, CPI may be used to evaluate the relative performances of several project alternatives or to evaluate the feasibility and acceptability of an individual alternative. In Figure 5.19, we present another cost-monitoring tool, referred to as a cost–control pie chart. The chart is used to track the percentage of the cost going into a specific component of a project. Control limits can be included in

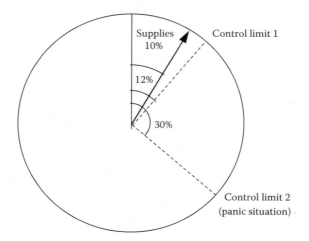

FIGURE 5.19 Cost–control pie chart.

the pie chart to identify costs that have become out of control. The example in Figure 5.19 shows that 10% of total cost is tied up in supplies. The control limit is located at 12% of total cost. Hence, the supplies expenditure is within control (so far, at least).

PROJECT BALANCE TECHNIQUE

One other approach to monitoring cost performance is the project balance technique. The technique helps in assessing the economic state of a project at a desired point in time in the life cycle of the project. It calculates the net cash flow of a project up to a given point in time. The project balance is calculated as follows:

$$B(i)_t = S_t - P(1+i)^t + \sum_{k=1}^{t} \text{PW}_{\text{income}}(i)_k$$

where
$B(i)_t$ = project balance at time t at an interest rate of $i\%$ per period
$\text{PW}_{\text{income}}(i)_k$ = present worth of net income from the project up to time k
P = initial cost of the project
S_t = salvage value at time t

The project balance at time t gives the net loss or net profit associated with the project up to that time.

COST AND SCHEDULE CONTROL SYSTEMS CRITERIA

Contract management involves the process by which goods and services are acquired, utilized, monitored, and controlled in a project. Contract management addresses the contractual relationships from the initiation of a project to the completion of the project (i.e., completion of services and/or hand over of deliverables). Some of the important aspects of contract management that STEP practitioners should be familiar, which include

STEP Cost Management

- Principles of contract law
- Bidding process and evaluation
- Contract and procurement strategies
- Selection of source and contractors
- Negotiation
- Worker safety considerations
- Product liability
- Uncertainty and risk management
- Conflict resolution

In 1967, the U.S. Department of Defense (DOD) introduced a set of 35 standards or criteria with which contractors must comply under cost or incentive contracts. The system of criteria is referred to as the Cost and Schedule Control Systems Criteria (C/SCSC). Although no longer in vogue, many government agencies still require compliance with modified and updated versions of C/SCSC, albeit under different "new" and trendy monikers. The primary goal of C/SCSC is to manage the risk of cost overrun to the government on major contracts. That goal is a desirable pursuit of any modern cost management and contract administration system although actual implementation is often lamentable. The C/SCSC system presents an integrated approach to cost and schedule management. This "integrated approach" is in agreement with the premise of STEP project management as presented in this book. C/SCSC has been widely used in major project undertakings. It is intended to facilitate greater uniformity and provide advance warning about impending schedule or cost overruns as well as performance risks. Some of the factors influencing schedule, performance, and cost problems are summarized below; with suggested lists of control actions:

Causes of schedule problems

- Delay of critical activities
- Unreliable time estimates
- Technical problems
- Precedence structure
- Change of due dates
- Bad time estimates
- Changes in management direction

Schedule control actions

- Use activity crashing
- Redesign tasks
- Revise milestones
- Update time estimates
- Change the scope of work
- Combine related activities
- Eliminate unnecessary activities (i.e., operate lean)

Causes of performance problems

- Poor quality
- Poor functionality

- Maintenance problems
- Poor mobility (knowledge transfer)
- Lack of training
- Lack of clear objectives

Performance control actions

- Use SMART (specific, measurable, aligned, realistic, timed) job objectives
- Use improved tools/technology
- Adjust project specifications
- Improve management oversight
- Review project priorities
- Modify project scope
- Allocate more resources
- Require higher level of accountability
- Improve work ethics (through training, mentoring, and education)

Causes of cost problems

- Inadequate budget
- Effects of inflation
- Poor cost reporting
- Increase in scope of work
- High overhead cost
- High labor cost

Cost control actions

- Reduce labor costs
- Use competitive bidding
- Modify work process
- Adjust work breakdown structure
- Improve coordination of project functions
- Improve cost estimation procedures
- Use less expensive raw materials
- Mitigate effects of inflationary trends (e.g., use of price hedging in procurement)
- Cut overhead costs
- Outsource work

The topics covered by C/SCSC or any of its modern derivates include cost estimating and forecasting, budgeting, cost control, cost reporting, earned value analysis, resource allocation and management, and schedule adjustments. There is no doubt that the contemporary evolution of cost management as presented in PMI's *PMBOK* was influenced by the foundational contents of C/SCSC. The important link between all of these developments is the dynamism of the relationship between performance, time, and cost, as was alluded to earlier in this book. Figure 5.20 illustrates an

STEP Cost Management

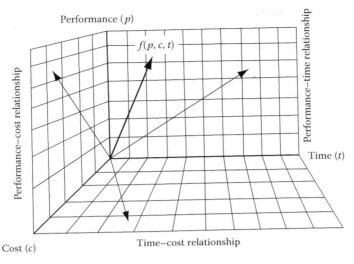

FIGURE 5.20 Cost–schedule–performance relationships.

example of the dynamism that exists in cost–schedule–performance relationships. The relationships represent a multi-objective problem. The resultant function, $f(p, c, t)$, in Figure 5.20 represents a vector of decision taking into account the relative nuances of project cost, schedule, and performance. Because performance, time, and cost objectives cannot be satisfied equally well, concessions or compromises need to be worked out in implementing C/SCSC or other project control criteria.

Another dimension of the performance–time–cost relationship is the U.S. Air Force's R&M 2000 standard, which addresses the reliability and maintainability of systems. R&M 2000 is intended to integrate reliability and maintainability into the performance, cost, and schedule management for government contracts. Together, C/SCSC, R&M 2000, and other recent project control guides constitute an effective template for industrial project planning, organizing, and control.

To comply with the ideals of cost management, contractors must use standardized planning and control methods based on *earned value*. Earned value refers to the actual dollar value of work performed at a given point in time compared to planned cost for the work. This is different from the conventional approach of measuring actual versus planned, which is explicitly forbidden by C/SCSC. In the conventional approach, it is possible to misrepresent the actual content (or value) of the work accomplished. The work rate analysis technique can be useful in overcoming the deficiencies of the conventional approach. C/SCSC is developed on a work content basis using the following factors:

- Actual cost of work performed (ACWP), which is determined on the basis of the data from cost accounting and information systems
- Budgeted cost of work scheduled (BCWS) or baseline cost determined by the costs of scheduled accomplishments
- Budgeted cost of work performed (BCWP) or earned value, the actual work of effort completed as of a specific point in time

The following equations can be used to calculate cost and schedule variances for a work package at any point in time.

$$\text{Cost variance} = \text{BCWP} - \text{ACWP}$$
$$\text{Percent cost variance} = 100 \times (\text{Cost variance}/\text{BCWP})$$
$$\text{Schedule variance} = \text{BCWP} - \text{BCWS}$$
$$\text{Percent schedule variance} = 100 \times (\text{Schedule variance}/\text{BCWS})$$
$$\text{ACWP and remaining funds} = \text{Target cost (TC)}$$
$$\text{ACWP} + \text{cost to complete} = \text{Estimated cost at completion (EAC)}.$$

The above characteristics of C/SCSC and R&M 2000 have undergone application modifications in recent years. Several new systems of cost control are now available in practice. The essential elements of cost control in any new approach are discussed in the section that follows.

ELEMENTS OF COST CONTROL

Cost control, in the context of cost management, refers to the process of regulating or rectifying cost attributes to bring them within acceptable levels. Because of the volatility and dynamism often encountered in STEPs, it is imperative to embrace the following project cost control practices as presented in *PMBOK*:

- Influence the factors that create changes to the cost baseline
- Ensure requested changes are agreed upon
- Manage the actual changes when and as they occur
- Assure that potential cost overruns do not exceed authorized funding (by period and in total)
- Monitor cost performance to detect and understand variances from the cost baseline
- Record all appropriate changes accurately against the cost baseline
- Prevent incorrect, inappropriate, or unapproved changes from being included in cost reports
- Inform appropriate stakeholders or approved changes
- Act to bring expected cost overruns within acceptable limits
- Use earned value technique (EVT) to track and rectify cost performance

CONTEMPORARY EARNED VALUE TECHNIQUE

This section details the elements of a contemporary EVT. EVT is used primarily for cost control purposes. The technique involves developing important diagnostic values for each schedule activity, work package, or control element. Although the definitions presented below are similar to those in the foregoing C/SCSC discussions, there are shades of differences that are important to highlight. The definitions according to PMI's *PMBOK* are summarized below:

Planned value (PV): This is the budgeted cost for the work scheduled to be completed on an activity or WBS element up to a given point in time.

Earned value (EV): This is the budgeted amount for the work actually completed on the schedule activity or WBS component during a given time period.

STEP Cost Management

Actual cost (AC): This is the total cost incurred in accomplishing work on the schedule activity or WBS component during a given time period. AC must correspond in definition, scale, units, and coverage to whatever was budgeted for PV and EV. For example, direct hours only, direct costs only, or all costs including indirect costs.

The PV, EV, and AC values are used jointly to provide performance measures of whether or not work is being accomplished as planned at any given point in time. The common measures of project assessment are cost variance (CV) and schedule variance (SV).

Cost variance (CV): This equals earned value minus actual cost. The cost variance at the end of the project will be the difference between the budget at completion (BAC) and the actual amount expended.

$$CV = EV - AC$$

Schedule variance (SV): This equals earned value minus planned value. Schedule variance will eventually become zero when the project is completed because all of the planned values will have been earned.

$$SV = EV - PV$$

Cost performance index (CPI): This is an efficiency indicator relating earned value to actual cost. It is the most commonly used cost-efficiency indicator. CPI value less than 1.0 indicates a cost overrun of the estimates. CPI value greater than 1.0 indicates a cost advantage (underrun) of the estimates.

$$CPI = \frac{EV}{AC}$$

Cumulative CPI (CPI^C): This is a measure that is widely used to forecast project costs at completion. It equals the sum of the periodic earned values (Cum. EV) divided by the sum of the individual actual costs (Cum. AC):

$$CPI^C = \frac{EV^C}{AC^C}$$

Schedule performance index (SPI): This is a measure that is used to predict the completion date of a project. It is used in conjunction with CPI to forecast project completion estimates:

$$SPI = \frac{EV}{PV}$$

Estimate to complete (ETC) *based on new estimate*: ETC equals the revised estimate for the work remaining as determined by the performing organization. This is an independent noncalculated estimate to complete for all the work remaining. It considers the performance or production of the resources to date. The calculation of ETC uses two alternate formulas based on earned value data.

ETC based on atypical variances: This calculation approach is used when current variances are seen as *atypical* and the expectations of the project team are that similar variances will *not* occur in the future:

$$ETC = BAC - EV^C$$

where BAC = budget at completion.

ETC based on typical variances: This calculation approach is used when current variances are seen as *typical* of what to expect in the future:

$$ETC = \frac{BAC - EV^C}{CPI^C}$$

Estimate at completion (EAC): This is a forecast of the most likely total value based on project performance. EAC is the projected or anticipated total final value for a schedule activity, WBS component, or project when the defined work of the project is completed. One EAC forecasting technique is based upon the performance organization providing an estimate at completion. Two other techniques are based on earned value data. The three calculation techniques are presented below. Each of the three approaches can be effective for any given project because it can provide valuable information and signal if the EAC forecasts are not within acceptable limits.

EAC using a new estimate: The approach calculates the actual costs to date plus a new ETC that is provided by the performing organization. This is most often used when past performance shows that the original estimating assumptions were fundamentally flawed or that they are no longer relevant due to a change in project operating conditions.

$$\mathbf{EAC = AC^C + ETC}$$

EAC using remaining budget: In this approach, EAC is calculated as cumulative actual cost plus the budget that is required to complete the remaining work where the remaining work is the budget at completion minus the earned value. This approach is most often used when current variances are seen as *atypical* and the project management team expectations are that similar variances will not occur in the future.

$$EAC = AC^C + (BAC - EV)$$

where (BAC − EV) = remaining project work = remaining PV.

EAC using cumulative CPI: In this approach, EAC is calculated as actual costs to date plus the budget that is required to complete the remaining project work, modified by a performance factor. The performance factor of choice is usually the cumulative CPI. This approach is most often used when current variances are seen as *typical* of what to expect in the future.

$$EAC = AC^C + \frac{(BAC - EV)}{CPI^C}$$

STEP Cost Management

Other important definitions and computational relationships among the earned value variables are

Earned → Budgeted cost of work actually performed
Planned → Budgeted cost of work scheduled
Actual → Cost of actual work performed
Ending CV = Budget at completion − Actual amount spent at the end
    = BAC − EAC
    = VAC (Variance at completion)
EAC = ETC + AC
    = (BAC − EV) + AC
    = AC + (BAC − EV)
ETC = EAC − AC
    = BAC − EV

Figure 5.21 illustrates the relationships among the earned value variables discussed above.

Activity-Based Costing

Activity-based costing (ABC) has emerged as an effective costing technique for industrial projects. The major motivation for ABC is that it offers an improved method to achieve enhancements in operational and strategic decisions. ABC offers a mechanism to allocate costs in direct proportion to the activities that are actually performed. This is an improvement over the traditional way of generically allocating costs to departments. It also improves the conventional approaches to allocating overhead costs. In general, ABC is a method for estimating the resources required to operate an organization's business activities, produce its products, and provide services to its clients.

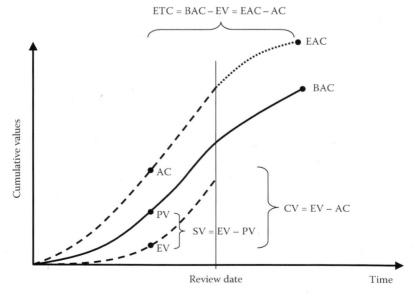

FIGURE 5.21 Graphical plot of earned value performance analysis.

The ABC methodology assigns resource costs through activities to the products and services provided to its customers. It is generally used as a tool for understanding product and customer costs with respect to project profitability. ABC is also frequently used to formulate strategic decisions such as product pricing, outsourcing, and process improvement efforts.

The use of PERT/CPM, precedence diagramming, the critical resource diagramming method, and work breakdown structure (WBS) can facilitate the decomposition or breakdown of a task to provide information for ABC. Some of the potential impacts of ABC on a production line include the following

- Identification and removal of unnecessary costs
- Identification of the cost impact of adding specific attributes to a product
- Indication of the incremental cost of improved quality
- Identification of the value-added points in a production process
- Inclusion of specific inventory carrying costs
- Provision of a basis for comparing production alternatives
- Ability to assess "what-if" scenarios for specific tasks

ABC is just one component of the overall activity-based management (ABM) in an organization, and thus has its limitations, as well. ABM involves a more global management approach to the planning and control of organizational endeavors. This requires consideration for product planning, resource allocation, productivity management, quality control, training, line balancing, value analysis, and a host of other organizational responsibilities. In the implementation of ABC, several issues must be considered:

- Level and availability of resources committed to developing activity-based information and cost
- Duration and level of effort needed to achieve ABC objectives
- Level of cost accuracy that can be achieved by ABC
- Ability to track activities based on ABC requirements
- Challenge of handling the volume of detailed information provided by ABC
- Sensitivity of the ABC system to changes in activity configuration

From ABM to ABC, there are both qualitative as well as quantitative aspects of tracking, managing, and controlling costs. Unfortunately, many attempts to use ABC often degenerate into conceptual arm-waving rather than real quantitative accountability. To be successful, the same SMART principle that was discussed previously can be applied for developing ABC strategies. Under ABM and ABC, cost tracking must satisfy the following SMART requirements:

Specific: Cost tracking must be specific so as to facilitate accountability

Measurable: Cost tracking must be measurable

Aligned: Cost tracking must be aligned with organization's goals

Realistic: Cost tracking must be realistic and within the organization's capability

Timed: Cost tracking must be timed in order to avoid ambiguities

Also, to increase the effectiveness of ABC, an organization should use parametric cost techniques, which utilize project characteristics (parameters) to develop mathematical models for cost management. In summary, STEP cost management requires more prudent approaches compared to conventional cost management practices. Frequent changes in science, technology, and engineering undertakings lead to dynamism of cost scenarios. Consequently, step-by-step tractable approaches must be used.

REFERENCES

Martin, H. L., *Techonomics: The Theory of Industrial Evolution*, Taylor & Francis/CRC Press, Boca Raton, FL, 2007.

PMI, *A Guide to the Project Management Body of Knowledge (PMBOK Guide)*, 3rd ed., Project Management Institute, Newtown Square, PA, 2004.

BIBLIOGRAPHY

Collins, J., *Good to Great*, HarperCollins Publishers, New York, 2001.

Dettmer, H. W., *Goldratt's Theory of Constraints: A Systems Approach to Continuous Improvement*, Quality Press, Milwaukee, WI, 1997.

Goldratt, E. M., *Critical Chain*, The North River Press, Great Barrington, MA, 1997.

Niven, P. R., *Balanced Scorecard: Step-by-Step: Maximizing Performance and Maintaining Results*, John Wiley, New York, 2002.

Woeppel, M. J., *Manufacturer's Guide to Implementing the Theory of Constraints*, St. Lucie Press, Boca Raton, FL, 2001.

6 STEP Quality Management

Good quality is everyone's responsibility; bad quality is everyone's fault.

–Adedeji Badiru, 1994

Project quality management is the next stage of the structural approach to project management in Project Management Body of Knowledge (PMBOK) guidelines (PMI, 2004). Quality management involves ensuring that the performance of a project conforms to specifications with respect to the requirements and expectations of the project stakeholders and participants. The objective of quality management is to minimize deviation from the actual project plans. Quality management must be performed throughout the life cycle of a project and not just by a final inspection of the product (Badiru and Ayeni, 1993).

QUALITY MANAGEMENT: STEP-BY-STEP IMPLEMENTATION

The quality management component of the PMBOK consists of the elements shown in the block diagram in Figure 6.1. The three elements in the block diagram are carried out across the process groups presented earlier in this book. The overlay of the elements and the process groups are shown in Table 6.1. Thus, under the knowledge area of quality management, the required steps are

- Step 1: Perform quality planning
- Step 2: Perform quality assurance
- Step 3: Perform quality control

Tables 6.2 through 6.4 present the inputs, tools, techniques, and outputs of each step.

Improvement programs have the propensity to drift into anecdotal, qualitative, and subjective processes. Having a quantifiable and measurable approach helps to overcome this deficiency. Figure 6.2 shows how operational efficiency transitions to effectiveness, quality, and then productivity.

SIX SIGMA AND QUALITY MANAGEMENT

The Six Sigma approach, which was originally introduced by Motorola's Government Electronics group, has caught on quickly in industry. Many major companies now embrace the approach as the key to high-quality industrial productivity. Six Sigma means six standard deviations from a statistical performance average. The Six Sigma

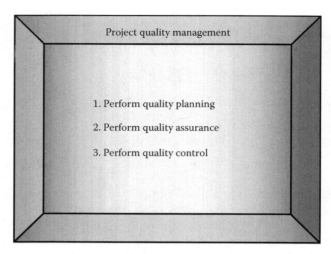

FIGURE 6.1 Block diagram of project quality management.

TABLE 6.1
Implementation of Project Quality Management across Process Groups

	Initiating	Planning	Executing	Monitoring and Controlling	Closing
Project Quality Management		Perform quality planning	Perform quality assurance	Perform quality control	

TABLE 6.2
Tools and Techniques for Quality Planning within Project Quality Management

STEP 1: Perform Quality Planning

Inputs	Tools and Techniques	Output(s)
Enterprise environmental factors	Cost/benefit analysis	Quality management plan
Organizational process assets	Benchmarking	Quality metrics
Project scope statement	Design of experiments	Quality check lists
Project management plan	Cost of quality (COQ) assessment	Process improvement plan
Other in-house (custom) factors of relevance and interest	Group decision techniques	Quality baseline
	Other in-house (custom) tools and techniques	Project management plan (updates)
		Other in-house outputs, reports, and data inferences of interest to the organization

STEP Quality Management

TABLE 6.3
Tools and Techniques for Quality Assurance within Project Quality Management

Step 2: Perform Quality Assurance

Inputs	Tools and Techniques	Output(s)
Quality management plan	Quality planning tools and techniques	Requested changes
Quality metrics		Recommended corrective actions
Process improvement plan	Quality audits	Organizational process assets (updates)
Work performance information	Process analysis	
Approved change requests	Quality control tools and techniques	Project management plan (updates)
Quality control measurements		
Implemented change requests	Other in-house (custom) tools and techniques	Other in-house outputs, reports, and data inferences of interest to the organization
Implemented corrective actions		
Implemented defect repair		
Implemented preventive repair		
Other in-house (custom) factors of relevance and interest		

TABLE 6.4
Tools and Techniques for Quality Control within Project Quality Management

STEP 3: Perform Quality Control

Inputs	Tools and Techniques	Output(s)
Quality management plan	Cause and effect diagram	Quality control measurements
Quality metrics	Control charts	Validated defect repair
Quality check lists	Flowcharting	Quality baseline (updates)
Organizational process assets	Histogram	Recommended corrective actions
Work performance information	Pareto chart	Recommended preventive actions
Approved change requests	Run chart	Requested changes
Deliverables	Scatter diagram	Recommended defect repair
Other in-house (custom) factors of relevance and interest	Statistical sampling	Organization process assets (updates)
	Quality inspection	Validated deliverables
	Defect repair review	Other in-house outputs, reports, and data inferences of interest to the organization
	Other in-house (custom) tools and techniques	

approach allows for no more than 3.4 defects per million parts in manufactured goods or 3.4 mistakes per million activities in a service operation. To appreciate the effect of the Six Sigma approach, consider a process that is 99% perfect. Such a process will produce 10,000 defects per million parts. With Six Sigma, the process will need to be 99.99966% perfect in order to produce only 3.4 defects per million.

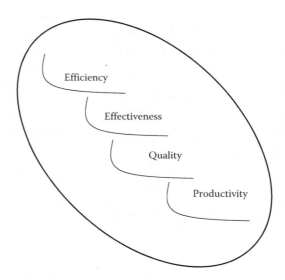

FIGURE 6.2 Foundations of STEP quality performance success.

TABLE 6.5
Interpretation of ±Sigma Intervals from Mean

Process Quality Range	Percentage Coverage	Interpretation of Standard
1 Sigma	68.26	Poor performance
2 Sigma	95.46	Below expectation
3 Sigma	99.73	Historical acceptable standard
4 Sigma	99.9937	Contemporary
6 Sigma	99.99999985	New competitive standard

Thus, Six Sigma is an approach that pushes the limit of perfection. Table 6.5 summarizes sigma ranges and process percentage coverage levels.

TAGUCHI LOSS FUNCTION

The philosophy of Taguchi loss function defines the concept of how deviation from an intended target creates a loss in the production process. Taguchi's idea of product quality analytically models the loss to the society from the time a product is shipped to customers. Taguchi loss function measures this conjectured loss with a quadratic function known as quality loss function (QLF), which is mathematically represented as shown below:

$$L(y) = k(y-m)^2$$

where
  k is a proportionality constant
  m is the target value
  y is the observed value of the quality characteristic of the product in question

STEP Quality Management

The quantity $(y - m)$ represents the deviation from the target. The larger the deviation, the larger is the loss to the society. The constant k can be determined if $L(y)$, y, and m are known. Loss, in the QLF concept, can be defined to consist of several components. Examples of loss are provided below:

- *Opportunity cost* of not having the service of the product due to its quality deficiency. The loss of service implies that something that should have been done to serve the society could not be done.
- *Time lost* in the search to find (or troubleshoot) the quality problem.
- *Time lost* (after finding the problem) in the attempt to solve the quality problem. The problem identification effort takes away some of the time that could have been productively used to serve the society. Thus, the society incurs a loss.
- *Productivity loss* that is incurred due to the reduced effectiveness of the product. The decreased productivity deprives the society of a certain level of service and, thereby, constitutes a loss.
- *Actual cost* of correcting the quality problem. This is, perhaps, the only direct loss that is easily recognized. But there are other subtle losses that the Taguchi method can help identify.
- *Actual loss* (e.g., loss of life) due to a failure of the product resulting from its low quality. For example, a defective automobile tire creates a potential for traffic fatality.
- *Waste* that is generated as a result of lost time and materials due to rework and other nonproductive activities associated with low quality of work.

IDENTIFICATION AND ELIMINATION OF SOURCES OF DEFECTS

The approach uses statistical methods to find problems that cause defects. For example, the total yield (number of nondefective units) from a process is determined by a combination of the performance levels of all the steps making up the process. If a process consists of 20 steps and each step is 98% perfect, then the performance of the overall process will be

$$(0.98)^{20} = 0.667608 \text{ (i.e., } 66.7608\%)$$

Thus, the process will produce 332,392 defects per million parts. If each step of the process is pushed to the Six Sigma limit, then the process performance will be

$$(0.9999966)^{20} = 0.999932 \text{ (i.e., } 99.9932\%)$$

Thus, the Six Sigma process will produce only 68 defects per million parts. This is a significant improvement over the original process performance. In many cases, it is not realistic to expect to achieve the Six Sigma level of production. But the approach helps to set a quality standard and provides a mechanism for striving to reach the goal. In effect, the Six Sigma process means changing the way workers perform their tasks so as to minimize the potential for defects.

The success of Six Sigma in industry ultimately depends on industry's ability to initiate and execute Six Sigma projects effectively. Thus, the project management approaches presented in this book are essential for realizing the benefits of Six Sigma. Project planning, organizing, team building, resource allocation, employee training, optimal scheduling, superior leadership, shared vision, and project control are all complementarily essential for implementing Six Sigma successfully. These success factors are not mutually exclusive. In many organizations, far too much focus is directed toward the statistical training for Six Sigma at the expense of proper project management development. This explains why many organizations have not been able to achieve the much-touted benefits of Six Sigma.

The success of the Toyota production system is not due to any special properties of the approach, but rather due to the consistency, persistence, and dedication of Toyota organizations in building their projects around all the essential success factors. Toyota focuses on changing the organizational mindset that is required in initiating and coordinating the success factors throughout the organization. Six Sigma requires the management of multiple projects with an identical mindset throughout the organization. The success of this requirement is dependent on proper application of project management tools and techniques.

ROLES AND RESPONSIBILITIES FOR SIX SIGMA

Human roles and responsibilities are crucial in executing Six Sigma projects. The different categories of team players are explained below:

Executive leadership: Develops and promulgates vision and direction. Leads change and maintains accountability for organizational results (on a full-time basis).

Employee group: Includes all employees, supports organizational vision, receives and implements Six Sigma specs, serves as points of total process improvement (TPM), exports mission statement to functional tasks, and deploys improvement practices (on full-time basis).

Six Sigma champion: Advocates improvement projects, leads business direction, and coordinates improvement projects (on a full-time basis).

Six Sigma project sponsor: Develops requirements, engages project teams, leads project scoping, and identifies resource requirements (on part-time basis).

Master belt: Trains and coaches black belts and green belts, leads large projects, and provides leadership (on full-time basis).

Black belt: Leads specific projects, facilitates troubleshooting, coordinates improvement groups, trains and coaches project team members (on full-time basis).

Green belt: Participates on black belt teams, leads small projects (on part-time project-specific basis).

Six Sigma project team members: Provide specific operational support, facilitate inward knowledge transfer, and link to functional areas (on part-time basis).

STATISTICAL TECHNIQUES FOR SIX SIGMA

Statistical process control (SPC) means controlling a process statistically. SPC originated from the efforts of the early quality control researchers. The techniques of SPC

STEP Quality Management

are based on basic statistical concepts normally used for statistical quality control. In a manufacturing environment, it is known that not all products are made exactly alike. There are always some inherent variations in units of the same product. The variation in the characteristics of a product provides the basis for using SPC for quality improvement. With the help of statistical approaches, individual items can be studied and general inferences can be drawn about the process or batches of products from the process. Since 100% inspection is difficult or impractical in many processes, SPC provides a mechanism to generalize concerning process performance. SPC uses random samples generated consecutively over time. The random samples should be representative of the general process. SPC can be accomplished through the following steps:

- Control charts ($\bar{X}$-chart, R-chart)
- Process capability analysis (nested design, C_p, C_{pk})
- Process control (factorial design, response surface)

CONTROL CHARTS

Two of the most commonly used control charts in industry are the X-bar charts ($\bar{X}$-charts) and the range charts (R-charts). The type of chart to be used normally depends on the kind of data collected. Data collected can be of two types: variable data and attribute data. The success of quality improvement depends on two major factors:

1. Quality of data available
2. Effectiveness of the techniques used for analyzing the data

Types of Data for Control Charts

Variable data: The control charts for variable data are listed below.

- Control charts for individual data elements (X)
- Moving range chart (MR-chart)
- Average chart ($\bar{X}$-chart)
- Range chart (R-chart)
- Median chart
- Standard deviation chart (σ-chart)
- Cumulative sum chart (CUSUM)
- Exponentially weighted moving average (EWMA)

Attribute data: The control charts for attribute data are listed below.

- Proportion or fraction defective chart (p-chart) (subgroup sample size can vary)
- Percent defective chart (100p-chart) (subgroup sample size can vary)
- Number defective chart (np-chart) (subgroup sample size is constant)
- Number defective (c-chart) (subgroup sample size = 1)
- Defective per inspection unit (u-chart) (subgroup sample size can vary)

The statistical theory useful to generate control limits is the same for all the above charts with the exception of EWMA and CUSUM.

X-Bar and Range Charts

The R-chart is a time plot useful in monitoring short-term process variations, while the X-bar chart monitors the longer term variations where the likelihood of special causes is greater over time. Both charts have control lines called upper and lower control limits as well as the central lines. The central line and control limits are calculated from the process measurements. They are not specification limits or a percentage of the specifications, or some other arbitrary lines based on experience. Therefore, they represent what the process is capable of doing when only common cause variation exists. If only common cause variation exists, then the data will continue to fall in a random fashion within the control limits. In this case, we say the process is in a state of statistical control. However, if a special cause acts on the process, one or more data points will be outside the control limits, so the process is not in a state of statistical control.

Data Collection Strategies

One strategy for data collection requires that about 20–25 subgroups be collected. Twenty to twenty-five subgroups should adequately show the location and spread of a distribution in a state of statistical control. If it happens that due to sampling costs or other sampling reasons associated with the process we are unable to have 20–25 subgroups, we can still use the available samples that we have to generate the trial control limits and update these limits as more samples are made available, because these limits will normally be wider than normal control limits and will therefore be less sensitive to changes in the process. Another approach is to use run charts to monitor the process until such time as 20–25 subgroups are made available. Then, control charts can be applied with control limits included on the charts. Other data collection strategies should consider the subgroup sample size as well as the sampling frequency.

Subgroup Sample Size

The subgroup samples of size n should be taken as n consecutive readings from the process and not random samples. This is necessary in order to have an accurate estimate of the process common cause variation. Each subgroup should be selected from some small period of time or small region of space or product in order to assure homogeneous conditions within the subgroup. This is necessary because the variation within the subgroup is used in generating the control limits. The subgroup sample size n can be between four and five samples. This is a good size that balances the pros and cons of using large or small sample size for a control chart as provided below.

Advantages of using small subgroup sample size

- Estimates of process standard deviation based on the range are as good and accurate as the estimates obtained from using the standard deviation equation, which is a complex hand calculation method.

- The probability of introducing special cause variations within a subgroup is very small.
- Range chart calculation is simple and easier to compute by hand on the shop floor by operators.

Advantages of using large subgroup sample size

- The central limit theorem supports the fact that the process average will be more normally distributed with larger sample size.
- If the process is stable, the larger the subgroup size, the better the estimates of process variability.
- A control chart based on larger subgroup sample size will be more sensitive to process changes.

The choice of a proper subgroup is very critical to the usefulness of any control chart. The following explains the importance of subgroup characteristics:

- If we fail to incorporate all common cause variations within our subgroups, the process variation will be underestimated, leading to very tight control limits. Then the process will appear to go out of control too frequently even when there is no existence of a special cause.
- If we incorporate special causes within our subgroups, then we will fail to detect special causes as frequently as expected.

Frequency of Sampling

The problem of determining how frequently one should sample depends on several factors. These factors include, but are not limited to the following.

- Cost of collecting and testing samples: The greater the cost of taking and testing samples, the less frequently we should sample.
- Changes in process conditions: The larger the frequency of changes to the process, the larger the sampling frequency. For example, if process conditions tend to change every 15 min, then sample every 15 min. If conditions change every 2 h, then sample every 2 h.
- Importance of quality characteristics: The more important the quality characteristic being charted is to the customer, the more frequently the characteristic will need to be sampled.
- Process control and capability: The more history of process control and capability, the less frequently the process needs to be sampled.

Stable Process

A process is said to be in a state of statistical control if the distribution of measurement data from the process has the same shape, location, and spread over time. In other words, a process is stable when the effects of all special causes have been removed from a process, so that the remaining variability is only due to common causes. Figure 6.3 shows an example of a stable distribution.

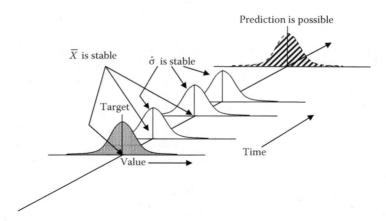

FIGURE 6.3 Stable distribution with no special causes.

Out-of-Control Patterns

A process is said to be unstable (not in a state of statistical control) if it changes from time to time because of a shifting average or shifting variability or a combination of shifting averages and variation. Figures 6.4 through 6.6 show examples of distributions from unstable processes.

Calculation of Control Limits

- Range (R)

This is the difference between the highest and lowest observations:

$$R = X_{highest} - X_{lowest}$$

- Center lines

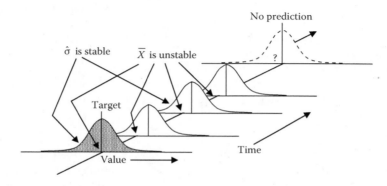

FIGURE 6.4 Unstable process average.

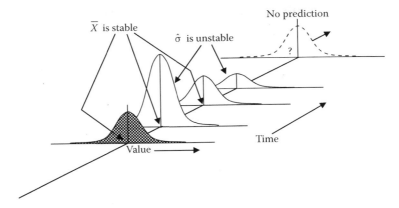

FIGURE 6.5 Unstable process variation.

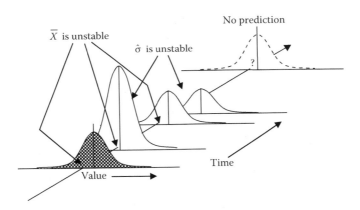

FIGURE 6.6 Unstable process average and variation.

Calculate $\overline{X}$ and $\overline{R}$:

$$\overline{X} = \frac{\sum X_i}{m}$$

$$\overline{R} = \frac{\sum R_i}{m}$$

where
  $\overline{X}$ is the overall process average
  $\overline{R}$ is the average range
  m is the total number of subgroups
  n is the within subgroup sample size

- Control limits based on R-chart

$$UCL_R = D_4 \bar{R}$$

$$LCL_R = D_3 \bar{R}$$

- Estimate of process variation

$$\hat{\sigma} = \frac{\bar{R}}{d_2}$$

- Control limits based on $\bar{X}$-chart

Calculate the upper and lower control limits for the process average:

$$UCL = \bar{X} + A_2 \bar{R}$$

$$LCL = \bar{X} - A_2 \bar{R}$$

Table 6.6 shows the values of d_2, A_2, D_3, and D_4 for different values of n. These constants are used for developing variable control charts.

Plotting Control Charts for Range and Average Charts

- Plot the range chart (R-chart) first.
- If R-chart is in control, then plot X-bar chart.
- If R-chart is not in control, identify and eliminate special causes, then delete points that are due to special causes, and recompute the control limits for the range chart. If process is in control, then plot X-bar chart.

TABLE 6.6
Table of Constants for Variables Control Charts

n	d_2	A_2	D_3	D_4
2	1.128	1.880	0	3.267
3	1.693	1.023	0	2.575
4	2.059	0.729	0	2.282
5	2.326	0.577	0	2.115
6	0.534	0.483	0	2.004
7	2.704	0.419	0.076	1.924
8	2.847	0.373	0.136	1.864
9	2.970	0.337	0.184	1.816
10	3.078	0.308	0.223	1.777
11	3.173	0.285	0.256	1.744
12	3.258	0.266	0.284	1.716

STEP Quality Management   169

- Check to see if X-bar chart is in control, if not search for special causes and eliminate them permanently.
- Remember to perform the eight trend tests.

Plotting Control Charts for Moving Range and Individual Control Charts

- Plot the moving range chart (MR-chart) first.
- If MR-chart is in control, then plot the individual chart (X).
- If MR-chart is not in control, identify and eliminate special causes, then delete special cause points, and recompute the control limits for the moving range chart. If MR-chart is in control, then plot the individual chart.
- Check to see if individual chart is in control, if not search for special causes from out-of-control points.
- Perform the eight trend tests.

Case Example: Plotting of Control Chart

An industrial engineer in a manufacturing company was trying to study a machining process for producing a smooth surface on a torque converter clutch. The quality characteristic of interest is the surface smoothness of the clutch. The engineer then collected four clutches every hour for 30 h and recorded the smoothness measurements in microinches. Acceptable values of smoothness lies between 0 (perfectly smooth) and 45 microinches. The data collected by the engineer are provided in Table 6.7. Histograms of the individual and average measurements are presented in Figure 6.7. The two histograms in the figure show that the hourly smoothness average ranges from 27 to 32 microinches, much narrower than the histogram of hourly individual smoothness, which ranges from 24 to 37 microinches. This is due to the fact that averages have less variability than individual measurements. Therefore, whenever we plot subgroup averages on an X-bar chart, there will always exist some individual measurements that will plot outside the control limits of an X-bar chart. The dot plots of the surface smoothness for individual and average measurements are shown in Figure 6.8.

The descriptive statistics for individual smoothness are presented below:

$N = 120$
MEAN = 29.367
MEDIAN = 29.00
TRMEAN = 29.287
STDEV = 2.822
SEMEAN = 0.258

The descriptive statistics for average smoothness are presented below:

$N = 30$
MEAN = 29.367
MEDIAN = 29.375
TRMEAN = 29.246
STDEV = 1.409
SEMEAN = 0.257

TABLE 6.7
Data for Control Chart Example

Smoothness (microinches)

Subgroup No.	I	II	III	IV	Average	Range
1	34	33	24	28	29.75	10
2	33	33	33	29	32.00	4
3	32	31	25	28	29.00	7
4	33	28	27	36	31.00	9
5	26	34	29	29	29.50	8
6	30	31	32	28	30.25	4
7	25	30	27	29	27.75	5
8	32	28	32	29	30.25	4
9	29	29	28	28	28.50	1
10	31	31	27	29	29.50	4
11	27	36	28	29	30.00	9
12	28	27	31	31	29.25	4
13	29	31	32	29	30.25	3
14	30	31	31	34	31.50	4
15	30	33	28	31	30.50	5
16	27	28	30	29	28.50	3
17	28	30	33	26	29.25	7
18	31	32	28	26	29.25	6
19	28	28	37	27	30.00	10
20	30	29	34	26	29.75	8
21	28	32	30	24	28.50	8
22	29	28	28	29	28.50	1
23	27	35	30	30	30.50	8
24	31	27	28	29	28.75	4
25	32	36	26	35	32.25	10
26	27	31	28	29	28.75	4
27	27	29	24	28	27.00	5
28	28	25	26	28	26.75	3
29	25	25	32	27	27.25	7
30	31	25	24	28	27.00	7
Total					881.00	172

Calculations

1. Natural limit of the process = $\bar{X} \pm 3s$ (based on empirical rule).
   s = estimated standard deviation of all individual samples
   Standard deviation (special and common), $s = 2.822$
   Process average, $\bar{X} = 29.367$
   Natural process limit = $29.367 \pm 3 (2.822) = 29.367 \pm 8.466$
   The natural limit of the process is between 20.90 and 37.83.

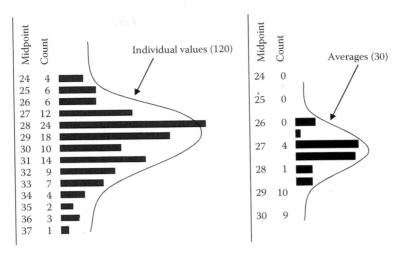

FIGURE 6.7 Histograms of individual measurements and averages for clutch smoothness.

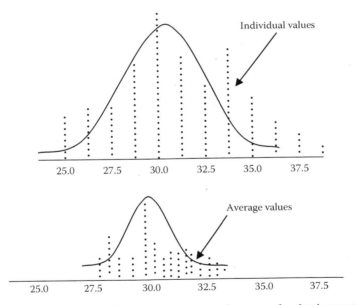

FIGURE 6.8 Dot plots of individual measurements and averages for clutch smoothness.

2. Inherent (common cause) process variability, $\hat{\sigma} = \bar{R}/d_2$
   $\bar{R}$ from the range chart = 5.83
   d_2 (for $n = 4$) = 2.059 (from Table 6.6)
   $\hat{\sigma} = \bar{R}/d_2 = 5.83/2.059 = 2.83$

Thus, the total process variation, s, is about the same as the inherent process variability. This is because the process is in control. If the process is out of control, the total standard deviation of all the numbers will be larger than $\bar{R}/d_2$.

3. Control limits for the range chart
Obtain constants D_3 and D_4 from Table 6.6 for $n = 4$.
   $D_3 = 0$
   $D_4 = 2.282$
   $\bar{R} = 172/30 = 5.73$
   $\text{UCL} = D_4^*\bar{R} = 2.282(5.73) = 16.16$
   $\text{LCL} = D_3^*\bar{R} = 0(5.73) = 0.0$

4. Control limits for the averages
Obtain constants A_2 from Table 6.6 for $n = 4$.
   $A_2 = 0.729$
   $\text{UCL} = \bar{X} + A_2(\bar{R}) = 29.367 + 0.729(5.73) = 33.54$
   $\text{LCL} = \bar{X} - A_2(\bar{R}) = 29.367 - 0.729(5.73) = 25.19$

5. Natural limit of the process = $\bar{X} \pm 3(\bar{R})/d_2 = 29.367 \pm 3(2.83) = 29.367 \pm 8.49$

The natural limit of the process is between 20.88 and 37.86, which is slightly different from ± 3s calculated earlier based on the empirical rule. This is due to the fact that $\bar{R}/d_2$ is used rather than the standard deviation of all the values. Again, if the process is out of control, the standard deviation of all the values will be greater than $\bar{R}/d_2$. The correct procedure is always to use $\bar{R}/d_2$ from a process that is in statistical control.

6. Comparison with specification
Since the specifications for the clutch surface smoothness is between 0 (perfectly smooth) and 45 microinches, and the natural limit of the process is between 20.88 and 37.86, then the process is capable of producing within the spec limits. Figure 6.9 presents the R and X-bar charts for clutch smoothness.

For this case example, the industrial engineer examined the above charts and concluded that the process is in a state of statistical control.

Process improvement opportunities
The industrial engineer realizes that if the smoothness of the clutch can be held below 15 microinches, then the clutch performance can be significantly improved. In this situation, the engineer can select key control factors to study in a two-level factorial or fractional factorial design.

Trend Analysis

After a process is recognized to be out of control, zone control charting technique is a logical approach to search for the sources of the variation problems. The following eight tests can be performed using MINITAB software or other statistical software tools. For this approach, the chart is divided into three zones. Zone A is between ± 3σ, zone B is between ± 2σ, and zone C is between ± 1σ.

Test 1

Pattern: One or more points falling outside the control limits on either side of the average. This is shown in Figure 6.10.

STEP Quality Management

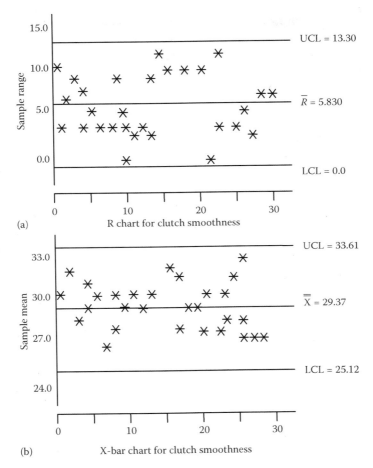

FIGURE 6.9 R and X-bar charts for clutch smoothness.

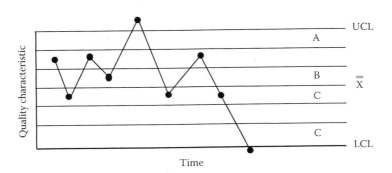

FIGURE 6.10 Test 1 for trend analysis.

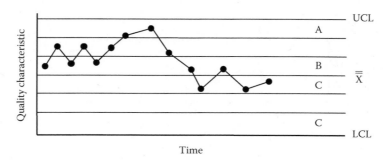

FIGURE 6.11 Test 2 for trend analysis.

Problem source: A sporadic change in the process due to special causes such as

- Equipment breakdown
- New operator
- Drastic change in raw material quality
- Change in method, machine, or process setting

Check: Go back and look at what might have been done differently before the out of control point signals.

Test 2

Pattern: A run of nine points on one side of the average as shown in Figures 6.6 through 6.11.

Problem source: This may be due to a small change in the level of process average. This change may be permanent at the new level.

Check: Go back to the beginning of the run and determine what was done differently at that time or prior to that time.

Test 3

Pattern: A trend of six points in a row either increasing or decreasing as shown in Figure 6.12.

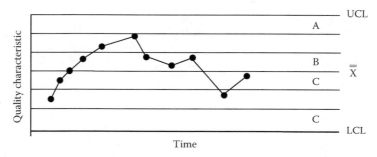

FIGURE 6.12 Test 3 for trend analysis.

STEP Quality Management

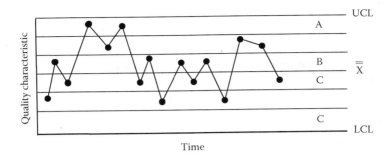

FIGURE 6.13 Test 4 for trend analysis.

Problem source: This may be due to the following:

- Gradual tool wear
- Change in characteristic such as gradual deterioration in the mixing or concentration of a chemical
- Deterioration of plating or etching solution in electronics or chemical industries

Check: Go back to the beginning of the run and search for the source of the run.

The above three tests are useful in providing good control of a process. However, in addition to the above three tests, some advanced tests for detecting out-of-control patterns can also be used. These tests are based on the zone control chart.

Test 4

Pattern: Fourteen points in a row alternating up and down within or outside the control limits as shown in Figure 6.13.

Problem source: This can be due to sampling variation from two different sources such as sampling systematically from high and low temperatures or lots with two different averages. This pattern can also occur if adjustment is being made all the time (over control).

Check: Look for cycles in the process, such as humidity or temperature cycles, or operator over control of process.

Test 5

Pattern: Two out of three points in a row on one side of the average in zone A or beyond. An example of this is presented in Figure 6.14.

Problem source: This can be due to a large, dramatic shift in the process level. This test sometimes provides early warning, particularly if the special cause is not as sporadic as in the case of Test 1.

Check: Go back one or more points in time and determine what might have caused the large shift in the level of the process.

Test 6

Pattern: Four out of five points in a row on one side of the average in zone B or beyond, as depicted in Figure 6.6 through 6.15.

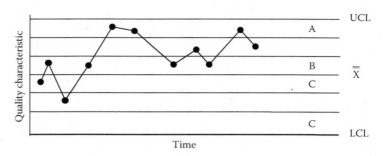

FIGURE 6.14 Test 5 for trend analysis.

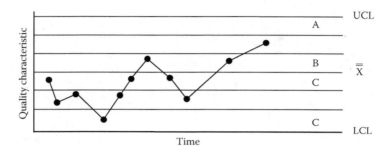

FIGURE 6.15 Test 6 for trend analysis.

Problem source: This may be due to a moderate shift in the process.

Check: Go back three or four points in time.

Test 7

Pattern: Fifteen points in a row on either side of the average in zone C as shown in Figure 6.16.

Problem source: This is due to the following:

- Unnatural small fluctuations or absence of points near the control limits
- At first glance may appear to be a good situation, but this is not a good control

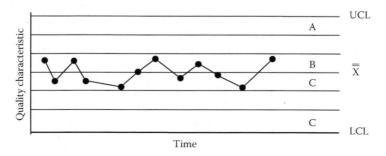

FIGURE 6.16 Test 7 for trend analysis.

STEP Quality Management

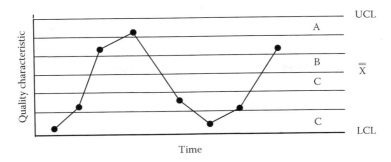

FIGURE 6.17 Test for trend analysis.

- Incorrect selection of subgroups. May be sampling from various subpopulations and combining them into a single subgroup for charting.
- Incorrect calculation of control limits.

Check: Look very close to the beginning of the pattern.

Test 8

Pattern: Eight points in a row on both sides of the center line with none in zone C. An example is shown in Figure 6.17.

Problem source: No sufficient resolution on the measurement system (see section on "Measurement System").

Check: Look at the range chart and see if it is in control.

PROCESS CAPABILITY ANALYSIS FOR SIX SIGMA

Industrial process capability analysis is an important aspect of managing industrial projects. The capability of a process is the spread that contains almost all values of the process distribution. It is very important to note that capability is defined in terms of a distribution. Therefore, capability can only be defined for a process that is stable (has distribution) with common cause variation (inherent variability). It cannot be defined for an out-of-control process (which has no distribution) with variation special to specific causes (total variability). Figure 6.18 shows a process capability distribution.

Capable Process (C_p)

A process is capable ($C_p \geq 1$) if its natural tolerance lies within the engineering tolerance or specifications. The measure of process capability of a stable process is $6\hat{\sigma}$, where $\hat{\sigma}$ is the inherent process variability that is estimated from the process. A minimum value of $C_p = 1.33$ is generally used for an ongoing process. This ensures a very low reject rate of 0.007% and therefore is an effective strategy for prevention of nonconforming items. C_p is defined mathematically as

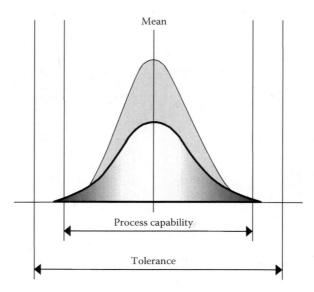

FIGURE 6.18 Process capability distribution.

$$C_p = \frac{\text{USL} - \text{LSL}}{6\hat{\sigma}}$$

$$= \frac{\text{Allowable process spread}}{\text{Actual process spread}}$$

where
  USL is the upper specification limit
  LSL is the lower specification limit
  C_p measures the effect of the inherent variability only

The analyst should use R-bar/d_2 to estimate $\hat{\sigma}$ from an R-chart that is in a state of statistical control, where R-bar is the average of the subgroup ranges and d_2 is a normalizing factor that is tabulated for different subgroup sizes (n). We do not have to verify control before performing a capability study. We can perform the study and then verify control after the study with the use of control charts. If the process is in control during the study, then our estimates of capabilities are correct and valid. However, if the process was not in control, we would have gained useful information as well as proper insights as to the corrective actions to pursue.

Capability Index (C_{pk})

Process centering can be assessed when a two-sided specification is available. If the capability index (C_{pk}) is equal to or greater than 1.33, then the process may be adequately centered. C_{pk} can also be employed when there is only one-sided specification. For a two-sided specification, it can be mathematically defined as

STEP Quality Management

$$C_{pk} = \text{minimum}\left\{\frac{USL - \bar{X}}{3\hat{\sigma}}, \frac{\bar{X} - LSL}{3\hat{\sigma}}\right\}$$

where $\bar{X}$ is the overall process average

However, for a one-sided specification, the actual C_{pk} obtained is reported. This can be used to determine the percentage of observations out of specification. The overall long-term objective is to make C_p and C_{pk} as large as possible by continuously improving or reducing process variability, $\hat{\sigma}$, for every iteration so that a greater percentage of the product is near the key quality characteristics target value. The ideal is to center the process with zero variability.

If a process is centered but not capable, one or several courses of action may be necessary. One of the actions may be that of integrating designed experiment to gain additional knowledge on the process and in designing control strategies. If excessive variability is demonstrated, one may conduct a nested design with the objective of estimating the various sources of variability. These sources of variability can then be evaluated to determine what strategies to use in order to reduce or permanently eliminate them. Another action may be that of changing the specifications or continuing production and then sorting the items. Three characteristics of a process can be observed with respect to capability, as summarized below. Figures 6.19 through 6.21 present the alternate characteristics.

1. Process may be centered and capable
2. Process may be capable but not centered
3. Process may be centered but not capable

Process capability example

Step 1: Using data for the specific process, determine if the process is capable.
Let us assume that the analyst has determined that the process is in a state

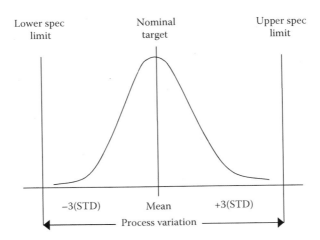

FIGURE 6.19 A process that is centered and capable.

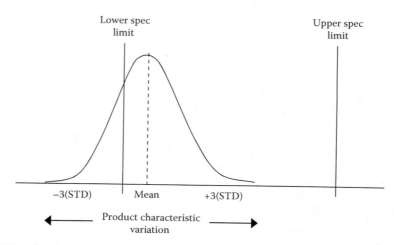

FIGURE 6.20 A process that is capable but not centered.

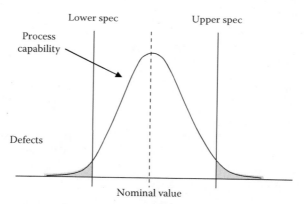

FIGURE 6.21 A process that is centered but not capable.

of statistical control. For this example, the specification limits are set at 0 (lower limit) and 45 (upper limit). The inherent process variability as determined from the control chart is

$$\hat{\sigma} = \frac{\bar{R}}{d_2} = \frac{5.83}{2.059} = 2.83$$

The capability of this process to produce within the specifications can be determined as

$$C_p = \frac{\text{USL} - \text{LSL}}{6\hat{\sigma}} = \frac{45 - 0}{6(2.83)} = 2.650$$

STEP Quality Management

The capability of the process, $C_p = 2.65 > 1.0$, indicating that the process is capable of producing clutches that will meet the specifications of between 0 and 45. The process average is 29.367.

Step 2: Determine if the process can be adequately centered. C_{pk} = minimum [C_l and C_u] can be used to determine if a process can be centered.

$$C_u = \frac{USL - \bar{X}}{3\hat{\sigma}} = \frac{45 - 29.367}{3(2.83)} = 1.84$$

$$C_l = \frac{\bar{X} - LSL}{3\hat{\sigma}} = \frac{29.367 - 0}{3(2.83)} = 3.46$$

Therefore, the capability index, C_{pk}, for this process is 1.84. Since C_{pk} = 1.84 is greater than 1.33, then the process can be adequately centered.

Possible Applications of Process Capability Index

The potential applications of process capability index are summarized below:

- Communication: C_p and C_{pk} have been used in industry to establish a dimensionless common language useful for assessing the performance of production processes. Engineering, quality, manufacturing, etc. can communicate and understand processes with high capabilities.
- Continuous improvement: The indices can be used to monitor continuous improvement by observing the changes in the distribution of process capabilities. For example, if there were 20% of processes with capabilities between 1 and 1.67 in a month, and some of these improved to between 1.33 and 2.0 the next month, then this is an indication that improvement has occurred.
- Audits: There are so many various kinds of audits in use today to assess the performance of quality systems. A comparison of in-process capabilities with capabilities determined from audits can help establish problem areas.
- Prioritization of improvement: A complete printout of all processes with unacceptable C_p or C_{pk} values can be extremely powerful in establishing the priority for process improvements.
- Prevention of nonconforming product: For process qualification, it is reasonable to establish a benchmark capability of C_{pk} = 1.33, which will make nonconforming products unlikely in most cases.

Potential Abuse of C_p and C_{pk}

In spite of its several possible applications, process capability index has some potential sources of abuse as summarized below:

- Problems and drawbacks: C_{pk} can increase without process improvement even though repeated testing reduces test variability. The wider the specifications, the larger the C_p or C_{pk}, but the action does not improve the process.
- Analysts tend to focus on number rather than on process.

- Process control: Analysts tend to determine process capability before statistical control has been established. Most people are not aware that capability determination is based on process common cause variation and what can be expected in the future. The presence of special causes of variation makes prediction impossible and capability index unclear.
- Nonnormality: Some processes result in nonnormal distribution for some characteristics. Since capability indices are very sensitive to departures from normality, data transformation may be used to achieve approximate normality.
- Computation: Most computer-based tools do not use $\bar{R}/d_2$ to calculate σ.

When analytical and statistical tools are coupled with sound managerial approaches, an organization can benefit from a robust implementation of improvement strategies. One approach that has emerged as a sound managerial principle is "lean," which has been successfully applied to many industrial operations.

LEAN PRINCIPLES AND APPLICATIONS

What is "Lean"? Lean means the identification and elimination of sources of *waste* in operations. Recall that Six Sigma involves the identification and elimination of source of *defects*. When Lean and Six Sigma are coupled, an organization can derive the double benefit of reducing waste and defects in operations, which leads to what is known as Lean–Six Sigma. Consequently, the organization can achieve higher product quality, better employee morale, better satisfaction of customer requirements, and more effective utilization of limited resources. The basic principle of "lean" is to take a close look at the elemental compositions of a process so that nonvalue-adding elements can be located and eliminated.

APPLYING KAIZEN TO A PROCESS

By applying the Japanese concept of *Kaizen*, which means "take apart and make better," an organization can redesign its processes to be lean and devoid of excesses. In a mechanical design sense, this can be likened to finite element analysis, which identifies how the component parts of a mechanical system fit together. It is by identifying these basic elements that improvement opportunities can be easily and quickly recognized. It should be recalled that the process of work breakdown structure in project management facilitates the identification of task-level components of an endeavor. Consequently, using a project management approach facilitates the achievement of the objectives of "lean." In the context of quality management, Figure 6.22 shows a process decomposition hierarchy that may help identify elemental characteristic that may harbor waste, inefficiency, and quality impedance. The functional relationships (f) are summarized as shown below:

Task = f(activity)
Subprocess = f(task)
Process = f(subprocess)
Quality system = f(process)

STEP Quality Management

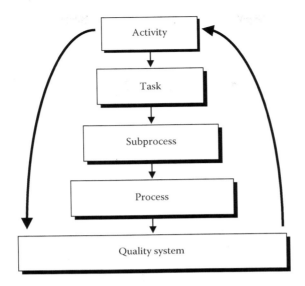

FIGURE 6.22 Hierarchy of process components.

Thus, quality improvement can be achieved by hierarchically improving a process and all the elements contained therein.

Fads come and go in industry. Over the years, we have witnessed the introduction and demise of many techniques that were hailed as the panacea of industry's ailments. Some of the techniques have survived the test of time because they do, indeed, hold some promise. Lean techniques appear to hold such promise, if it is viewed as an open-ended but focused application of the many improvement tools that have emerged over the years. The adoption of lean principles by the U.S. Air Force has given more credence to its application. The U.S. Air Force embarked on a massive endeavor to achieve widespread improvement in operational processes throughout the Air Force. The endeavor is called AFSO21 (Air Force Smart Operations for the Twenty-First Century or simply Air Force Smart Ops 21). This endeavor requires the implementation of appropriate project management practices at all levels. AFSO21 is a coordinated effort at achieving operational improvement in U.S. Air Force operations throughout the rank and file of the large organization. It is an integrative process of using Lean Principles, Theory of Constraints, Six Sigma, BPI, MBO, TQM, 6s, Project Management, and other classical management tools. However, the implementation of Lean principles constitutes about 80% of AFSO21 efforts. As a part of tools for lean practices and procedures, the following section presents lean task value rating system, which helps compare and rank elements of a process for retention, re-scoping, scaling, or elimination purposes.

LEAN TASK VALUE RATING SYSTEM

In order to identify value-adding elements of a lean project, the component tasks must be ranked and comparatively assessed. The method below applies relative ratings to tasks. It is based on the distribution of a total point system. The total points available to the composite process or project are allocated across individual tasks. The steps are explained below:

Steps:

1. Let T be the total points available to tasks.
2. $T = 100(n)$, where n = number of raters on the rating team.
3. Rate the value of each task on the basis of specified output (or quality) criteria on a scale of 0 to 100.
4. Let x_{ij} be the rating for task i by rater j.
5. Let m = number of tasks to be rated.
6. Organize the ratings by rater j as shown below:

   Rating for Task 1: x_{1j}
   Rating for Task 2: x_{2j}
   .
   .
   .
   Rating for Task m: x_{mj}
   Total rating points   100

7. Tabulate the ratings by the raters as shown in Table 6.8 and calculate the overall weighted score for each Task i from the expression below:

$$w_i = \frac{1}{n} \sum_{j=1}^{n} x_{ij}$$

The weighted score, w_i, is used to rank order the tasks to determine the relative value-added contributions of each task. Subsequently, using a preferred cut-off margin, the low or noncontributing activities can be slated for elimination.

In terms of activity prioritization, a comprehensive lean analysis can identify the important versus unimportant and urgent versus not urgent tasks. It is within the unimportant and not urgent quadrant that one will find "waste" task elements that should be eliminated. Using the familiar Pareto distribution format, Table 6.9 presents an example of task elements within a 20% waste elimination zone.

TABLE 6.8
Lean Task Rating Matrix

	Rating by Rater $j = 1$	Rating by Rater $j = 2$	...	...	Rating by Rater n	Total Pts. for Task i	w_i
Rating for Task $i = 1$							
Rating for Task $i = 2$							
...							
...							
Rating for Task m							
Total pts. from Rater j	100	100	...	...	100	$100n$	

**TABLE 6.9
Pareto Analysis of Unimportant
Process Task Elements**

	Urgent	Not Urgent
Important	20%	80%
Not Important	80%	20%

It is conjectured that activities that fall in the "not important" and "not urgent" zone run the risk of generating points of waste in any productive undertaking. That zone should be the first target of review for tasks that can be eliminated. Granted that there may be some "sacred cow" activities that an organization must retain for political, cultural, or regulatory reasons, attempts should still be made to categorize all task elements of a project. The long-established industrial engineering principle of time-and-motion studies is making a comeback due to the increased interest in eliminating waste in lean initiatives.

LEAN–SIX SIGMA WITHIN PROJECT MANAGEMENT

Lean and Six Sigma use analytical tools as the basis for pursuing their goals. But the achievement of those goals is predicated on having a structured approach to the activities of production. If proper project management is practiced at the outset on an industrial endeavor, it will pave the way for achieving Six Sigma results and realizing Lean outcomes. The key in any project endeavor is to have a structured design of the project so that diagnostic and corrective steps can easily be pursued. If the proverbial "garbage" is allowed to creep into a project, it would take much more time, effort, and cost to achieve a Lean–Six Sigma cleanup.

REFERENCES

Badiru, A. B. and B. J. Ayeni, *Practitioner's Guide to Quality and Process Improvement*, Chapman & Hall, London, 1993.

PMI, *A Guide to the Project Management Body of Knowledge (PMBOK Guide)*, 3rd ed., Project Management Institute, Newtown Square, PA, 2004.

7 STEP Human Resource Management

> It's all about collaborating with people, building trust and confidence, and making sure you take care of the followers.
>
> **–General Colin Powell**

> Whenever one person is found adequate to the discharge of a duty by close application thereto, it is worse executed by two persons, and scarcely done at all if three or more are employed therein.
>
> **–George Washington**

Project human resource management is the next stage of the structural approach to project management in *Project Management Body of Knowledge* (*PMBOK*) guidelines (PMI, 2004). The Colin Powell quote at the beginning of this chapter points out the importance of project leadership for human resources while the George Washington quote emphasizes the fact that there is a point of diminishing return for the application of human resources. It is a common practice to ask for additional human resources. But care must be exercised to determine where and when additional human resources are really needed and how they are deployed. A pertinent question to ask is: What incremental value or benefit is provided by the addition of one more unit of resource? This is of particular interest when allocating technical human resources in STEPs.

Human resources management provides the foundation for accomplishing project goals. Even in highly automated environments, human resources are still a key element in accomplishing goals and objectives. Human resources management involves the function of directing human resources throughout a project's life cycle. This requires the art and science of behavioral knowledge to achieve project goals. Employee involvement and empowerment are crucial elements of achieving the quality expectations of a project. The project manager is the key player in human resources management. Good leadership qualities and interpersonal skills are essential for dealing with both internal and external human resources associated with a project. The legal and safety aspects of employee welfare are important factors in human resources management. Human resource management is carried out to express and uphold an organization's standards and expectations in terms of the attributes of employees including the following:

- Desired competencies
- Current assignments

- Performance metrics
- Development plans
- Outline of future goals

Standard competencies of employees must be factored into in-house and customized reviews for project teams of technical professionals. Employee performance reviews should be tailored to employees' functional positions and accountability. Some standard competencies include the following:

- Achievement of focus
- Business and operational acumen
- Consultative personality
- Work design ability
- Continuous learning
- Customer focus
- Quality management
- Strategic thinking
- Team leadership
- Technology awareness
- Vision

Designing human systems interaction into a STEP environment ensures that human resources can integrate well with sophisticated science, technology, and engineering tools. Issues such as human effectiveness, performance excellence, occupational health, readiness, and safety are essential components of planning STEPs from a human resource management perspective.

AGING WORKFORCE IN SCIENCE, TECHNOLOGY, AND ENGINEERING

The issue of the aging workforce in science, technology, and engineering is very critical. Scientists and engineers who form the backbone of the advancements during and post–World War II are gradually phasing out of the STEP landscape. Managing and advancing contemporary science, technology, and engineering projects requires an infusion of a new crop of researchers, educators, and practitioners in those fields. Many nations are scrambling to find ways to train, acquire, or retain qualified STE professionals. Not only must the desired replenishment workforce be developed, it must also be preserved through the following strategies:

- Challenge the workforce to engage in higher level educational opportunities.
- Provide smooth avenues for the younger generation to follow the path of science, technology, and engineering.
- Create cohesive and progressive linkage between elementary education and higher education (e.g., K-12 to graduate studies education initiatives).
- Recognizing that science and technology are subject to dynamism and job uncertainties, create a guarantee of job security.

- Recognizing the high value of science, technology, and engineering professionals, institute programs that assure personal safety and security of the workforce.
- Recognizing that longevity is essential for getting the most out of the STE workforce, provide health care programs that extend and preserve the services of the workforce for as long as practicable.
- Diversify local availability of STE jobs to encourage recruitment, relocation, and local retention of the workforce.
- Encourage localities to provide physical infrastructure and favorable operating conditions to attract and retain science and technology industry.
- Create avenues for expanding job opportunities for the new and younger workforce so that "new" and "old" can coexist productively.
- Create tiered and mentoring workforce relationship such that age discrimination does not creep into workforce relationships.
- Take advantage of the wisdom and experience of the outgoing (retiring and departing) workforce in preparing the incoming (next generation) workforce.

KNOWLEDGE WORKERS IN STE WORK ENVIRONMENT

Knowledge workers will be the boon of the STE work environment of the future. While the brawn of yesteryears will still be needed in some quarters of the economy, the brain of the future is what will be needed to achieve and sustain technological advantage. A major shift will have to be institute to accommodate the needs of knowledge workers. While work-time accountability will still be needed for compensation purposes, the major shift will be to judge knowledge workers on the basis of their accomplishments rather than the number of hours spent directly on the job. Knowledge workers often work around the clock, even when they are not at work. Higher-level strategy formulation, rationalization, and pensive reflection on work actions most often occur away from the hustle and bustle of actual work environment. These off-site brain-intensive "work" translates directly or indirectly into work-site accomplishments. Many organizations already recognize this benefit and are already accounting for overall accomplishments in their compensation packages for employees. But this practice needs to spread to more organizations throughout the rank and file of the economy, even in industries that are not traditionally seen as being of high-tech caliber. The phrase below re-emphasizes the point of the unique asset that knowledge workers bring to the work environment:

> Evaluate knowledge workers on the basis of their overall accomplishments rather than how many office-hours it takes to achieve the accomplishments.

ELEMENTS OF HUMAN RESOURCE MANAGEMENT

Human resources are the basis of managing projects. Even highly automated systems must have human intervention at specific points to ensure overall process efficiency. Interhuman relationships must thus play a major role in an organization's strategy for managing human resources. Human divides do not mend easily. An organization must work hard to prevent a divide in the first place. This requires every organization to recognize and interface the following three major components of a project:

1. People (managers, team members, stakeholders, stockholders, vendors, suppliers, etc.)
2. Processes (work design, lean initiative, Six Sigma, business process re-engineering, etc.)
3. Tools (technology, widgets, facilities, information, etc.)

The success of any project is dependent on the human resources associated with linking its components. Human resources are distinguished from other resources because of the ability to learn, adapt to new project situations, and set goals. Human resources, technology resources, and management resources must coexist to pursue project goals. Managing human resources involves placing the right people with the right skills in the right jobs in the right environment. Good human resource management motivates workers to perform better. Both individual and organizational improvements are needed to improve overall quality by enriching jobs with the following strategies:

- Specify project goals in unambiguous terms
- Encourage and reward creativity on the job
- Eliminate mundane job control processes
- Increase accountability and responsibility for project results
- Define jobs in terms of manageable work packages that help identify line of responsibility
- Grant formal authority to make decisions at the task level
- Create advancement opportunities in each job
- Give challenging assignments that enable a worker to demonstrate his/her skill
- Encourage upward (vertical) communication of ideas
- Provide training and tools needed to get job done
- Maintain a stable management team

Several management approaches are used to manage human resources. Some of these approaches are formulated as direct responses to the cultural, social, family, or religious needs of workers. Examples of these approaches are

- Flexitime
- Religious holidays
- Half-time employment

These approaches can have a combination of several advantages. Some of the advantages are for the employer, while some are for the workers. The advantages are presented below:

- Low cost
- Cost savings on personnel benefits

- Higher employee productivity
- Less absenteeism
- Less work stress
- Better family/domestic situation, which may have positive effects on productivity

Workforce retraining is important for automation projects. Continuing education programs should be developed to retrain people who are only qualified to do jobs that do not require skilled humanpower. The retraining will create a ready pool of human resource that can help boost manufacturing output and competitiveness. Management stability is needed to encourage workers to adapt to the changes in industry. If management changes too often, workers may not develop a sense of commitment to the policies of management.

The major resource in any organization is humanpower both technical and nontechnical. People are the overriding factor in any project life cycle. Even in automated operations, the role played by whatever few people are involved can be very significant. Such operations invariably require the services of technical people with special managerial and professional needs. The high-tech manager in such situations would need special skills in order to discharge the managerial duties effectively. The manager must have auto-management skills that relate to the following:

- Managing self
- Being managed
- Managing others

Many of the managers who supervise technical people rise to the managerial posts from technical positions. Consequently, they often lack the managerial competence needed for the higher offices. In some cases, technical professionals are promoted to managerial levels and then transferred to administrative posts in functional areas different from their areas of technical competence. The poor managerial performance of these technical managers is not necessarily a reflection of poor managerial competence, but rather an indication of the lack of knowledge of the work elements in their surrogate function. Any technical training without some management exposure is, in effect, an incomplete education. Technical professionals should be trained for the eventualities of their professions.

In the transition from the technical to the management level, an individual's attention would shift from detail to overview, specific to general, and technical to administrative. Since most managerial positions are earned based on qualifications (except in aristocratic and autocratic systems), it is important to train technical professionals for possible administrative jobs. It is the responsibilities of the individual and the training institution to map out career goals and paths and institute-specific education aimed at the realization of those goals. One such path is outlined below.

1. *Technical professional.* This is an individual with practical and technical training and experience in a given field, such as industrial engineering. The individual must keep current in his/her area of specialization

through continuing education courses, seminars, conferences, and so on. The mentor program, which is now used in many large organizations, can be effectively utilized at this stage of the career ladder.
2. *Project manager.* This is an individual assigned the direct responsibility of supervising a given project through the phases of planning, organizing, scheduling, monitoring, and control. The managerial assignment may be limited to just a specific project. At the conclusion of the project, the individual returns to his/her regular technical duties. However, his/her performance on the project may help identify him/her as a suitable candidate for permanent managerial assignment later on.
3. *Group manager.* This is an individual who is assigned direct responsibility to plan, organize, and direct the activities of a group of people with a specific responsibility—for example, a computer data security advisory committee. This is an ongoing responsibility that may repeatedly require the managerial skills of the individual.
4. *Director.* An individual who oversees a particular function of the organization. For example, a marketing director has the responsibility of developing and implementing the strategy for getting the organization's products to the right market, at the right time, at the appropriate price, and in the proper quantity. This is a critical responsibility that may directly affect the survival of the organization. Only the individuals who have successfully proven themselves at the earlier career stages get the opportunity to advance to the director's level.
5. *Administrative manager.* This is an individual who oversees the administrative functions and staff of the organization. His/her responsibilities cut across several functional areas. He/she must have proven his/her managerial skills and diversity in previous assignment.

The above is just one of the several possible paths that can be charted for a technical professional as he/she gradually makes the transition from the technical ranks to the management level. To function effectively, a manager must acquire nontechnical background in various subjects. His/her experience, attitude, personality, and training will determine his/her managerial style. His/her appreciation of the human and professional needs of his subordinates will substantially enhance his/her managerial performance. Examples of subject areas in which a manager or an aspiring manager should get training include the ones outlined below.

1. Project management
    a. *Scheduling and budgeting*: Knowledge of project planning, organizing, scheduling, monitoring, and controlling under resource and budget restrictions.
    b. *Supervision*: Skill in planning, directing, and controlling the activities of subordinates.
    c. *Communication*: Skill of relating to others both within and outside the organization. This includes written and oral communication skills.

2. Personal and personnel management
    a. *Professional development*: Leadership roles played by participating in professional societies and peer recognition acquired through professional services.
    b. *Personnel development*: Skills needed to foster cooperation and encouragement of staff with respect to success, growth, and career advancement.
    c. *Performance evaluation*: Development of techniques for measuring, evaluating, and improving employee performance.
    d. *Time management*: Ability to prioritize and delegate activities as appropriate to maximize accomplishments within given time.
3. Operations management
    a. *Marketing*: Skills useful for winning new business for the organization or preserving existing market shares.
    b. *Negotiating*: Skills for moderating personnel issues, representing the organization in external negotiations, or administering company policies.
    c. *Estimating and budgeting*: Skills needed to develop reasonable cost estimates for company activities and the assignment of adequate resources to operations.
    d. *Cash flow analysis*: An appreciation for the time value of money, manipulations of equity and borrowed capitals, stable balance between revenues and expenditures, and maximization of returns on investments.
    e. *Decision analysis*: Ability to choose the direction of work by analyzing feasible alternatives.

A technical manager can develop the above skills through formal college courses, seminars, workshops, short courses, professional conferences, or in-plant company training. Several companies appreciate the need for these skills and are willing to bear the cost of furnishing their employees with the means of acquiring the skills. Many of the companies have custom formal courses, which they contract out to colleges to teach for their employees. This is a unique opportunity for technical professionals to acquire managerial skills needed to move up the company ladder.

Technical people have special needs. Unfortunately, some of these needs are often not recognized by peers, superiors, or subordinates. Inexperienced managers are particularly prone to the mistake of not distinguishing between technical and nontechnical professional needs. In order to perform more effectively, a manager must be administratively adaptive. He/she must understand the unique expectations of technical professionals in terms of professional preservation, professional peers, work content, hierarchy of needs, and the technical competence or background of their managers. Maslow's hierarchy of needs presents the organizational theory of how individual workers behave and respond to stimuli in the project environment.

Maslow's hierarchy of needs cover the basic elements summarized below:

1. *Physiological needs*: The needs for the basic necessities of life, such as food, water, housing, and clothing (survival needs). This is the level where access to wages is most critical. Biological needs fall into this category also.

2. *Safety needs*: The needs for security, stability, and freedom from threat of physical harm. Desire for safe working environment.
3. *Social needs*: The needs for social acceptance (sense of belonging), friends, love, affection, and association. Industrial outsourcing may bring about better economic outlook that may enable each individual to be in a better position to meet his or her social needs.
4. *Esteem needs*: The needs for accomplishment, respect, recognition, attention, self-respect, autonomy, and appreciation. These needs are important not only at the individual level, but also at the organizational level.
5. *Self-actualization needs*: These are the needs for self-fulfillment and self-improvement. They also involve the stage of opportunity to grow professionally. Industrial outsourcing may create opportunities for individuals to assert themselves socially and economically.

In addition to Abraham Maslow's hierarchy of needs, the following motivation theories are also essential.

- Theory X and Theory Y (presented by Douglas McGregor)
- Motivation–hygiene factors (presented by Frederick Herzberg)

Theory X can be effectively utilized in developing project teams. It has the following doctrines:

- Workers inherently dislike work, and whenever possible, will attempt to avoid it.
- Since workers dislike work, they must be coerced, controlled, cajoled, coaxed, enticed, persuaded, or threatened with punishment to achieve desired goals.
- Workers will evade and shift responsibilities and seek formal direction whenever possible.
- Most workers place security above all other factors associated with work and will display little ambition or self-motivation.

Theory Y can be embraced to take advantage of positive self-direction and self-actuating nature of workers to achieve project goals with little or no external prompting. The basic doctrines of Theory Y are

- Workers can view work as being as natural as personal normal pursuits such as recreation.
- A worker who is committed to the objectives of a project will exhibit self-direction, self-actuation, and self-control to get the job done.
- The average worker readily accepts and even seeks responsibilities to get the job done.
- Creativity, the ability to make good decisions, permeates the organization and not necessarily limited to the select few in management.

Motivation–hygiene factors also present constructive attributes of work that facilitate the achievement of project objectives.

- Motivation factors include achievement, recognition, job growth, work design itself, increased responsibility, career advancement.
- Hygiene factors include company policy, supervision practices, interpersonal relationships, work conditions, salary compensation, bonus program, personal life, professional status, and job security.

Professional preservation refers to the desire of a technical professional to preserve his/her identification with a particular job function. In many situations, the preservation is not possible due to a lack of humanpower to fill specific job slots. It is common to find people trained in one technical field holding assignments in other fields. An incompatible job function can easily become the basis for insubordination, egotism, and rebellious attitudes. While it is realized that in any job environment there will sometimes be the need to work outside one's profession, every effort should be made to match the surrogate profession as close as possible. This is primarily the responsibility of the human resources manager.

After a personnel team has been selected in the best possible manner, a critical study of the job assignments should be made. Even between two dissimilar professions, there may be specific job functions that are compatible. These should be identified and used in the process of personnel assignment. In fact, the mapping of job functions needed for an operation can serve as the basis for selecting a project team. In order to preserve the professional background of technical workers, their individualism must be understood. In most technical training programs, the professional is taught how to operate in the following ways:

1. Make decisions based on the assumption of certainty of information
2. Develop abstract models to study the problem being addressed
3. Work on tasks or assignments individually
4. Quantify outcomes
5. Pay attention to exacting details
6. Think autonomously
7. Generate creative insights to problems
8. Analyze systems operatability rather than profitability

However, in the business environment, not all of the above characteristics are desirable or even possible. For example, many business decisions are made with incomplete data. In many situations, it is unprofitable to expend the time and efforts to seek perfect data. As another example, many operating procedures are guided by company policies rather than creative choices of employees. An effective manager should be able to spot cases where a technical employee may be given room to practice his professional training. The job design should be such that the employee can address problems in a manner compatible with his professional training.

Professional peers. In addition to having professionally compatible job functions, technical people like to have other project team members to whom they can relate technically. A project team consisting of members from diversely unrelated technical fields can be a source of miscommunication, imposition, or introversion. The lack of a professional associate on the same project can cause a technical person to exhibit one or more of the following attitudes:

1. Withdraw into a shell and contribute very little to the project by holding back ideas that he/she feels the other project members cannot appreciate.
2. Exhibit technical snobbery and hold the impression that only he/she has the know-how for certain problems.
3. Straddle the fence on critical issues and develop no strong conviction for project decisions.

Providing an avenue for a technical "buddy system" to operate in an organization can be very instrumental in ensuring congeniality of personnel teams and in facilitating the eventual success of project endeavors. The manager in conjunction with the selection committee (if one is used) must carefully consider the mix of the personnel team on a given project. If it is not possible or desirable to have more than one person from the same technical area on the project, an effort should be made to provide as good a mix as possible. It is undesirable to have several people from the same department taking issues against the views of a lone project member from a rival department. Whether it is realized or not, admitted or not, there is a keen sense of rivalry among technical fields. Even within the same field, there are subtle rivalries between specific functions. It is important not to let these differences carry over to a project environment.

Work content. With the advent of new technology, the elements of a project task will need to be designed to take advantage of new developments. Technical professionals have a sense of achievement relative to their expected job functions. They will not be satisfied with mundane project assignments that will bring forth their technical competence. They prefer to claim contribution mostly where technical contribution can be identified. The project manager will need to ensure that the technical people of a project have assignments for which their background is really needed, It will be counterproductive to select a technical professional for a project mainly on the basis of personality. An objective selection and appropriate assignment of tasks will alleviate potential motivational problems that could develop later in the project.

Hierarchy of needs. Recalling Maslow's hierarchy of needs, the needs of a technical professional should be more critically analyzed. Being professionals, technical people are more likely to be higher up in the needs hierarchy. Most of their basic necessities for a good life would already have been met. Their prevailing needs will tend to involve esteem and self-actualization. As a result, by serving on a project team, a technical professional may have expectations that cannot usually be quantified in monetary terms. This is in contrast to nontechnical people who may look forward to overtime pay or other monetary gains that may result from being on the project. Technical professionals will generally look forward to one or several of the following opportunities:

1. *Professional growth and advancement*: Professional growth is a primary pursuit of most technical people. For example, a computer professional has to be frequently exposed to challenging situations that introduce new

technology developments and enable him to keep abreast of his field. Even occasional drifts from the field may lead to the fear of not keeping up and being left behind. The project environment must be reassuring to the technical people with regard to the opportunities for professional growth in terms of developing new skills and abilities.

2. *Technical freedom*: Technical freedom, to the extent permissible within the organization, is essential for the full utilization of a technical background. A technical professional will expect to have the liberty of determining how best the objective of his assignment can be accomplished. One should never impose a work method on a technical professional with the assurance that "this is the way it has always been done and will continue to be done!" If the worker's creative input to the project effort is not needed, then there is no need having him or her on the team in the first place.

3. *Respect for personal qualities*: Technical people have profound personal feelings despite the mechanical or abstract nature of their job functions. They will expect to be respected for their personal qualities. In spite of frequently operating in professional isolation, they do engage in interpersonal activities. They want their nontechnical views and ideas to be recognized and evaluated based on merit. They do not want to be viewed as "all technical." An appreciation for their personal qualities gives them the sense of belonging and helps them to become productive members of a project team.

4. *Respect for professional qualification*: A professional qualification usually takes several years to achieve and is not likely to be compromised by any technical professional. Technical professionals cherish the attention they receive due to their technical background. They expect certain preferential treatments. They like to make meaningful contributions to the decision process. They take approval of their technical approaches for granted. They believe they are on a project because they are qualified to be there. The project manager should recognize these situations and avoid the bias of viewing the technical person as being conceited.

5. *Increased recognition*: Increased recognition is expected as a by-product of a project effort. The technical professional, consciously or subconsciously, views his participation in a project as a means of satisfying one of his higher-level needs. He/she expects to be praised for the success of his/her efforts. He/she looks forward to being invited for subsequent technical endeavors. He/she savors hearing the importance of his contribution being related to his/her peers. Without going to the extreme, the project manager can ensure the realization of the above needs through careful comments.

6. *New and rewarding professional relationship*: New and rewarding professional relationships can serve as a bonus for a project effort. Most technical developments result from joint efforts of people that share closely allied interests. Professional allies are most easily found through project groups. A true technical professional will expect to meet new people with whom he/she can exchange views, ideas, and information later on. The project atmosphere should, as a result, be designed to be conducive to professional interactions.

Quality of leadership. The professional background of the project leader should be such that he/she commands the respect of technical subordinates. The leader must be reasonably conversant with the base technologies involved in the project. He/she must be able to converse intelligently on the terminologies of the project topic and be able to convey the project ideas to upper management. This serves to give him/her technical credibility. If technical credibility is lacking, the technical professionals on the project might view him/her as an ineffective leader. They will consider it impossible to serve under a manager to whom they cannot relate technically.

In addition to technical credibility, the manager must also possess administrative credibility. There are routine administrative matters that are needed to ensure a smooth progress for the project. Technical professionals will prefer to have those administrative issues successfully resolved by the project leader so that they can concentrate their efforts on the technical aspects. The essential elements of managing a group of technical professionals involve identifying the unique characteristics and needs of the group and then developing the means of satisfying those unique needs.

Recognizing the peculiar characteristics of technical professionals is one of the first steps in simplifying project management functions. The nature of manufacturing and automation projects calls for the involvement of technical human resources. Every manager must appreciate the fact that the cooperation or the lack of cooperation from technical professionals can have a significant effect on the overall management process. The success of a project can be enhanced or impeded by the management style utilized.

Work simplification. Work simplification is the systematic investigation and analysis of planned and existing work systems and methods for the purpose of developing easier, quicker, less fatiguing, and more economic ways of generating high-quality goods and services. Work simplification facilitates the content of workers, which invariably leads to better performance. Consideration must be given to improving the product or service, raw materials and supplies, the sequence of operations, tools, work place, equipment, and hand and body motions. Work simplification analysis helps in defining, analyzing, and documenting work methods.

HUMAN RESOURCE MANAGEMENT: STEP-BY-STEP IMPLEMENTATION

The human resource management component of the PMBOK consists of the elements shown in the block diagram in Figure 7.1. The four elements in the block diagram are carried out across the process groups presented earlier in this book. The overlay of the elements and the process groups are shown in Table 7.1. Thus, under the knowledge area of human resource management, the required steps are

- Step 1: Human resource planning
- Step 2: Acquire project team
- Step 3: Develop project team
- Step 4: Manage project team

Human resource planning involves identifying and documenting project roles, responsibilities, and creating staffing management plan. Availability, cost, and competence are essential attributes to be evaluated at this stage. Acquire project team involves

STEP Human Resource Management

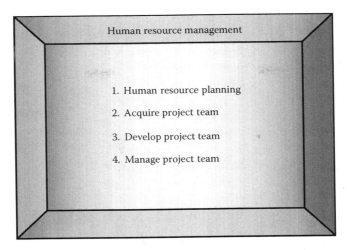

FIGURE 7.1 Block diagram of project human resource management.

TABLE 7.1
Implementation of Human Resource Management across Process Groups

	Initiating	Planning	Executing	Monitoring and Controlling	Closing
Project Human Resource Management		Human resource planning	Acquire project team Develop project team	Manage project team	

obtaining the human resources needed to complete the project. Acquiring human resources involves negotiation as well as possibility of using virtual teams. Develop project team involves improving the competencies and interaction of team members to enhance project performance. Developing human resources requires effective communication, motivation, problem solving, work facilitation, and influencing. Manage project team involves tracking team member performance, providing feedback, resolving conflicts, and coordinating changes to enhance project performance. Managing human resources implies team building and implementing project team ground rules. Tables 7.2 through 7.5 present the inputs, tools and techniques, and outputs of each step.

MANAGING HUMAN RESOURCE PERFORMANCE

Connecting with the employee is a basic requirement of managing human resource performance. Unbiased leadership means not judging others based on one's own values. As a leader, a project manager cannot be a good communicator if he/she is

TABLE 7.2
Tools and Techniques for Human Resource Planning within Project Human Resource Management

STEP 1: Human Resource Planning

Inputs	Tools and Techniques	Output(s)
Enterprise environmental factors	Organizational charts	Roles and responsibilities
Organizational process assets	Team networking	Project organization charts
Project management plan	Group dynamics	Staffing outline
Activity resource requirements	Organizational theory	Management plan
Other in-house (custom) factors of relevance and interest	Other in-house (custom) tools and techniques	Other in-house outputs, reports, and data inferences of interest to the organization

TABLE 7.3
Tools and Techniques for Acquiring Project Team within Project Human Resource Management

STEP 2: Acquire Project Team

Inputs	Tools and Techniques	Output(s)
Enterprise environmental factors	Preassignment	Project staff assignments
Organizational process assets	Negotiation	Resource availability database
Roles and responsibilities	Acquisition	Staffing management plan (updates)
Project organization charts	Virtual teaming	
Staffing management plan	Staff exchange programs	Other in-house outputs, reports, and data inferences of interest to the organization
Other in-house (custom) factors of relevance and interest	Colocation programs	
	Other in-house (custom) tools and techniques	

TABLE 7.4
Tools and Techniques for Developing Project Team within Project Human Resource Management

STEP 3: Develop Human Resource

Inputs	Tools and Techniques	Output(s)
Project staff assignments	General management skills	Team performance assessment
Staffing management plan	Training	Other in-house outputs, reports, and data inferences of interest to the organization
Resource availability	Team-building exercises	
Other in-house (custom) factors of relevance and interest	Ground rules formulation	
	Colocation strategies	
	Recognition and awards	
	Other in-house (custom) tools and techniques	

TABLE 7.5
Tools and Techniques for Managing Project Team within Project Human Resource Management

STEP 4: Manage Human Resource

Inputs	Tools and Techniques	Output(s)
Organizational process assets	Triple C Model (communication, cooperation, coordination)	Requested changes
Project staff assignments		Record of corrective actions
Roles and responsibilities		Record of preventive actions
Project organization chart	Hierarchy of needs	Organizational process assets
Staffing management plan	Theory X and Theory Y	Project management plan (updates)
Team performance assessment	Motivation–hygiene factors	
	Management by objective	Other in-house outputs, reports, and data inferences of interest to the organization
Performance reports	Management by exception	
Other in-house (custom) factors of relevance and interest	Observational programs	
	Staff conversation and dialogue techniques	
	Conflict management	
	Issue log	
	Other in-house (custom) tools and techniques	

not a good listener. The LINK (Look, Inquire, Note, Know) concept for connecting with employees requires that the leader exhibit empathy for the employee's specific needs. LINK is presented diagrammatically in Figure 7.2.

Managing human resource performance requires the following:

1. Managing employee information
2. Setting and managing goals
3. Documenting ongoing performance events
4. Developing employee performance improvement and advancement and strategies

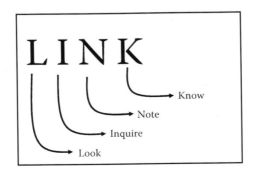

FIGURE 7.2 LINK concept of connecting with employees.

Negligence in managing employee information will most certainly lead to inaccuracies that will adverse affect how effectively performance can be managed or rectified.

QUANTITATIVE MODELING OF WORKER ASSIGNMENT

Operations research techniques are frequently used to enhance resource allocation decisions. One common resource allocation tool is the resource assignment algorithm, which can be applied to enhance human resource management. Suppose that there are n tasks that must be performed by n workers. The cost of worker i performing task j is c_{ij}. It is desired to assign workers to the tasks in a fashion that minimizes the cost of completing the tasks. This problem scenario is referred to as the *assignment problem*. The technique for finding the optimal solution to the problem is called the *assignment method*. Like the transportation method, the assignment method is an iterative procedure that arrives at the optimal solution by improving on a trial solution at each stage of the procedure. Conventional CPM and PERT can be used in controlling projects to ensure that the project will be completed on time; but both techniques do not consider the assignment of resources to the tasks that make up a project. The assignment method can be used to achieve an optimal assignment of resources to specific tasks in a project. Although the assignment method is cost-based, task duration can be incorporated into the modeling in terms of time–cost relationships. Of course, task precedence requirements and other scheduling constraints of the tasks will be factored into the computational procedure. The objective is to minimize the total cost of assigning workers to tasks. The formulation of the assignment problem is as follows:

Let
$x_{ij} = 1$ if worker i is assigned to task j, where $i, j = 1, 2, \ldots, n$
$x_{ij} = 0$ if worker i is not assigned to task j
c_{ij} is the cost of worker I performing task j

$$\text{Minimize:} z = \sum_{i=1}^{n}\sum_{j=1}^{n} c_{ij} x_{ij}$$

$$\text{Subject to:} \sum_{j=1}^{n} x_{ij} = 1, \quad i = 1, 2, \ldots, n$$

$$\sum_{i=1}^{n} x_{ij} = 1, \quad j = 1, 2, \ldots, n$$

$$x_{ij} \geq 0, \quad i, j = 1, 2, \ldots, n$$

The above formulation is a transportation problem with $m = n$ and all supplies and demands (sources to targets) are equal to 1. Note that we have used the nonnegativity constraint, $x_{ij} \geq 0$, instead of the integer constraint, $x_{ij} = 0$ or 1. However, the solution of the model will still be integer valued. Hence, the assignment problem is a special case of the transportation problem with $m = n$; $S_i = 1$ (supplies); and $D_j = 1$ (demands). Conversely, the transportation problem can also be viewed as a special case of the assignment problem. The basic requirements of an assignment problem are as follows:

1. There must be two or more tasks to be completed.
2. There must be two or more resources that can be assigned to the tasks.
3. The cost of using any of the resources to perform any of the tasks must be known.
4. Each resource is to be assigned to one and only one task.

If the number of tasks to be performed is greater than the number of workers available, we will need to add dummy workers to balance the problem formulation. Similarly, if the number of workers is greater than the number of tasks, we will need to add dummy tasks to balance the formulation. If there is no problem of overlapping, a worker's time may be split into segments so that the worker can be assigned more than one task. In this case, each segment of the worker's time will be modeled as a separate resource in the assignment problem. Thus, the assignment problem can be extended to consider partial allocation of resource units to multiple tasks.

Although the assignment problem can be formulated for and solved by the simplex method or the transportation method, a more efficient algorithm is available specifically for the assignment problem. The method, known as the Hungarian method, is a simple iterative technique. Details of the assignment problem and its solution techniques can be found in operations research textbooks. As an example, suppose that five workers are to be assigned to five tasks on the basis of the cost matrix presented in Table 7.6. Task 3 is a machine-controlled task with a fixed cost of $800.00 regardless of which worker it is assigned to. Using the data, we obtain the assignment solution presented in Table 7.7, which indicates the following:

$$x_{15} = 1$$

$$x_{23} = 1$$

TABLE 7.6
Cost Matrix for Worker Assignment Problem

Worker	Task 1	Task 2	Task 3	Task 4	Task 5
1	300	200	800	500	400
2	500	700	800	1250	700
3	300	900	800	1000	600
4	400	300	800	400	400
5	700	350	800	700	900

TABLE 7.7
Solution to Worker Assignment Problem

Worker	Task 1	Task 2	Task 3	Task 4	Task 5
1	0	0	0	0	1
2	0	0	1	0	0
3	1	0	0	0	0
4	0	0	0	1	0
5	0	1	0	0	0

$$x_{31} = 1$$

$$x_{44} = 1$$

$$x_{52} = 1$$

Thus, the minimum total cost is given by

$$\text{TC} = c_{15} + c_{23} + c_{31} + c_{44} + c_{52} = (400 + 800 + 300 + 400 + 350) = \$2250.00$$

The technique of work rate analysis can be used to determine the cost elements that go into an assignment problem. The solution of the assignment problem can then be combined with the technique of critical resource diagramming. This combination of tools and techniques can help enhance human resource management decisions from a quantitative modeling perspective.

RESOURCE WORK RATE ANALYSIS

When resources work concurrently at different work rates, the amount of work accomplished by each may be computed by the technique of work rate analysis. The critical resource diagram and the resource schedule chart provide information to identify when, where, and which resources work concurrently. The general relationship between work, work rate, and time can be expressed as shown below:

$$w = rt$$

where
  w is the amount of actual work accomplished (expressed in appropriate units, such as miles of road completed, lines of computer code typed, gallons of oil spill cleaned, units of widgets produced, and surface area painted)

r is the rate at which the work is accomplished
t is the total time required to accomplish the work

It should be noted that work rate can change due to the effects of learning curves. In the examples that follow, it is assumed that work rates remain constant for at least the duration of the work being analyzed. Work is defined as a physical measure of accomplishment with uniform density (i.e., homogeneous). For example, a computer programming task may be said to be homogeneous if one line of computer code is as complex and desirable as any other line of code in the program. Similarly, cleaning one gallon of oil spill is as good as cleaning any other gallon of oil spill within the same work environment. The production of one unit of a product is identical to the production of any other unit of the product. If uniform work density cannot be assumed for the particular work being analyzed, then the relationship presented above will need to be modified. If the total work to be accomplished is defined as one whole unit, then the tabulated relationship below will be applicable for the case of a single resource performing the work:

Human resource	Work rate	Time	Work done
Worker A	$1/x$	t	1.0

The variable $1/x$ is the amount of work accomplished per unit time. For a single resource to perform the whole unit of work, we must have the following:

$$\frac{1}{x}(t) = 1.0$$

This means that the absolute magnitude of x must equal the magnitude of t. For example, if Worker A is to complete one work unit in 30 min, it must work at the rate of 1/30 of work per unit time. If the magnitude of x is greater than the magnitude of t, then only a fraction of the required work will be accomplished. The information about the proportion of work completed may be useful for resource planning and productivity measurement purposes. In the case of multiple resources performing the work simultaneously, the work relationships are shown in Table 7.8.

TABLE 7.8
Multiple Resource Work Rate Data Table

Resource Type, i	Work Rate, r_i	Time, t_i	Work Done, w_i
RES 1	r_1	t_1	$(r_1)(t_1)$
RES 2	r_2	t_2	$(r_2)(t_2)$
...	...	...	...
RES n	r_n	t_n	$(r_n)(t_n)$
		Total	1.0

For multiple resources, we have the work rate formulation as follows:

$$\sum_{i=1}^{n} r_i t_i = 1.0$$

where
   n is the number of different resource types
   r_i is the work rate of resource type i
   t_i is the work time of resource type i

The expression indicates that even though the multiple resources may work at different rates, the sum of the total work they accomplished together must equal the required whole unit. For partial completion of work, the expression becomes

$$\sum_{i=1}^{n} r_i t_i = p$$

where p is the proportion of the required work actually completed. Suppose that RES 1, working alone, can complete a job in 50 min. After RES 1 has been working on the job for 10 min, RES 2 was assigned to help RES 1 in completing the job. Both resources working together finished the remaining work in 15 min. It is desired to determine the work rate of RES 2. The amount of work to be done is 1.0 whole unit. The work rate of RES 1 is 1/50 of work per unit time. Therefore, the amount of work completed by RES 1 in the 10 min it worked alone is (1/50)(10) = 1/5 of the required work. This may also be expressed in terms of percent completion or earned value using earned value technique. The remaining work to be done is 4/5 of the total work. The two resources working together for 15 min yield the tabulated result in Table 7.9.

Thus, we have $15/50 + 15(R_2) = 4/5$, which yields $r_2 = 1/30$ for the work rate of RES 2. This means that RES 2, working alone, could perform the job in 30 min. In this example, it is assumed that both resources produce identical quality of work. If quality levels are not identical for multiple resources, then the work rates may be adjusted to account for the different quality levels or a quality factor may be introduced into the analysis. The relative costs of the different resource types needed to perform the required work may be incorporated into the analysis by adding columns for worker pay rate and worker total cost to the layout in Table 7.8.

As another illustrative example, suppose that the work rate of RES 1 is such that it can perform a certain task in 30 days. It is desired to add RES 2 to the task so that

TABLE 7.9
Tabulation of Multiple Resource Work Rate Analysis

Resource Type, i	Work Rate, r_i	Time, t_i	Work Done, w_i
RES 1	1/50	15	15/50
RES 2	r_2	15	$15(r_2)$
		Total	1.0

STEP Human Resource Management

the completion time of the task could be reduced. The work rate of RES 2 is such that it can perform the same task alone in 22 days. If RES 1 has already worked 12 days on the task before RES 2 comes in, find the completion time of the task. It is assumed that RES 1 starts the task at time 0. As usual, the amount of work to be dome is 1.0 whole unit (i.e., the full task). The work rate of RES 1 is 1/30 of the task per unit time and the work rate of RES 2 is 1/22 of the task per unit time. The amount of work completed by RES 1 in the 12 days it worked alone is $(1/30)(12) = 2/5$ (or 40%) of the required work. Therefore, the remaining work to be done is 3/5 (or 60%) of the full task. Let T be the time for which both resources work together. The two resources working together to complete the task yield the following relationships:

Resource Type, i	Work Rate, r_i	Time, t_i	Work Done, w_i
RES 1	1/30	T	$T/30$
RES 2	1/22	T	$T/22$
		Total	3/5

Thus, we have $T/30 + T/22 = 3/5$, which yields $T = 7.62$ days. Consequently, the completion time of the task is $(12 + T) = 19.62$ days from time zero. The results of this example are summarized in the resource schedule charts in Figure 7.3. It is assumed that both resources produce identical quality of work and that the respective work rates remain consistent. As mentioned earlier, the respective costs of the different types may be incorporated into the work rate analysis.

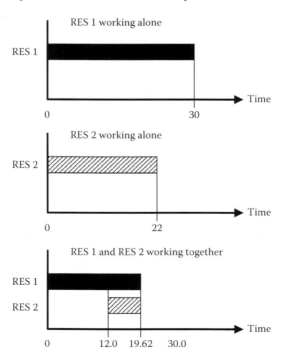

FIGURE 7.3 Resource schedule chart for work rate analysis.

MODEL FOR TECHNICAL HUMAN RESOURCE TRAINING

Technical training should be performance-based and designed in a way that allows workers to be trained better, faster, and cheaper. Industry is continually in a state of flux trying to adapt to the dynamics of the STE market place and ever-changing technology. This focus on the continuous improvement of an organization is based on a primary foundation of the organization's ability to promote cultural and technological changes. This ability of organizations to convince their workforce to change is a key indicator of the organization's success in the market place. Historically, the motivation of the workforce to change has come in a variety of approaches including management proclamation to focus on educating the workforce in key technical areas and the versatility of the workforce to move into higher positions. The increasingly competitive market provides incentives for improving the quality of training while lowering the overall time and cost. Education and training, designed to enhance the skills, knowledge, and competencies of a project team, can be achieved through various means including the following:

1. Classroom
2. Self-study
3. Working meetings
4. Team workshops
5. Group exercises

Sawhney et al. (2004) present a training model for technical professionals. There is a good correlation between formal education and training and higher organizational performance. It is well publicized that executives indicate that a cross-functional workforce is the most effective strategy to achieve organizational goals. The development of such a cross-functional workforce requires formal training, not only in areas of business operations but also for the area of project management. Since training is such a capital-intensive endeavor with no tangible revenue, justification of training is often difficult to ascertain. There are two aspects to the justification: effectiveness and efficiency of the training. Effectiveness refers to the benefits that the organization will receive by training the workforce to meet organizational objectives. Training efficiency is a function of determining the resources required for the training and subsequently providing the resources at the right time. The focus of the approach presented here is on variables that influence the learning process to enhance training effectiveness and efficiency.

In practice, there is a perceived lack of general awareness and, subsequently, a structured and consistent process to ensure training efficiency. A review and adaptation of what is available in the training literature leads to the summarized process below:

Step 1: Training assessment
Step 2: Training planning
Step 3: Training management
Step 4: Training risk management

STEP Human Resource Management

The objective of Step 1 is to assess the alignment of the training program to the organizational strategic goals. The objective of Step 2 is to create a plan that assists training personnel to become more systematic and organized in determining the resource requirement for the development of the training program. This plan focuses on the analysis, design, and development of the training material. Similarly the objective of Step 3 is to define plan for defining the resource requirements for monitoring, tracking, managing, evaluation, and documentation associated with training programs. Finally, the objective of Step 4 is to highlight the risks associated with the training program. Table 7.10 summarizes the characteristics of training concerns

TABLE 7.10
Summary of Training Concerns and Remedies

Process Step	Potential Failure Mode	Potential Failure Effects	Potential Causes	Current Process Controls	Actions Recommended	Responsible Party
People	Employees not motivated to train	Training is ineffective, does not produce results	Communication of reason and objective training is not viewed as priority	Informal communication	Formal communication of training	HR
	Employees pulled out of training by immediate needs	Training is ineffective, does not produce results	Production needs take priority	None	Management edict on training priority	HR
Planning	Incomplete training	Delayed results	Insufficient resources	None	Preplanning communication	HR
	Greater time required than anticipated	Increased cost	Improper planning	None	Preplanning communication	HR
		Delay in results	Fighting fires			
Implementation	Workforce did not learn	Failure in education change	Instructor material resources	Review by HR, domain	Stricter review process	HR
	Workforce did not implement	Failure in achieving result	Time confusion of responsibility	None	Management program, preplanning	Mgmt
Integration	Lack of system wide results	Inconsistent implementation	Resistance to change, Unions	None	Communication, preplanning	HR
Sustainability	Implementation is poor	System is worse than before	Management, resources, training	None	Communication, preplanning	Mgmt

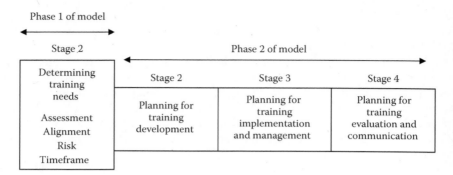

FIGURE 7.4 Proposed framework for training model.

with respect to the four-step process outlined above. The contents of the table lead to the need to develop a model that allows human resources personnel and other personnel in charge of training to plan an efficient training program.

A CONCEPTUAL APPROACH

A framework for the proposed model is presented in Figure 7.4. The framework is based on the planning for the various stages of a training program: needs assessment, developing the training, delivering the training, and subsequent evaluation and implementation in the organization.

Every concept needs a transformation mechanism that proves the value of the concept in practice. The purpose of the transformation mechanism in this case is to convert the framework in Figure 7.4 into a model that allows personnel responsible for training to properly determine required resources. The inspiration for this transformation mechanism is based on a combination of two models: Systematic Project Management Approach and NIOSH's 4-Stage Training Intervention Effectiveness Research Model (Sawhney et al., 2004). The contribution of each model is illustrated in the composite training model presented in Figure 7.5. The proposed model has two phases. The objective of phase 1 is to ensure that the training is in line with the organizations objectives, expectations, and resources. The objective of phase 2 is to provide personnel in charge of training the ability to forecast resource requirements and subsequently to garner these resources to assure the efficiency of the training. These resources are for developing the training program, conducting the training, and performing subsequent evaluation and implementation.

EXTENDED TRAINING MODEL

The extended model consists of two separate but complementary phases as outlined in Figure 7.6. Phase 1 of the extended model is associated with stage 1 of Figure 7.5, the alignment and risk associated with the training program. This phase consists of an analysis designed explicitly to guide training personnel to evaluate the value of the proposed training to the organization. This analysis is in the form of a worksheet,

STEP Human Resource Management

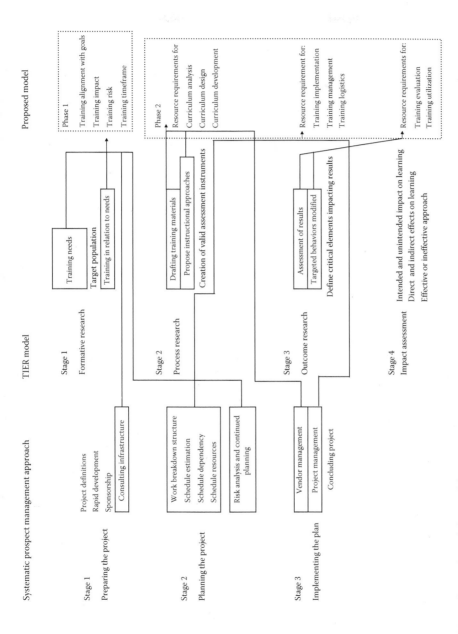

FIGURE 7.5 Composite training model.

Suggested Training	Training Alignment with Organizational Objectives	Alignment Rating (1–10) A	Training Impact on Workforce				Impact Rating (0–9) B	Risk Associated with Training				Risk Rating (1–10) C	Timeframe for Training Completion	Time Rating (1–10) D	Training Rating TR
			b_1	b_2	b_3	b_4		c_1	c_2	c_3	c_4				
			Tech	Manage	Safety	Cultural		Mgmt Com	Res Avail	Lack of Infra	Wkforce Accept				
Training Prog 1	Creates teamwork	8	5	8	0	9	5.5	6	5	4	2	2	Medium	5	440.0
Training Prog 2	Eliminates product waste	7	8	6	2	9	6.3	8	3	6	1	1	Long term	3	132.3
Training Prog 3	Teaches computer skills	5	5	9	0	0	3.5	4	8	8	6	4	Short term	8	560.0
		10 = High alignment					9 = High impact					10 = Low risk		10 = Short timeframe	

FIGURE 7.6 Worksheet for Stage 1 of the extended training model.

STEP Human Resource Management 213

as presented in Figure 7.6, allowing training personnel to consider the value of the training in terms of four distinct dimensions:

1. Alignment of training to organization's strategic goals
2. Identification of areas of impact to targeted workforce
3. Risk associated with not obtaining desired results from the training program
4. Timeframe required for obtaining the desired results

The following is a description of the steps required to fill and interpret the worksheet.

Step 1. *Alignment rating*: List the objectives of the training program. Next compare these objectives to the organization's strategic objectives. The user indicates the alignment of the training objectives with the organization's strategic objectives via an alignment rating. This range of the alignment rating is between 1 and 10. A rating of 1 indicates that the training does not support the organization's strategic objectives and a rating of 10 indicates a perfect correlation between training and organizational objectives. Any training with a low rating should be questioned for validity.

Step 2. *Impact rating*: Training provides content and information for the workforce in any of the following domain subsets: technical, managerial, safety, or cultural knowledge. Each training program needs to be evaluated in terms of the impact it makes in each of these categories. This range of impact rating is between 0 and 9. A rating of 0 indicates no impact while a rating of 9 indicates a significant impact. Not necessarily is each category equally important. Therefore, the worksheet allows the user to weigh each of the four categories in importance. This defined weighting scheme must remain constant for evaluation of all the training programs. The overall impact rating is then the weighted average of the four categories.

Step 3. *Risk rating*: There are four primary reasons that training programs do not provide the anticipated results to the organization. These include the lack of management commitment, lack of resources, lack of appropriate infrastructure, and the lack of workforce acceptance for the training. Every training program is rated for each category. This range of risk rating is between 1 and 10. A rating of 1 indicates extreme risk while a rating of 10 indicates no risk. The overall risk rating is the minimum rating value in each category because it is the constraint that detracts from the success of the training program.

Step 4. *Timeframe rating*: The timeframe rating allows one to anticipate the results of the training program to impact the organization. This range of the timeframe rating is between 1 and 10. A rating of 1 at this step indicates extremely long-term impact while a rating of 10 indicates immediate results.

Step 5. *Overall training rating*: The overall training rating is determined by multiplying the above four ratings. The ratings range from 0 to 9000. Statistical concepts can be utilized to further decompose the range and interpret the results. The computational process is summarized below:

A = Alignment rating number

$$B = \frac{w_{b1} b_1 + w_{b2} b_2 + w_{b3} b_3 + w_{b4} b_4}{\text{Weighted average}}$$

$C = \text{Min}(c_1, c_2, c_3, c_4)$
$D = $ Time rating number
$TR = W_A A \times W_B B \times W_C C \times W_D D$

This section presents phase 2 of the extended training model. The objective of this phase is to plan the resource requirements for stages 2, 3, and 4 of Figure 7.4. A proper planning will increase the probability of acquiring required resources at the right time. The foundation for phase 2 of the model is a component of a model used for instructional intervention (training). Specifically, the model exhibits three specific characteristics:

1. Utilization of work breakdown structure (WBS) to develop a precedence chart of the various levels of tasks within a training program
2. Defining time requirements for every task within each level of the WBS
3. Defining the dependency relationships between all tasks within each level of the WBS

The three types of dependency relationships are finish-to-start, start-to-start, and finish-to-finish. Further, for each dependency type there could be a lag, which is defined as a passage of time before the subsequent task can be initiated. Figure 7.7 presents the relationships. There are several assumptions defined below that are key to the model.

1. Tasks at any level are arranged chronologically from left to right such that the tasks can be analyzed left to right on any level.
2. Tasks at a lower level have to be completed before the tasks at a higher level can be initiated.
3. Dependency relationships between tasks at all levels are defined.
4. Values in parenthesis are necessary times (weeks) for each task.
5. Lag time is assumed to be 1 week.

Figure 7.7 illustrates that by utilizing the WBS structure, training will be completed in 18 weeks. This is compared to the traditional calculation of 14 weeks. This additional 4 weeks can be the difference between the successful completion of the training and the perceived failure of the training. This 4-week difference is due to the precedence requirements that training managers typically ignore. However, even the correction of this discrepancy does not provide resource details and the timings. The details of resource requirements and the timing of the resource requirements can be determined if a resource requirement planning (RRP) record is to be imposed for each WBS task in Figure 7.7. This is similar to the classic material requirement planning (MRP) record imposed on the bill of material (BOM). Such an RRP record is presented in Figure 7.8.

STEP Human Resource Management

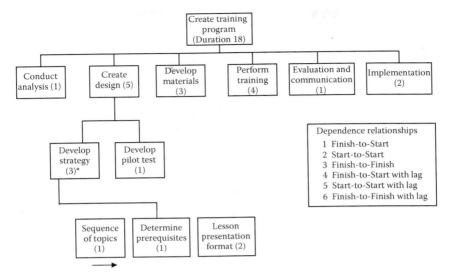

FIGURE 7.7 Stage 2: Precedence-based WBS.

Task/responsible resource			Time bucket					
Period (week)	1	2	3	4	5	6	7	8
Task time requirement (h)			10	10	20			
Precedence time adjustment				10	10	20		
Resource functionality adjustments				12	12	24		
Current resource commitment				10	10	10	10	
Resource misalignment				−2	−2	−14	10	−8
Current resource reallocation				−2	−4	−18	−8	
Total resource load				10	22	22	34	10

FIGURE 7.8 Resource requirement planning (RRP) record.

Each record is associated with a specific task from the WBS and is assigned a responsible department or individual. The record consists of a time bucket corresponding to the length of the training program. The following are brief explanations of each row of the RRP record.

1. Task time requirement (row 1): This is the estimated time required during each period to complete the task by the responsible department or individual. The time will include any delays due to contracting all or part of the work.

2. Precedence time adjustment (row 2): This row shifts the row 1 time requirements to appropriate time buckets to accommodate precedence and time lag constraints.
3. Resource functionality adjustment (row 3): This row increases or decreases the times in row 2 based on the historical evidence of the responsible department's meeting task obligations.
4. Current resource commitment (row 4): This row indicates the time commitment by the responsible department or individual. The time commitment may not be uniform over the training period.
5. Resource misalignment (row 5): This row balances the time requirements against the committed time for every period. The misalignment for each period is calculated as follows: row 4–row 3. The misalignments for each period are summed to indicate one of two problems: the incorrect allocation of resources (a positive summed value) or lack of resources (a negative summed number).
6. Current resource reallocation (row 6): This row balances the time requirements with the committed time based on the assumption that resources not utilized can be utilized in the subsequent period. The balance is calculated as follows: (row 6 from previous period + row 4) − row 3.
7. Total resource load (row 7): This row adds up all the time requirements for each period for all tasks associated with the responsible resource. Figure 7.8 assumes an additional 12, 2, 34, and 10 h of time required from other lower level tasks.

Figure 7.9 is an example of the mechanics of phase 2 of the proposed model. This example is based on a subset of the WBS in Figure 7.7. The model first analyzes the "develop strategy" task (task 1). This task influences the pilot study task (task 2). The time requirements task 2 are shifted from periods 1, 2, and 3 to periods 5, 6, and 7 because of the precedence relationship (finish-to-start with a lag) between task 1 and 2. Similarly, the time requirements of "create design" task (task 3) is moved because all the work in tasks 1 and 2 have to be completed before task 3 can be initiated. Further, the time requirements for task 2 have increased because historically it has taken about 1.5 times the given time to accomplish the task. The resource requirements and misalignments are presented for each task. Note that as we go from task 1 to 2 to 3 the cumulative resource requirements are identified because all three tasks are performed by the same responsible resource.

The models presented here have the potential to allow training managers to rank training programs including comparison of multiple programs as to their value to an organization. For example, the model has the ability to differentiate between training employees in both Lean and Six Sigma as compared to an integrated Lean-Sigma training program. The model further attempts to identify the resources required for any training program. For example, it has the ability to be able to distinguish between resource requirements of Lean and Six Sigma as compared to an integrated Lean-Sigma. The models can be adapted for customized implementation within STEP management processes.

STEP Human Resource Management

Task 3: Create design
Responsible resources; training

Period	1	2	3	4	5	6	7	8	9	10	11	12
TTR		10	20			40						
PTA								10	20			40
RFA (1.0)								10	20			40
CRC								20	20	20	20	
RM								10	0	20	20	
CRR								10	10	30	50	−40
TRL				30	40	10	30	45	60	10	20	10
												40

Task 2: Pilot strategy
Responsible resources; training

Period	1	2	3	4	5	6	7	8	9	10	11	12
TTR	20	30	40									
PTA					20	30	40					
RFA (1.5)					30	45	60					
CRC	40	40	40									
RM	40	40	40		−30	−45	−60					−15
CRR												
TRL	30	30	40	10	30	45	60					

Task 1: Develop strategy
Responsible resources; training

Period	1	2	3	4	5	6	7	8	9	10	11	12
TTR		30	40	10								
PTA		30	40	10								
RFA (1.0)		30	40	10								
CRC		30	20	20								
RM		10	−20	10								0
CRR		10	−10	0								
TRL		30	40	10								

FIGURE 7.9 Sample of Phase 2 of the proposed model.

REFERENCES

PMI, *A Guide to the Project Management Body of Knowledge (PMBOK Guide)*, 3rd ed., Project Management Institute, Newtown Square, PA, 2004.

Sawhney, R., A. B. Badiru, and A. Niranjan, A model for integrating and managing resources for technical training programs, in *Internet Economy: Opportunities and Challenges for Developed and Developing Regions of the World*, Y. A. Hosni and T. M. Khalil (Eds.), Elsevier, Boston, MA, 2004, pp. 337–351.

8 STEP Communications Management

Communication is the root of everything else.

–Adedeji Badiru, 2008

As the original quote at the beginning of this chapter suggests, communication is vital to everything else in a project. Any successful project manager would spend 90% of his or her time on communication activities. This is a vital function that is even more crucial in STEPs. Communications management refers to the functional interface between individuals and groups within the project environment. This involves proper organization, routing, and control of information needed to facilitate work. Good communication is in effect when there is a common understanding of information between the communicator and the target. Communications management facilitates unity of purpose in the project environment. The success of a project is directly related to the effectiveness of project communication. From the author's experience, most project problems can be traced to a lack of proper communication. Communication is achieved through a variety of means beyond verbal exchanges (Mooz et al, 2003; PMI, 2004). Telling, showing, and direct involvement are all effective modes of communication. A Chinese proverb says,

Tell me, and I forget;
Show me, and I remember;
Involve me, and I understand.

The project team should employ all possible avenues to get project information across to everyone.

COMMUNICATIONS MANAGEMENT: STEP-BY-STEP IMPLEMENTATION

The communications management component of the *Project Management Body of Knowledge* (*PMBOK*) consists of the elements shown in the block diagram in Figure 8.1. The four elements in the block diagram are carried out across the process groups presented earlier in this book. The overlay of the elements and the process groups are shown in Table 8.1. Thus, under the knowledge area of communications management, the required steps are

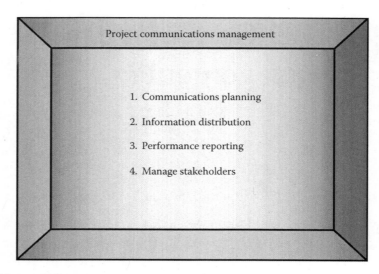

FIGURE 8.1 Block diagram of project communications management.

TABLE 8.1
Implementation of Communications Management across Process Groups

	Initiating	Planning	Executing	Monitoring and Controlling	Closing
Project Communications Management		Communications planning	Information distribution	Performance reporting Manage stakeholders	

Step 1: Communications planning
Step 2: Information distribution
Step 3: Performance reporting
Step 4: Manage stakeholders

Tables 8.2 through 8.5 present the inputs, tools, techniques, and outputs of each step of communications management. Communications planning involves determining the information and communication needs of the stakeholders regarding who needs what information, when, where, and how. Information distribution involves making the needed information available to project stakeholders in a timely manner and in appropriate dosage. Performance reporting involves collecting and disseminating performance information, which includes status reporting, progress measurement, and forecasting. Managing stakeholders involves managing communications to satisfy the requirements of the stakeholders so as to resolve issues that develop.

STEP Communications Management

TABLE 8.2
Tools and Techniques for Communications Planning within Project Communications Management

Step 1: Communications Planning

Inputs	Tools and Techniques	Output(s)
Enterprise environmental factors	Communications requirement analysis	Communications Management plan
Organizational process assets	Communications technology	Other in-house outputs, reports, and data inferences of interest to the organization
Project scope statement	Communications responsibility matrix	
Project constraints and assumptions	Collaborative alliance	
Other in-house (custom) factors of relevance and interest	Other in-house (custom) tools and techniques	

TABLE 8.3
Tools and Techniques for Information Distribution within Project Communications Management

Step 2: Information Distribution

Inputs	Tools and Techniques	Output(s)
Communication management plan	Communication modes and skills	Organizational process assets (updates)
Personnel distribution list	Social networking	Other in-house outputs, reports, and data inferences of interest to the organization
Other in-house (custom) factors of relevance and interest	Influence networking	
	Meetings and dialogues	
	Communication relationships	
	Information gathering and retrieval systems	
	Information distribution methods	
	Lessons learned	
	Best practices	
	Information exchange	
	Other in-house (custom) tools and techniques	

COMPLEXITY OF MULTIPERSON COMMUNICATION

Communication complexity increases with an increase in the number of communication channels. It is one thing to wish to communicate freely, but it is another thing to contend with the increased complexity when more people are involved. The statistical formula of combination can be used to estimate the complexity of communication as a function of the number of communication channels or number of

TABLE 8.4
Tools and Techniques for Performance Reporting within Project Communications Management

Step 3: Performance Reporting

Inputs	Tools and Techniques	Output(s)
Performance measurements	Performance information gathering and compilation	Forecasts
Forecasted completion		Requested changes
Quality control measurements	Status review meetings	Recommended corrective actions
Project performance measurement baseline	Time reporting systems	Organizational process assets
	Cost reporting systems	Other in-house outputs, reports, and data inferences of interest to the organization
Approved change requests	Other in-house (custom) tools and techniques	
List of deliverables		
Other in-house (custom) factors of relevance and interest		

TABLE 8.5
Tools and Techniques for Managing Stakeholders within Project Communications Management

STEP 4: Manage Stakeholders

Inputs	Tools and Techniques	Output(s)
Communications management plan	Communications methods	Conflict resolution report
	Issue logs	Approved change requests
Organizational process assets	Other in-house (custom) tools and techniques	Approved corrective actions
Other in-house (custom) factors of relevance and interest		Organizational process assets (updates)
		Other in-house outputs, reports, and data inferences of interest to the organization

participants. The combination formula is used to calculate the number of possible combinations of r objects from a set of n objects. This is written as

$$_nC_r = \frac{n!}{r![n-r]!}$$

In the case of communication, for illustration purposes, we assume communication is between two members of a team at a time. That is, combination of 2 from n team members. That is, number of possible combinations of 2 members out of a team of n people. Thus, the formula for communication complexity reduces to the expression below after some of the computation factors cancel out:

STEP Communications Management

$$_nC_2 = \frac{n(n-1)}{2}$$

In a similar vein, Badiru (2008) introduced a formula for cooperation complexity based on the statistical concept of permutation. Permutation is the number of possible arrangements of k objects taken from a set of n objects. The permutation formula is written as

$$_nP_k = \frac{n!}{(n-k)!}$$

Thus, for the number of possible permutations of 2 members out of a team of n members is estimated as

$$_nP_2 = n(n-1)$$

Permutation formula is used for cooperation because cooperation is bidirectional. Full cooperation requires that if A cooperates with B, then B must cooperate with A. But, A cooperating with B does not necessarily imply B cooperating with A. In notational form, that is

$$A \rightarrow B \text{ does not necessarily imply } B \rightarrow A$$

Figure 8.2 shows an example of communication channels in a project network. Figure 8.3 shows the relative plots of communication complexity and cooperation complexity as function of project team size, n. It is seen that complexity increases rapidly as the number of communication participants increases. Coordination complexity is even more exponential as the number of team members increases. Interested readers can derive their own coordination complexity formula based on

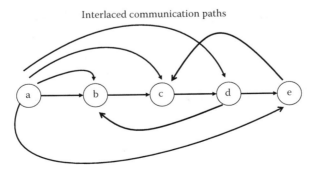

FIGURE 8.2 Example of communication channels in a STEP network.

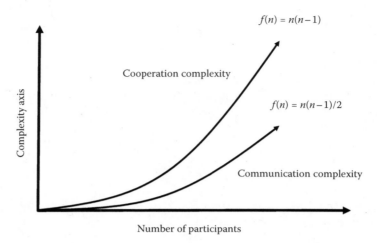

FIGURE 8.3 Plots of communication and cooperation complexities.

the standard combination and permutation formulas or other statistical measures. The complexity formulas indicate a need for a more structured approach to implementing the techniques of project management. The communications templates and guidelines presented in this chapter are useful for general management of STEPs. Each specific project implementation must adapt the guidelines to the prevailing scenario and constraints of a project.

COMMUNICATING THROUGH TRIPLE C MODEL

Badiru (2008) presents the Triple C model as an effective tool for achieving communication, cooperation, and coordination in complex project environment. The Triple C model states that project management can be enhanced by implementing it within the following integrated and hierarchical processes:

- Communication
- Cooperation
- Coordination

The model facilitates a systematic approach to project planning, organizing, scheduling, and control. The Triple C model requires communication to be the first and foremost function in the project endeavor. The model explicitly provides an avenue to address questions such as the following:

When will the project be accomplished?
Which tools are available for the project?
What training is needed for the project execution?
What resources are available for the project?
Who will be members of the project team?

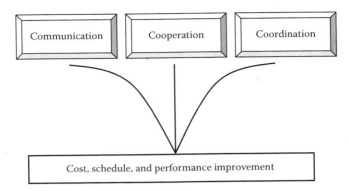

FIGURE 8.4 Triple C for planning, scheduling, and control.

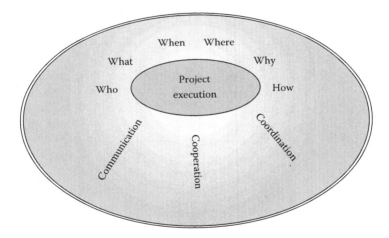

FIGURE 8.5 Triple C for who, what, why, when, where, and how.

Figure 8.4 illustrates the three elements of the Triple C model with respect to cost, schedule, and performance improvement goals.

Figure 8.5 presents how the basic questions of what, who, why, how, where, and when revolve around the Triple C model. It highlights what must be done and when. It can also help to identify the resources (personnel, equipment, facilities, etc.) required for each effort in the project. It points out important questions such as

- Does each project participant know what the objective is?
- Does each participant know his or her role in achieving the objective?
- What obstacles may prevent a participant from playing his or her role effectively?

Triple C can mitigate disparity between idea and practice because it explicitly solicits information about the critical aspects of a project. The different types of communication, cooperation, and coordination are outlined below.

Types of communication

- Verbal
- Written
- Body language
- Visual tools (e.g., graphical tools)
- Sensual (use of all five senses: sight, smell, touch, taste, hearing—olfactory, tactile, auditory)
- Simplex (unidirectional)
- Half-duplex (bidirectional with time lag)
- Full-duplex (real-time dialogue)
- One-on-one
- One-to-many
- Many-to-one

Types of cooperation

- Proximity
- Functional
- Professional
- Social
- Romantic
- Power influence
- Authority influence
- Hierarchical
- Lateral
- Cooperation by intimidation
- Cooperation by enticement

Types of coordination

- Teaming
- Delegation
- Supervision
- Partnership
- Token passing
- Baton hand-off

TYPICAL TRIPLE C QUESTIONS

Questioning is the best approach for getting information for effective project management. Everything should be questioned. By upfront questions, we can preempt and avert project problems later on. Typical questions to ask under Triple C approach are

- What is the purpose of the project?
- Who is in charge of the project?

STEP Communications Management

- Why is the project needed?
- Where is the project located?
- When will the project be carried out?
- How will the project contribute to increased opportunities for the organization?
- What is the project designed to achieve?
- How will the project affect different groups of people within the organization?
- What will be the project approach or methodology?
- What other groups or organizations will be involved (if any)?
- What will happen at the end of the project?
- How will the project be tracked, monitored, evaluated, and reported?
- What resources are required?
- What are the associated costs of the required resources?
- How do the project objectives fit the goal of the organization?
- What respective contribution is expected from each participant?
- What level of cooperation is expected from each group?
- Where is the coordinating point for the project?

TRIPLE C COMMUNICATION

Communication makes working together possible. The communication function of project management involves making all those concerned become aware of project requirements and progress. Those who will be affected by the project directly or indirectly, as direct participants or as beneficiaries, should be informed as appropriate regarding the following:

- Scope of the project
- Personnel contribution required
- Expected cost and merits of the project
- Project organization and implementation plan
- Potential adverse effects if the project should fail
- Alternatives, if any, for achieving the project goal
- Potential direct and indirect benefits of the project

The communication channel must be kept open throughout the project life cycle. In addition to internal communication, appropriate external sources should also be consulted. The project manager must

- Exude commitment to the project
- Utilize the communication responsibility matrix
- Facilitate multichannel communication interfaces
- Identify internal and external communication needs
- Resolve organizational and communication hierarchies
- Encourage both formal and informal communication links

When clear communication is maintained between management and employees and among peers, many project problems can be averted. Project communication may be carried out in one or more of the following formats:

- One-to-many
- One-to-one
- Many-to-one
- Written and formal
- Written and informal
- Oral and formal
- Oral and informal
- Nonverbal gestures

Good communication is affected when what is implied is perceived as intended. Effective communications are vital to the success of any project. Despite the awareness that proper communications form the blueprint for project success, many organizations still fail in their communications functions. The study of communication is complex. Factors that influence the effectiveness of communication within a project organization structure include the following.

1. *Personal perception.* Each person perceives events on the basis of personal psychological, social, cultural, and experimental background. As a result, no two people can interpret a given event the same way. The nature of events is not always the critical aspect of a problem situation. Rather, the problem is often the different perceptions of the different people involved.
2. *Psychological profile.* The psychological makeup of each person determines personal reactions to events or words. Thus, individual needs and level of thinking will dictate how a message is interpreted.
3. *Social environment.* Communication problems sometimes arise because people have been conditioned by their prevailing social environment to interpret certain things in unique ways. Vocabulary, idioms, organizational status, social stereotypes, and economic situation are among the social factors that can thwart effective communication.
4. *Cultural background.* Cultural differences are among the most pervasive barriers to project communications, especially in today's multinational organizations. Language and cultural idiosyncrasies often determine how communication is approached and interpreted.
5. *Semantic and syntactic factors.* Semantic and syntactic barriers to communications usually occur in written documents. Semantic factors are those that relate to the intrinsic knowledge of the subject of the communication. Syntactic factors are those that relate to the form in which the communication is presented. The problems created by these factors become acute in situations where response, feedback, or reaction to the communication cannot be observed.
6. *Organizational structure.* Frequently, the organization structure in which a project is conducted has a direct influence on the flow of information

though only a few days...

STEP Communications Management   229

and, consequently, on the effectiveness of communication. Organization hierarchy may determine how different personnel levels perceive a given communication.

7. *Communication media.* The method of transmitting a message may also affect the value ascribed to the message and consequently how it is interpreted or used. The common barriers to project communications are

- Inattentiveness
- Lack of organization
- Outstanding grudges
- Preconceived notions
- Ambiguous presentation
- Emotions and sentiments
- Lack of communication feedback
- Sloppy and unprofessional presentation
- Lack of confidence in the communicator
- Lack of confidence by the communicator
- Low credibility of communicator
- Unnecessary technical jargon
- Too many people involved
- Untimely communication
- Arrogance or imposition
- Lack of focus

Some suggestions on improving the effectiveness of communication are presented next. The recommendations may be implemented as appropriate for any of the forms of communications listed earlier. The recommendations are for both the communicator and the audience.

1. Never assume that the integrity of the information sent will be preserved as the information passes through several communication channels. Information is generally filtered, condensed, or expanded by the receivers before relaying it to the next destination. When preparing a communication that needs to pass through several organization structures, one safeguard is to compose the original information in a concise form to minimize the need for recomposition of the project structure.
2. Give the audience a central role in the discussion. A leading role can help make a person feel a part of the project effort and responsible for the projects' success. He or she can then have a more constructive view of project communication.
3. Do homework and think through the intended accomplishment of the communication. This helps eliminate trivial and inconsequential communication efforts.
4. Carefully plan the organization of the ideas embodied in the communication. Use indexing or points of reference whenever possible. Grouping ideas into related chunks of information can be particularly effective. Present the

short messages first. Short messages help create focus, maintain interest, and prepare the mind for the longer messages to follow.
5. Highlight why the communication is of interest and how it is intended to be used. Full attention should be given to the content of the message with regard to the prevailing project situation.
6. Elicit the support of those around you by integrating their ideas into the communication. The more people feel they have contributed to the issue, the more expeditious they are in soliciting the cooperation of others. The effect of the multiplicative rule can quickly garner support for the communication purpose.
7. Be responsive to the feelings of others. It takes two to communicate. Anticipate and appreciate the reactions of members of the audience. Recognize their operational circumstances and present your message in a form they can relate to.
8. Accept constructive criticism. Nobody is infallible. Use criticism as a springboard to higher communication performance.
9. Exhibit interest in the issue in order to arouse the interest of your audience. Avoid delivering your messages as a matter of a routine organizational requirement.
10. Obtain and furnish feedback promptly. Clarify vague points with examples.
11. Communicate at the appropriate time, at the right place, to the right people.
12. Reinforce words with positive action. Never promise what cannot be delivered. Value your credibility.
13. Maintain eye contact in oral communication and read the facial expressions of your audience to obtain real-time feedback.
14. Concentrate on listening as much as speaking. Evaluate both the implicit and explicit meanings of statements.
15. Document communication transactions for future references.
16. Avoid asking questions that can be answered yes or no. Use relevant questions to focus the attention of the audience. Use questions that make people reflect upon their words, such as, "How do you think this will work?" compared to "Do you this will work?"
17. Avoid patronizing the audience. Respect their judgment and knowledge.
18. Speak and write in a controlled tempo. Avoid emotionally charged voice inflections.
19. Create an atmosphere for formal and informal exchange of ideas.
20. Summarize the objectives of the communication and how they will be achieved.

SMART COMMUNICATION

The key to getting everyone on board with a project is to ensure that task objectives are clear and comply with the principle of SMART as outlined below:

Specific: Task objective must be specific

Measurable: Task objective must be measurable

Aligned: Task objective must be achievable and aligned with overall project goal
Realistic: Task objective must be realistic and relevant to the organization
Timed: Task objective must have a time basis

If a task has the above intrinsic characteristics, then the function of communicating the task will more likely lead to personnel cooperation. Figure 8.6 shows an example of a design of a communication responsibility matrix. A communication responsibility matrix shows the linking of sources of communication and targets of communication. Cells within the matrix indicate the subject of the desired communication. There should be at least one filled cell in each row and each column of the matrix. This assures that each individual of a department has at least one communication source or target associated with him or her. With a communication responsibility matrix, a clear understanding of what needs to be communicated to whom can be developed. Communication in a project environment can take any of several forms. The specific needs of a project may dictate the most appropriate mode. Three popular computer communication modes are discussed next in the context of communicating data and information for project management.

Simplex communication. This is a unidirectional communication arrangement in which one project entity initiates communication to another entity or individual within the project environment. The entity addressed in the communication does not have mechanism or capability for responding to the communication. An extreme example of this is a one-way, top-down communication from top management to the project personnel. In this case, the personnel have no communication access or input to top management. A budget-related example is a case where top management allocates budget to a project without requesting and reviewing the actual needs of the project. Simplex communication is common in authoritarian organizations.

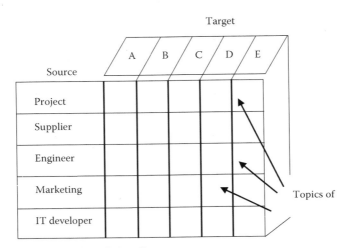

A: Project manager B: Supplier

FIGURE 8.6 Triple C communication matrix.

Half-duplex communication. This is a bidirectional communication arrangement whereby one project entity can communicate with another entity and receives a response within a certain time lag. Both entities can communicate with each other but not at the same time. An example of half-duplex communication is a project organization that permits communication with top management without a direct meeting. Each communicator must wait for a response from the target of the communication. Request and allocation without a budget meeting is another example of half-duplex data communication in project management.

Full-duplex communication. This involves a communication arrangement that permits a dialogue between the communicating entities. Both individuals and entities can communicate with each other at the same time or face-to-face. As long as there is no clash of words, this appears to be the most receptive communication mode. It allows participative project planning in which each project personnel has an opportunity to contribute to the planning process.

Each member of a project team needs to recognize the nature of the prevailing communication mode in the project. Management must evaluate the prevailing communication structure and attempt to modify it if necessary to enhance project functions. An evaluation of who is to communicate with whom about what may help improve the project data/information communication process. A communication matrix may include notations about the desired modes of communication between individuals and groups in the project environment.

TRIPLE C COOPERATION

The cooperation of the project personnel must be explicitly elicited. Merely voicing consent for a project is not enough assurance of full cooperation. The participants and beneficiaries of the project must be convinced of the merits of the project. Some of the factors that influence cooperation in a project environment include personnel requirements, resource requirements, budget limitations, past experiences, conflicting priorities, and lack of uniform organizational support. A structured approach to seeking cooperation should clarify the following:

- Cooperative efforts required
- Precedents for future projects
- Implication of lack of cooperation
- Criticality of cooperation to project success
- Organizational impact of cooperation
- Time frame involved in the project
- Rewards of good cooperation

Cooperation is a basic virtue of human interaction. More projects fail due to a lack of cooperation and commitment than any other project factors. To secure and retain the cooperation of project participants, you must elicit a positive first reaction to the project. The most positive aspects of a project should be the first items of project communication. For project management, there are different types of cooperation that should be understood.

STEP Communications Management

Functional cooperation. This is cooperation induced by the nature of the functional relationship between two groups. The two groups may be required to perform related functions that can only be accomplished through mutual cooperation.

Social cooperation. This is the type of cooperation effected by the social relationship between two groups. The prevailing social relationship motivates cooperation that may be useful in getting project work done.

Legal cooperation. Legal cooperation is the type of cooperation that is imposed through some authoritative requirement. In this case, the participants may have no choice other than to cooperate.

Administrative cooperation. This is cooperation brought on by administrative requirements that make it imperative that two groups work together on a common goal.

Associative cooperation. This type of cooperation may also be referred to as collegiality. The level of cooperation is determined by the association that exists between two groups.

Proximity cooperation. Cooperation due to the fact that two groups are geographically close is referred to as proximity cooperation. Being close makes it imperative that the two groups work together.

Dependency cooperation. This is cooperation caused by the fact that one group depends on another group for some important aspect. Such dependency is usually of a mutual two-way nature. One group depends on the other for one thing while the latter group depends on the former for some other thing.

Imposed cooperation. In this type of cooperation, external agents must be employed to induced cooperation between two groups. This is applicable for cases where the two groups have no natural reason to cooperate. This is where the approaches presented earlier for seeking cooperation can became very useful.

Lateral cooperation. Lateral cooperation involves cooperation with peers and immediate associates. Lateral cooperation is often easy to achieve because existing lateral relationships create an environment that is conducive for project cooperation.

Vertical cooperation. Vertical or hierarchical cooperation refers to cooperation that is implied by the hierarchical structure of the project. For example, subordinates are expected to cooperate with their vertical superiors.

Irrespective of the type of cooperation available in a project environment, the cooperative forces should be channeled toward achieving project goals. Documentation of the prevailing level of cooperation is useful for winning further support for a project. Clarification of project priorities will facilitate personnel cooperation. Relative priorities of multiple projects should be specified so that a priority to all groups within the organization is clearly understood by everyone. Some guidelines for securing cooperation for most projects are

- Establish achievable goals for the project
- Clearly outline the individual commitments required
- Integrate project priorities with existing priorities
- Eliminate the fear of job loss due to industrialization
- Anticipate and eliminate potential sources of conflict

- Use an open-door policy to address project grievances
- Remove skepticism by documenting the merits of the project

Commitment. Cooperation must be supported with commitment. To cooperate is to support the ideas of a project. To commit is to willingly and actively participate in project efforts again and again through the thick and thin of the project. Provision of resources is one way in which management can express commitment to a project.

TRIPLE C COORDINATION

After the communication and cooperation functions have successfully been initiated, the efforts of the project personnel must be coordinated. Coordination facilitates harmonious organization of project efforts. The construction of a responsibility chart can be very helpful at this stage. A responsibility chart is a matrix consisting of columns of individual or functional departments and rows of required actions. Cells within the matrix are filled with relationship codes that indicate who is responsible for what. Table 8.6 illustrates an example of a responsibility matrix for the planning of a seminar program. The matrix helps avoid neglecting crucial communication requirements and obligations. It can help resolve questions such as

TABLE 8.6
Example of Responsibility Matrix for Project Coordination

	Person Responsible				Status of Task			
TASKS	Staff A	Staff B	Staff C	Mgr	31-Jan	15-Feb	28-Mar	21-Apr
Brainstorming meeting	R	R	R	R	D			
Identify speakers				R		O		
Select seminar location	I	R	R			O		
Select banquet location	R	R				D		
Prepare publicity materials		C	R	I	O	O	D	
Draft brochures		C	R					D
Develop schedule			R			L	L	
Arrange for visual aids			R		L	L	L	
Coordinate activities			R				L	
Periodic review of tasks	R	R	R	S				D
Monitor progress of program	C	R	R			O	L	
Review program progress	R				O	O	L	L
Closing arrangements	R							L
Post-program review and evaluation	R	R	R	R			D	

Notes: Responsibility codes: R, responsible; I, inform; S, support; C, consult. Task codes: D, done; O, on track; L, late.

- Who is to do what?
- How long will it take?
- Who is to inform whom of what?
- Whose approval is needed for what?
- Who is responsible for which results?
- What personnel interfaces are required?
- What support is needed from whom and when?

CONFLICT RESOLUTION USING TRIPLE C APPROACH

Conflicts can and do develop in any work environment. Conflicts, whether intended or inadvertent, prevent an organization from getting the most out of the work force. When implemented as an integrated process, the Triple C model can help avoid conflicts in a project. When conflicts do develop, it can help in resolving the conflicts. The key to conflict resolution is open and direct communication, mutual cooperation, and sustainable coordination. Several sources of conflicts can exist in projects. Some of these are discussed below.

Schedule conflict. Conflicts can develop because of improper timing or sequencing of project tasks. This is particularly common in large multiple projects. Procrastination can lead to having too much to do at once, thereby creating a clash of project functions and discord among project team members. Inaccurate estimates of time requirements may lead to infeasible activity schedules. Project coordination can help avoid schedule conflicts.

Cost conflict. Project cost may not be generally acceptable to the clients of a project. This will lead to project conflict. Even if the initial cost of the project is acceptable, a lack of cost control during implementation can lead to conflicts. Poor budget allocation approaches and the lack of a financial feasibility study will cause cost conflicts later on in a project. Communication and coordination can help prevent most of the adverse effects of cost conflicts.

Performance conflict. If clear performance requirements are not established, performance conflicts will develop. Lack of clearly defined performance standards can lead each person to evaluate his or her own performance based on personal value judgments. In order to uniformly evaluate quality of work and monitor project progress, performance standards should be established by using the Triple C approach.

Management conflict. There must be a two-way alliance between management and the project team. The views of management should be understood by the team. The views of the team should be appreciated by management. If this does not happen, management conflicts will develop. A lack of a two-way interaction can lead to strikes and industrial actions, which can be detrimental to project objectives. The Triple C approach can help create a conductive dialogue environment between management and the project team.

Technical conflict. If the technical basis of a project is not sound, technical conflict will develop. New industrial projects are particularly prone to technical conflicts because of their significant dependence on technology. Lack of a comprehensive technical feasibility study will lead to technical conflicts. Performance requirements and systems specifications can be integrated through the Triple C approach to avoid technical conflicts.

Priority conflict. Priority conflicts can develop if project objectives are not defined properly and applied uniformly across a project. Lack of a direct project definition can lead each project member to define his or her own goals, which may be in conflict with the intended goal of a project. Lack of consistency of the project mission is another potential source of priority conflicts. Overassignment of responsibilities with no guidelines for relative significance levels can also lead to priority conflicts. Communication can help defuse priority conflict.

Resource conflict. Resource allocation problems are a major source of conflict in project management. Competition for resources, including personnel, tools, hardware, software, and so on, can lead to disruptive clashes among project members. The Triple C approach can help secure resource cooperation.

Power conflict. Project politics lead to a power play that can adversely affect the progress of a project. Project authority and project power should be clearly delineated. Project authority is the control that a person has by virtue of his or her functional post. Project power relates to the clout and influence, which a person can exercise due to connections within the administrative structure. People with popular personalities can often wield a lot of project power in spite of low or nonexistent project authority. The Triple C model can facilitate a positive marriage of project authority and power to the benefit of project goals. This will help define clear leadership for a project.

Personality conflict. Personality conflict is a common problem in projects involving a large group of people. The larger the project, the larger the size of the management team needed to keep things running. Unfortunately, the larger management team creates an opportunity for personality conflicts. Communication and cooperation can help defuse personality conflicts. In summary, conflict resolution through Triple C can be achieved by observing the following guidelines:

1. Confront the conflict and identify the underlying causes.
2. Be cooperative and receptive to negotiation as a mechanism for resolving conflicts.
3. Distinguish between proactive, inactive, and reactive behaviors in a conflict situation.
4. Use communication to defuse internal strife and competition.
5. Recognize that short-term compromise can lead to long-term gains.
6. Use coordination to work toward a unified goal.
7. Use communication and cooperation to turn a competitor into a collaborator.

It is the little and often neglected aspects of a project that lead to project failures. Several factors may constrain the project implementation. All the relevant factors can be evaluated under the Triple C model right from the project initiation stage.

APPLICATION OF TRIPLE C TO STEPS

Having now understood the intrinsic elements of Triple C, we can see how and where it could be applicable to the steps of project management. Communication explains project scope and requirements through the stages of planning, organizing, scheduling, and control. Cooperation is required to get human resource buy-in and stakeholder endorsement across all facets of planning, organizing, scheduling, and control. Coordination facilitates adaptive interfaces over all the elements of planning, organizing, scheduling, and control. The Triple C model should be implemented as an iterative loop process that moves a project through the communication, cooperation, and coordination functions.

DMAIC AND TRIPLE C

Many organizations now explore Six Sigma DMAIC (define, measure, analyze, improve, and control) methodology and associated tools to achieve better project performance. Six Sigma means six standard deviations from a statistical performance average. The Six Sigma approach allows for no more than 3.4 defects per million parts in manufactured goods or 3.4 mistakes per million activities in a service operation. To explain the effect of the Six Sigma approach, consider a process that is 99% perfect. That process will produce 10,000 defects per million parts. With Six Sigma, the process will need to be 99.99966% perfect in order to produce only 3.4 defects per million. Thus, Six Sigma is an approach that moves a process toward perfection. Six Sigma, in effect, reduces variability among products produced by the same process. By contrast, Lean approach is designed to reduce/eliminate waste in the production process.

Six Sigma provides a roadmap for the five major steps of DMAIC, which are applicable to the planning and control steps of project management. We cannot improve what we cannot measure. Triple C provides a sustainable approach to obtain cooperation and coordination for DMAIC during improvement efforts. DMAIC requires project documentation and reporting, which coincide with project control requirements.

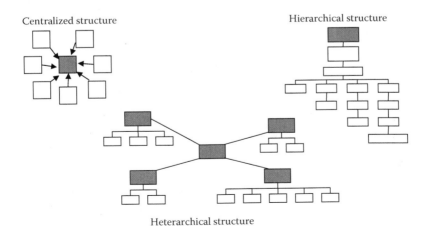

FIGURE 8.7 Team-building organizational structures for effective communication.

A summary of lessons to be inferred from a Triple C approach are

- Use proactive planning to initiate project functions.
- Use preemptive planning to avoid project pitfalls.
- Use meetings strategically. Meeting is not *work*. Meeting should be done to facilitate work.
- Use project assessment to properly frame the problem, adequately define the requirements, continually ask the right questions, cautiously analyze risks, and effectively scope the project.
- Be bold to terminate a project when termination is the right course of action. Every project needs an exit plan. In some cases, there is victory in capitulation.

The sustainability of the Triple C approach is summarized below:

- For effective communication, create good communication channels.
- For enduring cooperation, establish partnership arrangements.
- For steady coordination, use a workable organization structure for communications team building using a combination of strategies as illustrated in Figure 8.7.

REFERENCES

Badiru, A. B. *Triple C Model of Project Management*, Taylor & Francis/CRC Press, Boca Raton, FL, 2008.

Mooz, H., K. Forsberg, and H. Cotterman, *Communicating Project Management*, John Wiley & Sons, Hoboken, NJ, 2003.

PMI, *A Guide to the Project Management Body of Knowledge (PMBOK Guide)*, 3rd ed., Project Management Institute, Newtown Square, PA, 2004.

9 STEP Risk Management

It is not the critic who counts. Not the man who points out how the strong man stumbled or where the doer of deeds could have done better. The credit belongs to the man who is actually in the arena, whose face is marred by dust and sweat and blood; who strives valiantly; who errs and comes short again and again; who knows the great enthusiasms, the great devotions; who spends himself in a worthy cause. Who, at the best, knows in the end the triumph of high achievement, and who at the worst, at least fails while daring greatly, so that his place shall never be with those timid souls who know neither victory nor defeat.

–Ted Roosevelt

The opening quote at the beginning of this chapter is about taking risk, venturing out and discovering what is out there, and exploring what exists within or outside the realm of possibility. Risk management is the process of identifying, analyzing, and recognizing the various risks and uncertainties that might affect a project. Change can be expected in any project environment. Change portends risk and uncertainty. Risk analysis outlines possible future events and their likelihood of occurrence. With the information from risk analysis, the project team can be better prepared for change with good planning and control actions. By identifying the various project alternatives and their associated risk, the project team can select the most appropriate courses of action.

Risk permeates every aspect of a project. In fact, each and every one of the other elements in the *Project Management Body of Knowledge* (*PMBOK*) is subject to some level of risk. Scope presents risks. Communication has risk components. Cost is subject to risk. Time has factors of risk and uncertainty. Quality variability contains a dimension of risk. Human resources pose operational risks. Procurement is subject to risk realities of the marketplace. Just as risk presents opportunities, it also poses threats. Thus, risk management is a crucial component of project management, particularly in science and technology based projects where system dynamics can disrupt operations in a flash.

RISK DEFINITION

PMI's *PMBOK* defines risk as "an uncertain event or condition that, if it occurs, has a positive or negative effect on at least one project objective, such as time, cost, scope, or quality." In risk management, it is assumed that there exist a number of possible future states of a variable. Each occurrence of the variable has a known or assumed probability of occurring. There are often interdependencies in factors associated with a risk event. Thus, quantitative assessment is often very complex. Once a risk occurs, it is no longer a risk; it is a fact. There are three elements of risk:

1. There is some future event that has not occurred yet.
2. There is some level of uncertainty associated with the event.
3. There is a consequence (positive or negative) emanating from the risk event.

Risk management is the process of identifying, analyzing, and recognizing the various risks and uncertainties that might affect a project. Figure 9.1 shows the life cycle of risk management. The purpose of risk management is to achieve one of the following:

- Maximize the probability and consequence of positive events
- Minimize the probability and consequence of negative events

There are three possible risk response behaviors for risk management:

1. Risk-averse behavior: conscious and deliberate attempt to avoid risk
2. Risk-seeking behavior: conscious and deliberate pursuit of risk, perhaps as a manifestation of the old West saying that "you cannot accumulate if you don't speculate"
3. Risk-neutral behavior: indifference to the presence or absence of risk

Typical recommended level of investment in risk management is around 5%–10% of total project budget. If there is no risk management plan in a project, then the project

FIGURE 9.1 Life cycle of risk management.

STEP Risk Management

is operating in a fire-fighting mode. This is an example of management by exception (MBE) in an organization that does not make contingency plans.

RISK MANAGEMENT: STEP-BY-STEP IMPLEMENTATION

The risk management component of the *PMBOK* consists of the elements shown in the block diagram in Figure 9.2. The six elements in the block diagram are carried out across the process groups presented earlier in this book. The overlay of the elements and the process groups are shown in Table 9.1. Thus, under the knowledge area of communications management, the required steps are

Step 1: Risk management planning
Step 2: Risk identification
Step 3: Qualitative risk analysis
Step 4: Quantitative risk analysis
Step 5: Risk response planning
Step 6: Risk monitoring and control

Tables 9.2 through 9.7 present the inputs, tools, techniques, and outputs of each step of risk management. It should be emphasized that risk itself is not identified in the risk management planning phase. The planning phase is used only to identify the processes that will be used to handle risk. Also, risk identification simply develops a list of risks; it does not rank or analyze the risks. Risk register, which is an output of risk identification, presents a list (preferably a spreadsheet) of risk events, their root causes, and associated responses. Risk rating matrix, which is

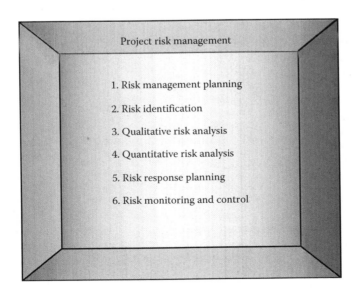

FIGURE 9.2 Block diagram of project risk management.

TABLE 9.1
Implementation of Project Risk Management across Process Groups

	Initiating	Planning	Executing	Monitoring and Controlling	Closing
Project Risk Management		Risk management planning Risk identification Qualitative risk analysis Quantitative risk analysis Risk response planning		Risk monitoring and control	

a tool for qualitative risk analysis, presents a matrix of risk events with respect to their respective probabilities of occurrence and impact levels. The probability ranges are presented as "near certainty," "highly likely," "likely," "unlikely," and "remote." The impact levels are presented as "negligible," "minor," "moderate," "serious," and "critical."

PROJECT DECISIONS UNDER RISK AND UNCERTAINTY

Traditional decision theory classifies decisions under three different influences:

- Decision under certainty: Made when possible event(s) or outcome(s) of a decision can be positively determined
- Decisions under risk: Made using information on the probability that a possible event or outcome will occur

TABLE 9.2
Tools and Techniques for Risk Planning within Project Risk Management

Step 1: Risk Planning

Inputs	Tools and Techniques	Output(s)
Enterprise environmental factors Organizational process assets Project scope statement Other in-house (custom) factors of relevance and interest	Communications requirement analysis Planning meetings and analysis Other in-house (custom) tools and techniques	Risk management plan Other in-house outputs, reports, and data inferences of interest to the organization

TABLE 9.3
Tools and Techniques for Risk Identification within Project Risk Management

Step 2: Risk Identification

Inputs	Tools and Techniques	Output(s)
Enterprise environmental factors	Documentation reviews	Risk register
Organizational process assets	Information-gathering techniques	Other in-house outputs, reports, and data inferences of interest to the organization
Project scope statement	Survey of subject matter experts to get risk information	
Risk management plan	Checklist analysis	
Project management plan	Assumptions analysis	
Other in-house (custom) factors of relevance and interest	Diagramming techniques	
	Other in-house (custom) tools and techniques	

- Decisions under uncertainty: Made by evaluating possible event(s) or outcome(s) without information on the probability that the event(s) or outcome(s) will occur

Many authors make a distinction between decisions under risk and under uncertainty. In the literature, decisions made under uncertainty are increasingly incorporating

TABLE 9.4
Tools and Techniques for Qualitative Risk Analysis within Project Risk Management

Step 3: Qualitative Risk Analysis

Inputs	Tools and Techniques	Output(s)
Organizational process assets	Critical incident safety management (CISM) for risk analysis	Risk register (updates)
Project scope statement	Risk rating matrix	Other in-house outputs, reports, and data inferences of interest to the organization
Risk management plan	Risk probability and impact assessment	
Risk register	Probability and impact matrix	
Other in-house (custom) factors of relevance and interest	Risk data quality assessment	
	Risk categorization	
	Risk urgency assessment	
	Other in-house (custom) tools and techniques	

TABLE 9.5
Tools and Techniques for Quantitative Risk Analysis within Project Risk Management

Step 4: Quantitative Risk Analysis

Inputs	Tools and Techniques	Output(s)
Organizational process assets	Data gathering and representation techniques	Risk register (updates)
Project scope statement		Other in-house outputs, reports, and data inferences of interest to the organization
Risk management plan	Quantitative risk analysis and modeling techniques	
Risk register		
Project management plan	Other in-house (custom) tools and techniques	
Other in-house (custom) factors of relevance and interest		

decisions made under risk, as defined above. In this book, no special distinction will be made between risk and uncertainty. Some of the chapters in this book contain a number of procedures to illustrate how project decisions may be made under uncertainty. Some of the parameters that normally change during a project life cycle include project costs, time requirements, and performance specifications. The uncertainties associated with these parameters are a concern for project managers. Cost, time, and performance must be managed throughout the project life cycle.

COST UNCERTAINTIES

In an inflationary economy, project costs can become very dynamic and intractable. Cost estimates include various tangible and intangible components of a project, such

TABLE 9.6
Tools and Techniques for Risk Response Planning within Project Risk Management

Step 5: Risk Response Planning

Inputs	Tools and Techniques	Output(s)
Risk management plan	Risk tolerance level	Risk register (updates)
Risk register	Risk-averse tendencies	Project management plan (updates)
Other in-house (custom) factors of relevance and interest	Strategies for negative risks or threats	Risk-related contractual agreements
	Strategies for positive risks or opportunities	Other in-house outputs, reports, and data inferences of interest to the organization
	Contingency strategy	
	Other in-house (custom) tools and techniques	

TABLE 9.7
Tools and Techniques for Risk Monitoring and Control within Project Risk Management

Step 6: Risk Monitoring and Control

Inputs	Tools and Techniques	Output(s)
Risk management plan	Risk reassessment	Risk register (updates)
Risk register	Risk mitigation method	Requested changes
Approved change requests	Risk audits	Recommended preventive actions
Work performance information	Variance and trend analysis	Organizational process assets (updates)
Performance reports	Technical performance measurement	Project management plan (updates)
Other in-house (custom) factors of relevance and interest	Reserve analysis	Other in-house outputs, reports, and data inferences of interest to the organization
	Status meetings	
	Other in-house (custom) tools and techniques	

as machines, inventory, training, raw materials, design, and personnel wages. Costs can change during a project for a number of reasons including

- External inflationary trends
- Internal cost adjustment procedures
- Modification of work process
- Design adjustments
- Changes in cost of raw materials
- Changes in labor costs
- Adjustment of work breakdown structure
- Cash flow limitations
- Effects of tax obligations

These cost changes and others combine to create uncertainties in the project's cost. Even when the cost of some of the parameters can be accurately estimated, the overall project cost may still be uncertain due to the few parameters that cannot be accurately estimated.

SCHEDULE UNCERTAINTIES

Unexpected engineering change orders (ECO) and other changes in a project environment may necessitate schedule changes, which introduce uncertainties to the project. The following are some of the reasons project schedules change:

- Task adjustments
- Changes in scope of work

- Changes in delivery arrangements
- Changes in project specification
- Introduction of new technology

PERFORMANCE UNCERTAINTIES

Performance measurement involves observing the value of parameter(s) during a project and comparing the actual performance, based on the observed parameter(s), to the expected performance. Performance control then takes appropriate actions to minimize the deviations between actual performance and expected performance. Project plans are based on the expected performance of the project parameters. Performance uncertainties exist when expected performance cannot be defined in definite terms. As a result, project plans require a frequent review.

The project management team must have a good understanding of the factors that can have a negative impact on the expected project performance. If at least some of the sources of deficient performance can be controlled, then the detrimental effects of uncertainties can be alleviated. The most common factors that can influence project performance include the following:

- Redefinition of project priorities
- Changes in management control
- Changes in resource availability
- Changes in work ethic
- Changes in organizational policies and procedures
- Changes in personnel productivity
- Changes in quality standards

To minimize the effect of uncertainties in project management, a good control must be maintained over the various sources of uncertainty discussed above. The same analytic tools that are effective for one category of uncertainties should also work for other categories.

RISK AND DECISION TREES

Decision tree analysis is used to evaluate sequential decision problems. In project management, a *decision tree* may be useful for evaluating sequential project milestones. A decision problem under certainty has two elements: action and consequence. The decision maker's choices are the actions while the results of those actions are the consequences. For example, in a CPM network planning, the choice of one task among three potential tasks in a given time slot represents a potential action. The consequences of choosing one task over another may be characterized in terms of the slack time created in the network, the cost of performing the selected task, the resulting effect on the project completion time, or the degree to which a specified performance criterion is satisfied.

If the decision is made under uncertainty, as in PERT network analysis, a third element, called an *event*, is introduced into the decision problem. Extending the

CPM task selection example to a PERT analysis, the actions may be defined as Select Task 1, Select Task 2, and Select Task 3. The durations associated with the three possible actions can be categorized as long task duration, medium task duration, and short task duration. The actual duration of each task is uncertain. Thus, each task has some probability of exhibiting long, medium, or short durations.

The events can be identified as weather incidents: rain or no rain. The incidents of rain or no rain are uncertain. The consequences may be defined as increased project completion time, decreased project completion time, and unchanged project completion time. These consequences are also uncertain due to the probable durations of the tasks and the variable choices of the decision maker. That is, the consequences are determined partly by choice and partly by chance. The consequences also depend on which event occurs—rain or no rain.

To simplify the decision analysis, the decision elements may be summarized by using a decision table. A *decision table* shows the relationship between pairs of decision elements. Table 9.8 shows the decision table for the task duration example discussed above. In the table, each row corresponds to an event and each column corresponds to an action. The consequences appear as entries in the body of the table. The consequences have been coded as I (increased), D (decreased), U (unchanged). Each event–action combination has a specific consequence associated with it.

In some decision problems, the consequences may not be unique. Thus, a consequence, which is associated with a particular event–action pair, may also be associated with another event–action pair. The actions included in the decision table are the only ones that the decision maker wishes to consider. Subcontracting and task elimination, for example, are other possible choices for the decision maker. The actions included in the decision problem are mutually exclusive and collectively exhaustive, so that exactly one will be selected. The events are also mutually exclusive and collectively exhaustive.

The decision problem can also be conveniently represented as a decision tree as shown in Figure 9.3. The tree representation is particularly effective for decision problems with choices that must be made at different times over an extended period. Resource allocation decisions, for example, must be made several times during the life cycle of a project. The choice of actions is shown as a decision junction with

---

TABLE 9.8
Decision Table for Task Selection

	Actions								
	Task 1			Task 2			Task 3		
Event	Long	Medium	Short	Long	Medium	Short	Long	Medium	Short
Rain	I	I	U	I	U	D	I	I	U
No rain	I	D	D	U	D	D	U	U	U

Note: I, increased project duration; D, decreased project duration; U, unchanged project duration.

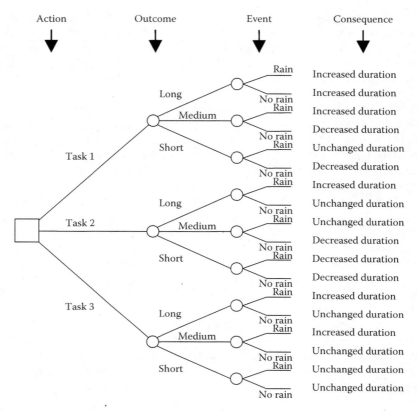

FIGURE 9.3 Decision tree for task selection example.

a separate branch for each action. The events are also represented by branches in separate fields.

To avoid confusion in very complex decision trees, the nodes for action forks are represented by squares, while the nodes for event junctions are represented by circles. The basic convention for constructing a tree diagram is that the flow should be chronological from left to right. The actions are shown on the initial junction because the decision must be made before the actual event is known. The events are shown as branches in the third-stage forks. The consequence resulting from an event–action combination is shown as the endpoint of the corresponding path from the root of the tree.

Figure 9.3 shows six paths leading to an increase in the project duration, five paths leading to a decrease in project duration, and seven paths leading to a unchanged project duration. The total number of paths is given by

$$P = \prod_{i=1}^{N} n_i$$

STEP Risk Management

where
P is the total number of paths in the decision tree
N is the number of decision stages in the tree
n_i is the number of branches emanating from each node in stage i

Thus, for the example in Figure 9.3, the number of paths is $P = (3)(3)(2) = 18$ paths. As mentioned previously, some of the paths, even though they are distinct, lead to identical consequences.

Probability values can be incorporated into the decision structure as shown in Figure 9.4. Note that the selection of a task at the decision node is based on choice rather than probability. In this example, we assume that the probability of having a particular task duration is independent of whether or not it rains. In some cases, the weather sensitivity of a task may influence the duration of the task. Also, the probability of rain or no rain is independent of any other element in the decision structure.

If the items in the probability tree are interdependent, then the appropriate conditional probabilities would need to be computed. This will be the case if the duration of a task is influenced by whether or not it rains. In such a case, the probability tree should be redrawn as shown in Figure 9.5, which indicates that the weather event

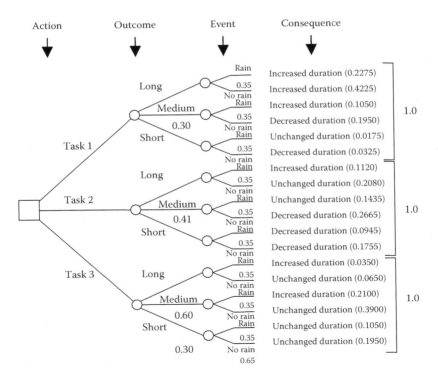

FIGURE 9.4 Probability tree diagram for task selection example.

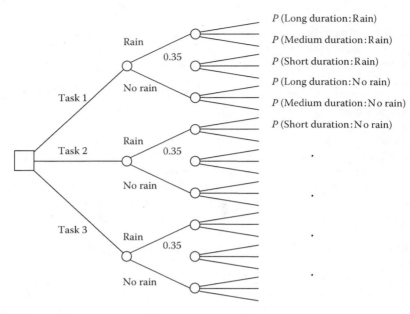

FIGURE 9.5 Probability tree for weather-dependent task durations.

will need to be observed first before the task duration event can be determined. For Figure 9.5, the conditional probability of each type of duration, given that it rains or it does not rain, will need to be calculated.

The respective probabilities of the three possible consequences are shown in Figure 9.4. The probability at the end of each path is computed by multiplying the individual probabilities along the path. For example, the probability of having an increased project completion time along the first path (Task 1, long duration, and rain) is calculated as

$$(0.65)(0.35) = 0.2275$$

Similarly, the probability for the second path (Task 1, long duration, and no rain) is calculated as

$$(0.65)(0.65) = 0.4225$$

The sum of the probabilities at the end of the paths associated with each action (choice) is equal to one as expected. Table 9.9 presents a summary of the respective probabilities of the three consequences based on the selection of each task. The probability of having an increased project duration when Task 1 is selected is calculated as

$$\text{Probability} = 0.2275 + 0.4225 + 0.105 = 0.755$$

TABLE 9.9
Probability Summary for Project Completion Time

	Selected Task					
Consequence	Task 1		Task 2		Task 3	
Increased duration	0.2275 +0.4225 +0.105	0.755	0.112	0.112	0.035 +0.21	0.245
Decreased duration	0.195 +0.0325	0.2275	0.2665 +0.0945 +0.1755	0.5635	0.0	0.0
Unchanged duration	0.0175	0.0175	0.208 +0.1435	0.3515	0.065 +0.39 +0.105 +0.195	0.755
Sum of probabilities		1.0		1.0		1.0

Likewise, the probability of having an increased project duration when Task 3 is selected is calculated as

$$\text{Probability} = 0.035 + 0.21 = 0.245$$

If the selection of tasks at the first node is probable in nature, then the respective probabilities would be included in the calculation procedure. For example, Figure 9.6 shows a case where Task 1 is selected 25% of the time, Task 2 45% of the time, and Task 3 30% of the time. The resulting ending probabilities for the three possible consequences have been revised accordingly. Note that all probabilities at the end of all the paths add up to one in this case. Table 9.10 presents the summary of the probabilities of the three consequences for the case of weather-dependent task durations.

TABLE 9.10
Summary for Weather-Dependent Task Durations

Consequence	Path Probabilities	Row Total
Increased duration	0.056875 + 0.105625 + 0.02625 + 0.0504 + 0.0105 + 0.063	0.312650
Decreased duration	0.04875 + 0.119925 + 0.042525 + 0.078975	0.290175
Unchanged duration	0.004375 + 0.008125 + 0.0936 + 0.064575 + 0.0195 + 0.117 + 0.0315 + 0.0585	0.397175
Column total		1.0

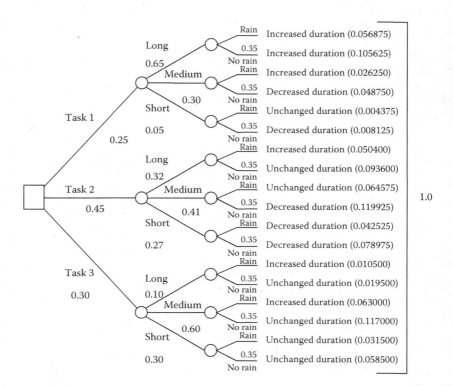

FIGURE 9.6 Modified probability tree for task selection example.

The examples presented above can be extended to other decision problems in project management, which can be represented in terms of decision tables and trees. For example, resource allocation decision problems under uncertainty can be handled by appropriate decision tree models.

10 STEP Procurement Management

If you don't speculate, you can't accumulate.

−Chinese proverb

Procurement management involves the process of acquiring the necessary equipment, tools, goods, services, and resources needed to successfully accomplish project goals. As the quote in the chapter opening suggests, it is alright to speculate about the needs of a project in order to acquire and accumulate the resources needed for the project. Procurement is also often called acquisition, purchasing, or contracting. This represents the process of acquiring (through contracting) products, results, or services for direct usage on a project. Recall that the end results of a project fall in three major categories of

- Products
- Services
- Results

Procurement is needed as a formal process of obtaining the above from a vendor or supplier whether the products or services are already in existence or must be newly designed, developed, tested, or demonstrated. Procurement involves all aspects of contract administration during the project life cycle (PMI, 2004). The buy, lease, or make options available to the project must be evaluated with respect to time, cost, and technical performance requirements. Contractual agreements, in written or unwritten (verbal) format, constitute the legal document that defines work obligation of each participant in a project. Procurement refers to the actual process of obtaining the needed services and resources. A contract, within the context of project procurement, is a mutually binding agreement that obligates the vendor to provide the specified products, services, or results and obligates the buyer to provide monetary return for the contract rendered. The procurement cycle occurs at the project–supplier interface and covers all processes necessary to ensure that materials are available for executing the project schedule. The supply chain networking becomes very essential during the procurement cycle.

Coordinated procurement is particularly crucial for science, technology, and engineering projects. Sourcing, within the procurement process, involves selection of suppliers, development of contracts, product design collaboration, materials supply, and evaluation of vendor performance. Figure 10.1 shows a typical procurement cycle suitable for adaptation for STEP project management. Just like any partnership

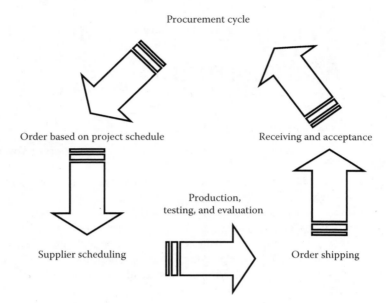

FIGURE 10.1   STEP procurement cycle.

relationship, the STEP project management team must cultivate, nurture, and sustain a positive alliance with vendors for the project and the alliance must center around the following dimensions of partnership:

- Project–vendor communication
- Project–vendor cooperation
- Project–vendor coordination

PROCUREMENT MANAGEMENT: STEP-BY-STEP IMPLEMENTATION

The procurement management component of the *Project Management Body of Knowledge* consists of the elements shown in the block diagram in Figure 10.2. The six elements in the block diagram are carried out across the process groups presented earlier in this book. The overlay of the elements and the process groups are shown in Table 10.1. Thus, under the knowledge area of communications management, the required steps are

    Step 1:   Plan purchases and acquisitions
    Step 2:   Plan contracting
    Step 3:   Request seller responses
    Step 4:   Select vendors (sellers)
    Step 5:   Contract administration
    Step 6:   Contract closure

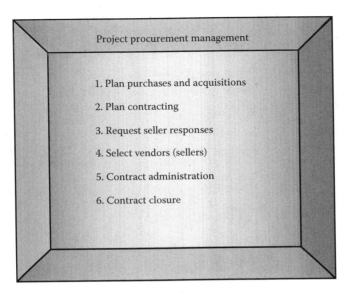

FIGURE 10.2 Block diagram of project procurement management.

TABLE 10.1
Implementation of Project Procurement Management across Process Groups

	Initiating	Planning	Executing	Monitoring and Controlling	Closing
Project Procurement Management		Plan purchases and acquisitions Plan contracting	Request seller responses Select sellers	Contract administration	Contract closure

Tables 10.2 through 10.7 present the inputs, tools, techniques, and outputs of each step in procurement management. Plan purchases and acquisitions constitute the process of identifying which components of a project to acquire through the procurement process. This involves the following queries:

- Whether or not to acquire the component?
- How to acquire the component?
- What to acquire?
- How much of the component to acquire?
- When to acquire the component?

Explanations of the entries in the step-by-step tables are provided below:

TABLE 10.2
Tools and Techniques for Purchases and Acquisitions within Project Procurement Management

Step 1: Plan Purchases and Acquisitions

Inputs	Tools and Techniques	Output(s)
Enterprise environmental factors	Make-or-buy analysis	Procurement management plan
Organizational process assets	Breakeven analysis	Contractor statement of work
Project scope statement	Expert judgment	Make or buy decision
WBS Dictionary	Contract type selection	Requested changes
Project management plan	Project selection criteria	Other in-house outputs, reports, and data inferences of interest to the organization
Other in-house (custom) factors of relevance and interest	Minimum revenue requirement analysis	
	Other in-house (custom) tools and techniques	

Enterprise environmental factors describe marketplace conditions, what is available, from whom, and in what quantity and quality.

Organizational process assets provide the formal and informal policies, procedures, guidelines, and management systems for the procurement management plan and contract type.

Project scope statement describes project boundaries, requirements, (e.g., safety clearance and permit), constraints (e.g., budget limitation), and assumptions (e.g., resource availability) related to the project scope.

Work breakdown structure (WBS) provides the relationship among project components and deliverables.

TABLE 10.3
Tools and Techniques for Contracting within Project Procurement Management

Step 2: Plan Contracting

Inputs	Tools and Techniques	Output(s)
Procurement management plan	Make-or-buy analysis	Procurement documents
Contract statement of work (CSOW)	Breakeven analysis	Evaluation criteria
Make-or-buy decisions	Contracting standard forms	CSOW updates
Project management plan	Contract administration planning	Other in-house outputs, reports, and data inferences of interest to the organization
Other in-house (custom) factors of relevance and interest	Expert judgment	
	Other in-house (custom) tools and techniques	

STEP Procurement Management

TABLE 10.4
Tools and Techniques for Requesting Vendors within Project Procurement Management

Step 3: Request Vendors

Inputs	Tools and Techniques	Output(s)
Procurement management plan	Bidder conferences	Qualified sellers list
Organizational process assets	Advertising	Procurement document package
Procurement documents	Broad-agency announcement	Proposals
Other in-house (custom) factors of relevance and interest	Development of qualified sellers list	Other in-house outputs, reports, and data inferences of interest to the organization
	Request for proposals (RFP)	
	Request for bids (RFB)	
	Invitation for bid (IFB)	
	Request for quotation (RFQ)	
	Invitation for negotiation (IFN)	
	Other in-house (custom) tools and techniques	

WBS dictionary identifies the deliverable with a description of work for each WBS component.

Project management plan provides overall plan and includes the procurement management plan including other considerations such as risk register for risks, owners, and risk responses, risk-related contractual agreements, insurance, activity resource requirements, project schedule, activity cost estimates, and cost baseline.

TABLE 10.5
Tools and Techniques for Selecting Vendors within Project Procurement Management

Step 4: Select Vendors

Inputs	Tools and Techniques	Output(s)
Procurement management plan	Weighting system	Selected vendors
Organizational process assets	Independent estimates	Contract issuance
Procurement document package	Screening system	Contract management plan
Evaluation criteria	Contract negotiation	Resource availability
Proposals	Vendor rating system	Update procurement management plan
Qualified sellers list	Expert judgment	Requested changes
Project management plan	Proposal evaluation techniques	Other in-house outputs, reports, and data inferences of interest to the organization
Other in-house (custom) factors of relevance and interest	Multicriteria outsourcing techniques	
	Other in-house (custom) tools and techniques	

TABLE 10.6
Tools and Techniques for Contract Administration within Project Procurement Management

Step 5: Contract Administration

Inputs	Tools and Techniques	Output(s)
Contract	Contract change control system	Contract documentation
Contract management plan	Buyer-conducted performance review	Requested changes
Performance reports		Recommended corrective actions
Approved change requests	Inspection and audits	Organization process assets (updates)
Work performance information	Performance reporting	
Other in-house (custom) factors of relevance and interest	Payment system	Project management plan (updates)
	Claims administration	
	Records management system	Other in-house outputs, reports, and data inferences of interest to the organization
	Information technology	
	Other in-house (custom) tools and techniques	

Make-or-buy analysis is a general decision technique to determine whether a particular product or service can be produced more cost effectively organically by the project organization or purchased from an external source. The make-or-buy analysis reflects the interests and strategy of the project organization, the capability of the vendor organization, as well as the immediate needs of the project.

Expert judgment, in the context of procurement management, assesses the inputs and outputs needed for an effective procurement decision. Inputs would normally include the interests and oversight of other units within the project organization including such departments as legal, contracts, technical support, subject matter experts, and management preferences. In addition, inputs from external sources such as consultants,

TABLE 10.7
Tools and Techniques for Contract Closure within Project Procurement Management

Step 6: Contract Closure

Inputs	Tools and Techniques	Output(s)
Procurement management plan	Procurement audits	Closed contracts
Contract management plan	Records management systems	Lessons learned documentation
Contract documentation	Other in-house (custom) tools and techniques	Dissemination of project results
Contract closure procedure		Organizational process assets
Other in-house (custom) factors of relevance and interest		Other in-house outputs, reports, and data inferences of interest to the organization

STEP Procurement Management

regulatory requirements, professional organizations, technical associations, and industry groups are often instrumental in making procurement decisions.

Contract type selection helps to align procurement decisions with decision factors and project constraints such as cost, schedule, and performance expectations. The type of contract selected is based on the following:

- Overall cost
- Schedule compatibility
- Quality acceptance
- Degree of risk
- Product or service complexity (e.g., technical risk)
- Contractor's accountability, responsibility, and risk
- Concurrent contracts
- Outsourcing and subcontracting preferences
- Vendor's accounting system and reliability
- Urgency of need

Contracts fall into one of three major categories as explained below:

1. *Fixed price or lump sum contracts* have the following characteristics:
   a. Fixed total price for a well-defined product or service.
   b. If the product is not well defined, both the project and vendor are at risk.
   c. The simplest form of this is to use purchase order for a specified item, at a specified price, and for a specific date.
   d. Fixed price contracts may also include incentives for meeting project objectives.
2. *Time and materials contracts* have the following characteristics:
   a. This contains aspects of both cost reimbursable and fixed price contracts.
   b. It is often open-ended and full value is usually not defined at the time of award.
   c. Unit rates for this type of contract can be preset.
3. *Cost reimbursable contracts* have the following characteristics:
   a. This involves payment to the vendor for actual costs of product or service rendered.
   b. Costs are classified as direct costs or indirect costs. Direct costs are costs incurred exclusively for the purpose of the project. Indirect costs are overhead costs that are allocated to the project by the performing organization.

Cost-reimbursable contracts are further categorized into the following types:

- Cost-plus-fee (CPF) or cost-plus-percentage of cost (CPPC): In this case, the vendor is reimbursed for all allowable costs plus agreed fee at an agreed percentage of costs. Fee varies with actual costs.

- Cost-plus-fixed-fee (CPFF): In this case, the vendor is reimbursed for all allowable costs plus a fixed fee payment based on a percentage of the estimated project costs. Fee does not vary with actual costs.
- Cost-plus-incentive-fee (CPIF): In this case, the vendor is reimbursed for all allowable costs and a predetermined fee (incentive bonus) based on achieving certain performance objectives. Both the vendor and buyer could benefit from cost savings on the basis of a negotiated cost formula.

COMPLETION AND TERM CONTRACTS

A contract can be executed either as a *completion contract* or a *term contract*. In a completion contract, the contractor is required to deliver a definitive end product. The contract is complete upon delivery and formal customer acceptance. The final payment is made upon delivery. In a term contract, the contractor is required to deliver a specific "level of effort," where the effort is expressed in "person-days" over a specified period of time. The contractor is under no further obligation after the effort is performed. Final payment is not dependent upon technical accomplishment. Figure 10.3 shows the varying levels of risks associated with the different types of contract.

PROCUREMENT MANAGEMENT PLAN

A procurement management plan specifies how the remaining procurement processes will be managed. It may be formal or informal, highly detailed or broadly stated, based on the specific needs of the project. It is a subsidiary of the overall project plan.

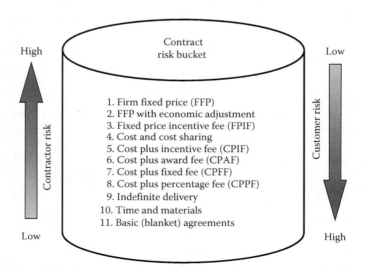

FIGURE 10.3 Risk levels for types of contracts.

CONTRACTOR STATEMENT OF WORK

The statement of work (SOW) describes the procurement item in sufficient detail to allow prospective vendors to determine if they are capable of providing the product or service. Each individual procurement item requires a separate statement of work. However, the multiple products or services may be grouped as one procurement item with a single SOW. Statement of work will often influence the development of additional contract evaluation criteria such as the following queries:

- Does the vendor demonstrate an understanding of the needs of the project? This can be evident in the contents of the proposal.
- What level of overall or life cycle cost is offered by the vendor? Will the selected vendor produce the lowest total cost including contract cost as well as operating cost?
- Does the vendor have adequate technical capability? Does the vendor currently have, or can be expected to acquire, the technical capabilities and knowledge needed by the project?
- Will the vendor's management approach ensure a successful execution of the project?
- Does the vendor have the financial status and capability adequate to execute the contract successfully and adequately?
- What certifications are available on the vendor's history, resources, and quality records?

ORGANIZATION PROCESS ASSETS

Organization process assets include historical lists of qualified vendors, past experience, and previous relationships. The list of preferred vendors is developed through some sort of rigorous methodology. Some quantitative methodologies are presented in this chapter. Bidder conferences, contractor conferences, vendor conferences, and pre-bid conferences are examples of meetings with prospective vendors prior to preparation of a proposal. The prospective vendors must have a clear understanding of the procurement process.

CONTRACT FEASIBILITY ANALYSIS

Procurement should be preceded by a formal feasibility analysis. The feasibility of a project can be ascertained in terms of technical factors, economic factors, or both. Some of the topics to be evaluated include contract responsibilities and authorities, applicable terms and laws, technical and business management approaches, and financing source. A complex procurement process may require an independent or external negotiation process. A feasibility study is documented with a report showing all the ramifications of the project and should be broken down into the following categories:

Technical feasibility. Technical feasibility refers to the ability of the process to take advantage of the current state of the technology in pursuing further improvement. The technical capability of the personnel as well as the capability of the available technology should be considered.

Managerial feasibility. Managerial feasibility involves the capability of the infrastructure of a process to achieve and sustain process improvement. Management support, employee involvement, and commitment are key elements required to ascertain managerial feasibility.

Economic feasibility. This involves the ability of the proposed project to generate economic benefits. A benefit–cost analysis and a breakeven analysis are important aspects of evaluating the economic feasibility of new science and technology projects. The tangible and intangible aspects of a project should be translated into economic terms to facilitate a consistent basis for evaluation.

Financial feasibility. Financial feasibility should be distinguished from economic feasibility. Financial feasibility involves the capability of the project organization to raise the appropriate funds needed to implement the proposed project. Project financing can be a major obstacle in large multiparty projects because of the level of capital required. Loan availability, credit worthiness, equity, and loan schedule are important aspects of financial feasibility analysis.

Cultural feasibility. Cultural feasibility deals with the compatibility of the proposed project with the cultural setup of the project environment. In labor-intensive projects, planned functions must be integrated with the local cultural practices and beliefs. For example, religious beliefs may influence what an individual is willing to do or not do.

Social feasibility. Social feasibility addresses the influences that a proposed project may have on the social system in the project environment. The ambient social structure may be such that certain categories of workers may be in short supply or nonexistent. The effect of the project on the social status of the project participants must be assessed to ensure compatibility. It should be recognized that workers in certain industries may have certain status symbols within the society.

Safety feasibility. Safety feasibility is another important aspect that should be considered in project planning. Safety feasibility refers to an analysis of whether the project is capable of being implemented and operated safely with minimal adverse effects on the environment. Unfortunately, environmental impact assessment is often not adequately addressed in complex projects. As an example, the North America Free Trade Agreement (NAFTA) between the United States, Canada, and Mexico was temporarily suspended in 1993 because of the legal consideration of the potential environmental impacts of the projects to be undertaken under the agreement.

Political feasibility. A politically feasible project may be referred to as a "politically correct project." Political considerations often dictate the direction for a proposed project. This is particularly true for large projects with national visibility that may have significant government inputs and political implications. For example, political necessity may be a source of support for a project regardless of the project's merits. On the other hand, worthy projects may face insurmountable opposition simply because of political factors. Political feasibility analysis requires an evaluation of the compatibility of project goals with the prevailing goals of the political system. In general, feasibility analysis for a project should include following items:

1. *Need analysis*: This indicates recognition of a need for the project. The need may affect the organization itself, another organization, the public, or the government. A preliminary study is conducted to confirm and evaluate the need. A proposal of how the need may be satisfied is then made. Pertinent questions that should be asked include the following:
   a. Is the need significant enough to justify the proposed project?
   b. Will the need still exist by the time the project is completed?
   c. What are alternate means of satisfying the need?
   d. What are the economic, social, environmental, and political impacts of the need?
2. *Process work*: This is the preliminary analysis done to determine what will be required to satisfy the need. The work may be performed by a consultant who is an expert in the project field. The preliminary study often involves system models or prototypes. For technology-oriented projects, artist conceptions and scaled-down models may be used for illustrating the general characteristics of a process. A simulation of the proposed system can be carried out to predict the outcome before the actual project starts.
3. *Engineering and design*: This involves a detailed technical study of the proposed project. Written quotations are obtained from suppliers and subcontractors as needed. Technology capabilities are evaluated as needed. Product design, if needed, should be done at this stage.
4. *Cost estimate*: This involves estimating project cost to an acceptable level of accuracy. Levels of around −5% to +15% are common at this level of a project plan. Both the initial and operating costs are included in the cost estimation. Estimates of capital investment, recurring, and nonrecurring costs should also be contained in the cost-estimate document. Sensitivity analysis can be carried out on the estimated cost values to see how sensitive the project plan is to changes in the project scenario.
5. *Financial analysis*: This involves an analysis of the cash flow profile of the project. The analysis should consider rates of return, inflation, sources of capital, payback periods, breakeven point, residual values, and sensitivity.
6. *Project impacts*: This portion of the feasibility study provides an assessment of the impact of the proposed project. Environmental, social, cultural, political, and economic impacts may be some of the factors that will determine how a project is perceived by the public. The value-added potential of the project should also be assessed.
7. *Conclusions and recommendations*: The feasibility study should end with the overall outcome of the project analysis. This may constitute either an endorsement or disapproval of the project.

CONTENTS OF PROJECT PROPOSAL

The project proposal should present a detailed plan for executing the proposed project. The proposal may be directed to a management team within the same organization or to an external organization. The proposal contents may be written in two parts: a technical section and a management section.

TECHNICAL SECTION OF PROJECT PROPOSAL

Project background
- Organization's expertise in the project area
- Project scope
- Primary objectives
- Secondary objectives

Technical approach
- Required technology
- Available technology
- Problems and their resolutions
- Work breakdown structure

Work statement
- Task definitions and list
- Expectations

Schedule
- Gantt charts
- Milestones
- Deadlines

Project deliverables

Value of the project
- Significance
- Benefit
- Impact

MANAGEMENT SECTION OF PROJECT PROPOSAL

Project staff and experience
- Personnel credentials

Organization
- Task assignment
- Project manager, liaison, assistants, consultants, etc.

Cost analysis
- Personnel cost
- Equipment and materials
- Computing cost
- Travel
- Documentation preparation
- Cost sharing
- Facilities cost

STEP Procurement Management

Delivery dates
- Specified deliverables

Quality control measures
- Rework policy

Progress and performance monitoring
- Productivity measurement

Cost-control measures
- Milestone analysis
- Cost benchmarks

A contract awarded following a successful feasibility analysis conveys a legal relationship subject to remedy through the legal system. The contract will spell out the statement of work, period of performance, pricing, product support, limitation of liability, incentives, insurance, subcontractor approval, termination, and disputes resolution strategy. Requested changes from selected vendors are incorporated into the contract through the integrated change control process and constitute a part of the overall project and procurement plans.

CONTRACT TEAMWORK AND COOPERATION

We can usually get a lot done when a cohesive team exists. Using the methodology of Triple C, we can improve communication, cooperation, and coordination to manage relationships between vendors and the project team. Figure 10.4 illustrates the merging of contract administration efforts using the Triple C framework. Managing contract relationships has the following attributes:

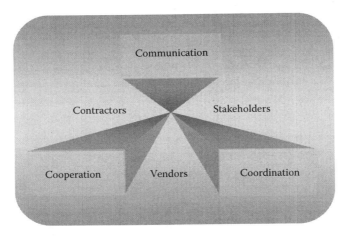

FIGURE 10.4 Triple C linkage of contract relationships.

- Ensures that the vendor's performance meets contractual requirements
- Manages the interfaces among the various providers
- Makes project team aware of project requirements
- Clarifies legal and regulatory requirements
- Facilitates across-the-board collaboration
- Enhances response to the triple constraints of time, cost, and quality
- Provides justification for contract changes

Changes are a fact of project management. There cannot be a workable permanent boilerplate statement in a contract. This is particularly true in science, technology, and engineering projects where project execution may be subject to frequent dynamic developments. Some of the possible reasons for contract changes include the following:

- Unrealistic performance expectations
- Lack of specificity in the contract
- Lack of measurable basis for work
- Work not aligned with core business goals
- Misinterpretation of contract
- Excessive inspection of work
- Knowledge uncertainties
- Changes in delivery schedule
- Improperly executed contract options
- Proprietary and nondisclosure disagreements
- Technological advances
- Budgetary changes

VENDOR RATING SYSTEM

A vendor must be committed to the producer; the producer must be committed to the vendor. Just as customers are expected to be involved in project success, so also should vendors be expected to be involved. Customer requirements should be relayed to vendors so that the goods and services they supply to the project will satisfy what is required to meet project requirements. Selected vendors may be certified based on their previous records of supplying high-quality products. A comprehensive program of vendor–producer commitment should hold both external vendors and internal project process jointly responsible for high-quality products, services, and results all through the project life cycle. The importance of vendor involvement is outlined below:

- Vendor and project team have a joint understanding of project requirements
- Skepticism about a vendor's supply is removed
- Excessive inspection of a vendor's supply is avoided
- Cost of inspecting a vendor's supply is reduced
- Vendors reduce their costs by reducing scrap, rework, and returns
- Vendor morale is improved by the feeling of participation in the project's mission

To facilitate vendor involvement, the producer may assign a liaison to work directly with the vendor in ensuring that the joint quality objectives are achieved. In some cases, the liaison will actually spend time in the vendor's plant. This physical presence helps to solidify the vendor–project relationship. Also, the technical and managerial capability of the producer can be made available to the vendor for the purpose of source quality improvement. Many large companies have arrangements whereby a team of technical staff is assigned to train and help vendors with their quality improvement efforts.

RATING PROCEDURE

A formal system for vendor rating can be useful in encouraging vendor involvement. Vendors who have been certified as supplying high-quality products will enjoy favorable prestige in an organization. Presented below is a simple but effective vendor rating system. The system is based on the opinion poll of a team of individuals.

REQUIREMENTS

1. Form a vendor quality rating team of individuals who are familiar with project operations and the vendor's products.
2. Determine the set of vendors to be included in the rating process.
3. Inform the vendors of the rating process.
4. Each member of the rating team should participate in the rating process.
5. Each member will submit an anonymous evaluation of each vendor based on specified quality criteria.
6. Develop a weighted evaluation of the vendors to arrive at overall relative weights.

COMPUTATION STEPS

1. Let T be the total points available to vendors.
2. Set $T = 100(n)$, where n is the number of individuals in the rating team.
3. Rate the performance of each vendor on the basis of specified quality criteria on a scale of 0–100.
4. Let x_{ij} be the rating for vendor i by team member j.
5. Let m be the number of vendors to be rated.
6. Organize the ratings by team member j as shown below:
   Rating for vendor 1 = x_{1j}
   Rating for vendor 2 = x_{2j}
   Rating for vendor 3 = x_{3j}
   $\vdots$
   Rating for vendor $m = x_{mj}$

   Total rating points (from team member j) = 100
7. Tabulate the team ratings as shown in Table 10.8 and calculate the overall weighted score for each vendor i using the following equation

TABLE 10.8
Layout of Vendor Rating Matrix

	Rating by Member $j=1$	Rating by Member $j=2$	...	Rating by Member $j=n$	Total Points for Vendor i	w_i
Rating for vendor $i=1$						w_1
Rating for vendor $i=2$						w_2
⋮						⋮
⋮						⋮
Rating for vendor $i=m$						w_m
Total points from j	100	100	...	100	$100n$	

$$w_i = \frac{1}{n}\sum_{j=1}^{n} x_{ij}$$

For the case of multiple vendors for the same item, the relative weights, w_i, may be used to determine the fraction of the total supply that should be obtained from each vendor. The fraction is calculated as follows:

$$F_i = w_i(\text{size of total order})$$

where F_i is the fraction of the total supply that should be obtained from vendor i. The size of the order may be expressed in terms of monetary currency or equivalent product units.

MULTICRITERIA VENDOR SELECTION TECHNIQUE

The vendor selection problem is very much like an outsourcing problem and they both can benefit from rigorous analytical selection tools and techniques. Some of the commonly used techniques for vendor selection include the following:

- Total cost approach: In this approach, the quoted price from each vendor is taken as the starting point and each constraint under consideration is replaced iteratively by a cost factor. The contract is awarded to the vendor with the lowest unit total cost.
- Multiattribute utility theory (MAUT): In this approach, multiple, and possibly conflicting, attributes are fed into a comprehensive mathematical model. This approach is useful for global contracting applications.
- Multiobjective programming: In this approach, flexibility and vendor inclusiveness are achieved by allowing a varying number of vendors into the solution such that suggested volume of allocation to each vendor is recommended by the mathematical model.

STEP Procurement Management

- Total cost of ownership: In this philosophy-based approach, the selection process looks beyond price of purchase to include other purchase-related costs. This is useful for demonstrating vendor buy-in and overall involvement in project success.
- Analytic hierarchy process: In this approach, pair-wise comparison of vendors is conducted in a stage-by-stage decision process. This is useful for cases where qualitative considerations are important for the decision process.

Wadhwa–Ravindran Vendor Selection Technique

Several other mathematical models are available in the literature. One comprehensive quantitative technique that uses multicriteria modeling is presented by Wadhwa and Ravindran (2007). They present a multicriteria formulation of the vendor selection problem with multiple buyers and multiple vendors under price discounts. This is applicable to cases where different divisions of an organization buy through one central purchasing department. The number of buyers in this scenario is set equal to the number of divisions buying through the central purchasing office. The model is also applicable for a case where the number of buyers is equal to one. The formulation considers the least restrictive case where any of the buyers can acquire one or more products from any vendors. The potential set of vendors chosen by an organization is constrained by the following:

- Quality level of the products from different vendors
- Lead time of the supplied products
- Production capacity of the vendors

The Wadhwa–Ravindran model helps any organization to select a subset of the most favorable vendors for various outsourced components and to determine the respective quantities to order from each of the chosen most favorable vendors with the objective of meeting project needs. The model uses the following notations:

I = set of products to be purchased
J = set of buyers who procure multiple units in order to fulfill some demand
K = potential set of vendors
M = set of incremental price breaks
p_{ikm} = cost of acquiring one unit of product i from vendor k at price level m
b_{ikm} = quantity at which incremental price breaks occurs for product i by vendor k
F_k = fixed ordering cost associated with vendor k
d_{ij} = demand of product i for buyer j
l_{ijk} = lead time of vendor k to produce and supply product i to buyer j
q_{ik} = quality that vendor k maintains for product i (measured in percent of defects)
L_{ij} = lead time that buyer j requires for product i
Q_j = minimum quality level that buyer j requires for all vendors to maintain (percent rejection)
CAP_k = production capacity of vendor k

N = maximum number of vendors that can be selected
X_{ijkm} = number of units of product i supplied by vendor k to buyer j at price level m
Z_k = decision variable denoting whether or not a particular vendor is chosen (1 or 0)
Y_{ijkm} = decision variable indicating whether or not price level m is used (1 or 0)

The objective of the model is to simultaneously minimize price, lead time, and rejects. The mathematical representations of these multiple objectives are presented below for price, lead time, and quality:

Total purchasing cost = Total variable cost + Total fixed cost

$$= \sum_i \sum_j \sum_k \sum_m P_{ikm} X_{ijkm} + \sum_k F_k Z_k$$

Total lead-time = Summation over all products, buyers, and vendors

$$= \sum_i \sum_j \sum_k \sum_m l_{ijk} X_{ijkm}$$

Quality = Sum of rejects over all products, buyers, and vendors

$$= \sum_i \sum_j \sum_k \sum_m q_{ijk} X_{ijkm}$$

The constraints in the model are expressed in terms of capacity constraint, demand constraint, maximum number of vendors, linearization, and non-negativity. These are expressed as follows:

Capacity constraint: $\sum_i \sum_j \sum_m X_{ijkm} \leq CAP_k Z_k \quad \forall k$

Demand constraint: $\sum_k \sum_m X_{ijkm} = d_{ij} \quad \forall i, j$

Maximum number of vendors: $\sum_k Z_k \leq N$

Because of price discounts, the objective function will be nonlinear. Linearizing constraints are needed to convert the nonlinear objective function to a linear function. These constraints are expressed as

$$X_{ijkm} \leq (b_{ikm} - b_{ikm-1}) * Y_{ijkm} \quad \forall i, j, k; \quad 1 \leq m \leq m_k$$

$$X_{ijkm} \geq (b_{ikm} - b_{ikm-1}) * Y_{ijkm+1} \quad \forall i, j, k; \quad 1 \leq m \leq m_k - 1$$

Note that price breaks occur at the following sequence of quantities:

$$0 = b_{i,k,0} < b_{i,k,1} < \cdots < b_{i,k,n}$$

The unit price of ordering X_{ijkm} units from vendor k at price level m is given by p_{ikm} if $b_{i,k,m-1} < X_{ijkm} \leq b_{i,k,m}$ ($1 \leq m \leq m[k]$).

The linearizing constraints force quantities in the discount range for a vendor to be incremental. Because the "quantity" is incremental, if the order quantity lies in discount interval m, i.e., $Y_{ijkm} = 1$, then the quantities in the interval 1 to $(m-1)$ should be at the maximum of those ranges The first of the two constraints also assures that a quantity in any range is no greater than the width of the range. The non-negativity and binary constraint is expressed as

$$X_{ijkm} \geq 0; \quad Z_k, Y_{ijkm} \in (0,1).$$

The above formulations present the general structure of the Wadhwa–Ravindran model. Interested readers should consult the Wadhwa and Ravindran (2007) for the full exposition of the model as well as a numeric example of the model. Several different methods are available for solving multiobjective optimization problems. Wadhwa and Ravindran cover the following solution methods:

1. Weighted objective method
2. Goal programming method
3. Compromise programming method

WEIGHTED OBJECTIVE METHOD

Weighing the objectives to obtain an efficient or Pareto-optimal solution is a common multiobjective solution technique. Under the weighted objective approach, the vendor selection problem is transformed to the following single-objective optimization problem:

$$\text{Min } w_1 \left[\sum_i \sum_j \sum_k \sum_m p_{ikm} X_{ijkm} \sum_k F_k \right] + w_2 \left[\sum_i \sum_j \sum_k \sum_m l_{ijk} X_{ijkm} \right]$$
$$+ w_3 \left[\sum_i \sum_j \sum_k \sum_m q_{ik} X_{ijkm} \right],$$

where w_1, w_2, and w_3 are the weights on each of the objectives. The optimal solution to the weighted problem is a noninferior solution to the multiobjective problem as long as all the weights are positive. The weights can be systematically varied to generate several efficient solutions. This is not generally a good method for finding an exact representation of the efficient set. It is often used to approximate the efficient solution set.

GOAL PROGRAMMING

Goal programming approach views a decision problem as a set of goals to be accomplished subject to a set of *soft constraints* representing the targets to be achieved. Typical optimization techniques assume that the decision constraints are *hard constraints* that cannot be violated. Goal programming relaxes that strict requirement by focusing on compromises that can be accommodated in favor of achieving an overall improvement in the set of goals. The compromises are modeled as deviations from the goals. Goal programming attempts to minimize the set of deviations from the specified goals. The goals are considered simultaneously, but they are weighted in accordance with their relative importance to the decision maker. Goal programming is a three-step approach.

Step 1: The decision maker provides the goals and targets to be achieved for each objective. Because the goals are not hard constraints, some of the goals may not be achievable. Let us consider an objective f_i with a target value of b_i. The goal constraint is written as:

$$f_i(x) + d_i^- - d_i^+ = b_i$$

where
d_i^- = underachievement of goal
d_i^+ = overachievement of goal

Step 2: The decision maker provides his/her preference on achieving the goals. This can be done as ordinal (preemptive rank order), cardinal (absolute weights) or hybrid measure.

Step 3: Find a solution that will come as close as possible to the stated goal in the specified preference order. As an illustration, preemptive weights are used in the model presented here. Priority order is assigned to the goals. Goals with higher priorities are satisfied before lower-priority goals are considered. For the example below, price is the highest priority goal, followed by lead time, and then quality. The formulation is represented as shown below:

$$\text{Min } Z = P_1 d_1^+ + P_2 d_2^+ + P_3 d_3^+$$

Subject to

$$\sum_i \sum_j \sum_k \sum_m P_{ikm} \cdot X_{ijkm} + \sum_k F_k \cdot Z_k + d_1^- - d_1^+; \text{ for price goal}; \quad \forall i,j$$

$$\sum_k \sum_m q_{ik} \cdot X_{ijkm} + d_2^- - d_2^+ \text{ for quality goal}; \quad \forall i,j$$

$$\sum_k \sum_m q_{ik} \cdot X_{ijkm} + d_2^- - d_2^+ \text{ for quality goal}; \quad \forall i,j$$

where $p_1, p_2,$ and p_3 are the preemptive priorities assigned to each criterion.

COMPROMISE PROGRAMMING

Compromise programming (CP) is an approach that sets the identification of an ideal solution as a point where each attribute under consideration achieves its optimum value and seeks a solution that is as close as possible to the ideal point. Comparative weights are used as measures of relative importance of the attributes in the CP model. Although weights representing relative importance are used as the preference structure in CP, the mathematical basis for applying CP is superior to conventional weighted-sum methods for locating efficient solutions, or the so-called Pareto points. Compromise programming is very useful for collective decision making, such as procurement selection. It is a methodology for approaching the *ideal solution* as closely as possible within the decision sphere. An ideal solution corresponds to the best value that can be achieved for each objective, ignoring other objectives, subject to the overall constraints. Since the objectives are conflicting, the ideal solution cannot be achieved; but it can be approached as closely as possible. "Closeness," in this regard, is represented by a distance metric, L_p, defined as follows:

$$L_p = \left[\sum_{i=1}^{k} \lambda_i^p (f_i - f_i^*)^p \right]^{1/p} \quad \text{for } p = 1, 2, \ldots, \infty$$

where the variables, $f_1, f_2, \ldots, f_k$, are the different objectives. The factor, $f_i^* = \min(f_i)$, ignoring other criteria, is called the ideal value for the ith objective. The weights given to the various criteria are the λ_i values. In general, using w_i's as the relative weights, we have the following relationship:

$$\lambda_i = \frac{w_i}{f_i^*}$$

A *compromise solution* is identified as any point that minimizes the L_p function for the following conditions:

$$\lambda_i > 0$$
$$\sum \lambda_i = 1$$
$$1 \leq p \leq \infty$$

The compromise solution is always nondominated in the optimization sense. As p increases, larger deviations are assigned higher weights. For $p = \infty$, the largest of the deviations completely dominates the distance determination. For the vendor selection application, the CP approach will proceed as follows:

Step 1: Obtain the ideal solution by optimizing the problem separately for each objective. The ideal values for each of the three objectives price, lead time, and quality are denoted, respectively, by p_i^*, l_i^*, and q_i^*.

Step 2: Obtain compromise solution by using an appropriate distance measure.

Thus, we have the following mathematical expression for the vendor selection problem:

$$\text{Min} \left[w_1 \left\{ \frac{\left(\sum_i \sum_j \sum_k \sum_m p_{ijkm} * X_{ijkm} + \sum_k F_k * Z_k \right) - p_i^*}{p_i^*} \right\}^p + w_2 \left\{ \frac{\left(\sum_i \sum_j \sum_k \sum_m l_{ijk} * X_{ijkm} \right) - l_i^*}{l_i^*} \right\}^p + w_3 \left\{ \frac{\left(\sum_i \sum_j \sum_k \sum_m q_{ik} * X_{ijkm} \right) - q_i^*}{q_i^*} \right\}^p \right]^{1/p}$$

Typical values used for p are 1, 2, and ∞. By changing the value of the parameter p, different efficient solutions can be obtained from the above expression. LINGO optimization software can be used to solve the compromise programming formulation.

INVENTORY ANALYSIS AND PROCUREMENT

Inventory management is one quantitative approach to managing scope, cost, and schedule as a part of the project management knowledge areas. Inventoried items are an important component of any procurement management effort. Consequently, inventory management strategies should be developed for effective procurement management. Tracking activities is analogous to tracking inventory items. The important aspects of inventory management for procurement management include the following:

- Ability to satisfy work demands promptly by supplying materials from stock
- Availability of bulk rates for purchases and shipping
- Possibility of maintaining more stable and level resource or workforce

The following presents some basic inventory control models.

ECONOMIC ORDER QUANTITY MODEL

The economic order quantity (EOQ) model determines the optimal order quantity based on purchase cost, inventory carrying cost, demand rate, and ordering cost. The objective is to minimize the total relevant costs of inventory. For the formulation of the model, the following notations are used:

Q is the replenishment order quantity (in units)
A is the fixed cost of placing an order

v is the variable cost per unit of the item to be inventoried
r is the inventory carrying charge per dollar of inventory per unit time
D is the demand rate of the item
TRC is the total relevant costs per unit time

Figure 10.5 shows the basic inventory pattern with respect to time. One complete cycle starts from a level of Q and ends at zero inventory.

The total relevant cost for order quantity Q is given by the expression below. Figure 10.6 shows the costs as functions of replenishment quantity.

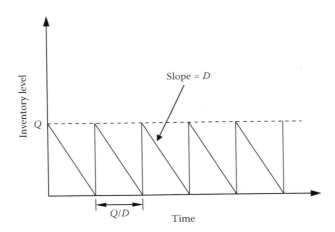

FIGURE 10.5 Basic inventory pattern.

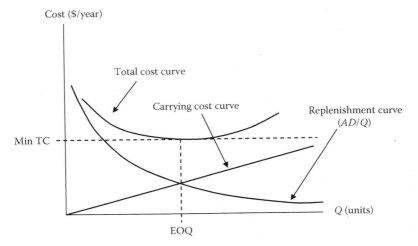

FIGURE 10.6 Inventory costs as functions of replenishment quantity.

$$\text{TRC}(Q) = \frac{Qvr}{2} + \frac{AD}{Q}$$

When the TRC(Q) function is optimized with respect to Q, we obtain the expression for the EOQ,

$$\text{EOQ} = \sqrt{\frac{2AD}{vr}}$$

which represents the minimum total relevant costs of inventory. The above formulation assumes that the cost per unit is constant regardless of the order quantity. In some cases, quantity discounts may be applicable to the inventory item. The formulation for quantity discount situation is presented below.

QUANTITY DISCOUNT

A quantity discount may be available if the order quantity exceeds a certain level. This is referred to as the single breakpoint discount. The unit cost is represented as shown below. Figure 10.7 presents the price breakpoint for a quantity discount.

$$v = \begin{cases} v_0, & 0 \leq Q < Q_b \\ v_0(1-d), & Q_b \leq Q \end{cases}$$

where
    v_0 is the basic unit cost without discount
    d is the discount (in decimals) and d is applied to all units when $Q \leq Q_b$
    Q_b is the breakpoint

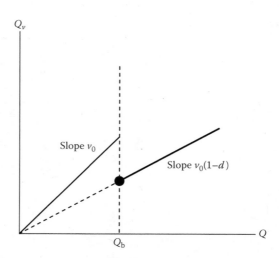

FIGURE 10.7 Price breakpoint for quantity discount.

CALCULATION OF TOTAL RELEVANT COST

For $0 \leq Q < Q_b$ we obtain

$$\text{TRC}(Q) = \left(\frac{Q}{2}\right)v_0 r + \left(\frac{A}{Q}\right)D + Dv_0$$

For $Q_b \leq Q$ we have

$$\text{TRC}(Q)_{\text{discount}} = \left(\frac{Q}{2}\right)v_0(1-d)r + \left(\frac{A}{Q}\right)D + Dv_0(1-d)$$

Note that for any given value of Q, $\text{TRC}(Q)_{\text{discount}} < \text{TRC}(Q)$. Therefore, if the lowest point on the $\text{TRC}(Q)_{\text{discount}}$ curve corresponds to a value of $Q^* > Q_b$ (i.e., Q is valid), then set $Q_{\text{opt}} = Q^*$.

EVALUATION OF THE DISCOUNT OPTION

The trade-off between extra carrying costs and the reduction in replenishment costs should be evaluated to see if the discount option is cost justified. A reduction in replenishment costs can be achieved by two strategies:

1. Reduction in unit value
2. Fewer replenishments per unit time

Case a: If reduction in acquisition costs is greater than extra carrying costs, then set $Q_{\text{opt}} = Q_b$.

Case b: If reduction in acquisition costs is greater than extra carrying costs, then set $Q_{\text{opt}} =$ EOQ with no discount.

Case c: If Q_b is relatively small, then set $Q_{\text{opt}} =$ EOQ with discount. The three cases are illustrated in Figure 10.8.

Based on the three cases shown in Figure 10.8, the optimal order quantity, Q_{opt}, can be found as follows.

Step 1: Compute EOQ when d is applicable:

$$\text{EOQ}(\text{discount}) = \sqrt{\frac{2AD}{v_0(1-d)r}}$$

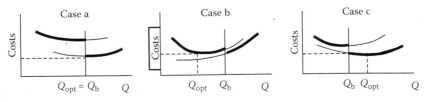

FIGURE 10.8 Cost curves for discount options.

Step 2: Compare EOQ(d) with Q_b:
If EOQ(d) $\geq Q_b$, set Q_{opt} = EOQ(d)
If EOQ(d) $< Q_b$, go to Step 3.

Step 3: Evaluate TRC for EOQ and Q_b:

$$\text{TRC}(\text{EOQ}) = \sqrt{2ADv_0 r} + Dv_0$$

$$\text{TRC}(Q_b)_{discount} = \left(\frac{Q_b}{2}\right)v\,(1-d)r + \left(\frac{A}{Q_b}\right)D + Dv_0(1-d)$$

If $\text{TRC}(\text{EOQ}) < \text{TRC}(Q_b)$, set Q_{opt} = EOQ (no discount):

$$\text{EOQ (no discount)} = \sqrt{\frac{2AD}{v_0 r}}$$

If $\text{TRC}(\text{EOQ}) > \text{TRC}(Q_b)$, set $Q_{opt} = Q_b$

The following example illustrates the use of quantity discount. Suppose $d = 0.02$ and $Q_b = 100$ for the three items shown in Table 10.9.

Item 1 (Case a):
Step 1: EOQ (discount) = 19 units < 100 units
Step 2: EOQ (discount) < Q_b, go to Step 3
Step 3: TRC values

$$\text{TRC}(\text{EOQ}) = \sqrt{2(1.50)(416)(14.20)(0.24)} + 416(14.20)$$
$$= \$5972.42/\text{year}$$

$$\text{TRC}(Q_b) = \frac{100(14.20)(0.98)(0.24)}{2} + \frac{(1.50)(416)}{100} + 416(14.20)(0.98)$$
$$= \$5962.29/\text{year}$$

Since TRC(EOQ)>TRC(Q_b), set Q_{opt} = 100 units.

Item 2 (Case b):
Step 1: EOQ (discount) = 21 units < 100 units
Step 2: EOQ (discount) < Q_b, go to Step 3
Step 3: TRC values

TABLE 10.9
Items Subject to Quantity Discount

Item	D (Units/Year)	v_0 ($/Unit)	(A) ($)	r ($/$/Year)
Item 1	416	14.20	1.50	0.24
Item 2	104	3.10	1.50	0.24
Item 3	4160	2.40	1.50	0.24

$$\text{TRC}(\text{EOQ}) = \sqrt{2(1.50)(104)(3.10)(0.24)} + 104(3.10)$$
$$= \$337.64/\text{year}$$
$$\text{TRC}(Q_b) = \frac{100(3.10)(0.98)(0.24)}{2} + \frac{(1.50)(104)}{100} + 104(3.10)(0.98)$$
$$= \$353.97/\text{year}$$

$\text{TRC}(\text{EOQ}) < \text{TRC}(Q_b)$, set $Q_{\text{opt}} = \text{EOQ}$ (without discount):

$$\text{EOQ} = \sqrt{\frac{2(1.50)(104)}{3.10(0.24)}}$$
$$= 20 \text{ units}$$

Item 3 (Case c):
Step 1:   Compute EOQ (discount)

$$\text{EOQ}(\text{discount}) = \sqrt{\frac{2(1.50)(4160)}{2.40(0.98)(0.24)}}$$
$$= 149 \text{ units} > 100 \text{ units}$$

Step 2:   EOQ (discount) > Q_b. Set $Q_{\text{opt}} = 149$ units.

SENSITIVITY ANALYSIS

Sensitivity analysis involves a determination of the changes in the values of a parameter that will lead to a change in the dependent variable. It is a process for determining how wrong a decision will be if some or any of the assumptions on which the decision is based prove to be incorrect. For example, a "decision" may be dependent on the changes in the values of a particular parameter, such as inventory cost. The cost itself may in turn depend on the values of other parameters, as shown below:

$$\text{Subparameter} \rightarrow \text{Main parameter} \rightarrow \text{Decision}$$

It is of interest to determine what changes in parameter values can lead to changes in a decision. With respect to inventory management, we may be interested in the cost impact of the deviation of actual order quantity from the EOQ. The sensitivity of cost to departures from EOQ is analyzed as presented below:
Let p represent the level of change from EOQ:

$$|p| \leq 1.0$$
$$Q' = (1-p)\text{EOQ}$$

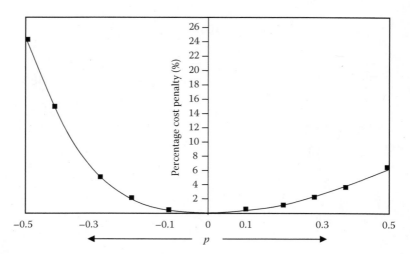

FIGURE 10.9 Sensitivity analysis based on PCP.

Percentage cost penalty (PCP) is defined as follows:

$$\text{PCP} = \frac{\text{TRC}(Q') - \text{TRC}(\text{EOQ})}{\text{TRC}(\text{EOQ})}(100)$$

$$= 50\left(\frac{p^2}{1+p}\right)$$

A plot of the cost penalty is shown in Figure 10.9. It is seen that the cost is not very sensitive to minor departures from EOQ. We can conclude that changes within 10% of EOQ will not significantly affect the total relevant cost.

REFERENCES

PMI, *A Guide to the Project Management Body of Knowledge (PMBOK Guide)*, 3rd ed., Project Management Institute, Newtown Square, PA, 2004.

Wadhwa, Vijay, and A. Ravi Ravindran, Vendor selection in outsourcing, *Computers & Operations Research*, 34, 2007, 3725–3737.

BIBLIOGRAPHY

Badiru, Adedeji B., *Industry's Guide to ISO 9000*, John Wiley & Sons, New York, NY, 1995.

Badiru, Adedeji B., Abi Badiru, and Ade Badiru, *Industrial Project Management*, Taylor & Francis/CRC Press, Boca Raton, FL, 2007.

Chopra, Sunil, and Peter Meindl, *Supply Chain Management: Strategy, Planning, and Operation*, 2nd ed., Pearson Prentice-Hall, Upper Saddle River, NJ, 2004.

11 STEP Case Study: Space Shuttle *Challenger*

Dare to dream and achieve the impossible; let lessons of the past guide the future.

—Adedeji Badiru, 2008

Catastrophes sometimes have a way of catalyzing positive changes and improvement. What is most important is to have the fortitude to move forward and institute improvement changes. Disasters often eradicate complacency. The original quote opening this chapter emphasizes the importance of learning from the past to dream about the future. If we consider the lackadaisical response to Hurricane Katrina in August 2005 to the well-coordinated evaluation response of August 2008, we would see a marked difference that confirms that the disaster of 2005 taught us lessons that served us well in 2008 and possibly in future years.

Despite the many years since the Space Shuttle *Challenger* accident occurred on January 28, 1986, it still offers a classic case study of what can happen in any complex science, technology, and engineering project (STEP) and how to mitigate similar problems in the future (Badiru, 1996). Many of the points of failure in the project can never be overemphasized because the mistakes are being repeated in many large STEPs. More recent accidents point to the need to re-visit the case study again and again to re-emphasize the importance of coordinated project management. Now that NASA is working toward phasing out the space shuttle in 2010 to be replaced by a new generation of space vehicles, named Orion, contemporary knowledge-based project management practices must be instituted.

Orion is expected to carry a new generation of explorers back to the moon and later to Mars. Orion will succeed the space shuttle as NASA's primary vehicle for human space exploration. Orion's first flight with astronauts onboard is planned for no later than 2014 to the International Space Station. Its first flight to the moon is planned for no later than 2020. These are tough targets in terms of science, technology, and engineering challenges to be overcome. STEP project management practices are needed to accomplish these targets within budget and on schedule with the satisfactory performance level. Already, there are concerns emerging in the scientific community regarding the guiding vision for Orion as well as the science and technology priorities heading up the ambitious project. The priority conflicts and concerns can be resolved, or at least, mitigated through structural project management. Figure 11.1 shows the design profile of the crew module of Orion Space Vehicle. Figure 11.2 illustrates a collection of next-generation space vehicles.

FIGURE 11.1 Orion space vehicle crew module. (Courtesy of NASA pictures.)

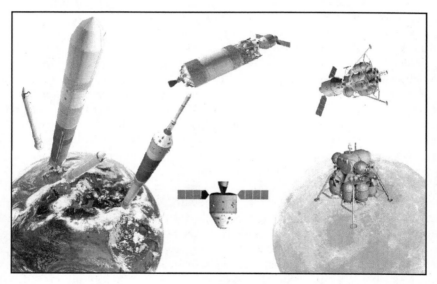

FIGURE 11.2 Collection of next generation space vehicles. (Courtesy of NASA pictures.)

CASE BACKGROUND

This case study illustrates some of the problems involved in managing high-technology projects. The case study involves the events leading to the Space Shuttle *Challenger* accident that occurred on January 28, 1986. Although the accident happened, several years ago, it continues to serve as an excellent example of a high-tech project failure. Its repercussions are still being felt in the space programs around the world. Readers should pay particular attention to the problems in the planning, conceptualization, technology assessment, communication, cooperation, and coordination processes of the *Challenger* mission. Every aspect of a STEP methodology can be extracted from the detailed account of mission and the ensuing investigations.

The case account presented here is based on the report of the Presidential Commission that investigated the Space Shuttle *Challenger* accident. The Commission was chaired by William P. Rogers. The report was published in June 1986 and is available to the public as a matter of public record. As much as possible, the testimony statements are recounted in their original presentation formats and sentence structure in order to capture and preserve the essence of the accident investigation proceedings. Having the shuttle accident as a case study in a book helps to increase public awareness and provides a formal setting for a critical analysis of a complex project management process. The full Presidential Commission report (in five volumes) is overwhelming with pieces of managerial details logged in various sections. Thus, a case study provides a unified and condensed presentation of the management aspects of the accident. The case study can serve as a format for the managerial analysis of any complex system involving science, technology, and engineering interfaces.

FOUNDATION FOR LESSONS LEARNED

On September 30, 1986, B. I. Edelson, associate administrator for space science and applications, wrote an open letter to the science community requesting thoughts and advice on problems of the Space Program in the wake of the *Challenger* accident. This case study is one of the many responses that came out of that call. Readers can develop good managerial analyses (or advice) from the case study to form a guideline for enhancing future operations not only for the space program, but also for other science, technology, and engineering endeavors. Most of the materials in the case study are direct excerpts from the executive summary of the Presidential Commission report. The testimonies, in particular, are reproduced as published (verbatim) in the report in order to preserve the integrity of the original commission hearing. The quoted testimonies are indented and printed in small print to distinguish them from other text.

Points for discussion or reader analysis are printed in italics at the appropriate places in the case study. It is recommended that readers participate in the discussion process so as to develop competence in analyzing managerial decision problems. The scenarios of shuttle management should serve as models for readers to use in evaluating problems in their own operating environments. Efforts have been made to include as much of the testimony as possible to allow readers to have enough information to conduct the recommended managerial analysis.

SPACE SHUTTLE MISSION BACKGROUND AND AUTHORIZATION

The space shuttle concept originated in the 1960s during the development of Apollo lunar landing spacecraft. The objective was to achieve economical access to space by using a reusable launch system. In September 1969, a Space Task Group offered a choice of three long-range plans:

1. A program costing $80 to $10 billion per year involving a manned Mars expedition, a space station in lunar orbit, and a 50-person Earth-orbiting station serviced by a reusable ferry, or space shuttle.
2. An intermediate program that would include the Mars mission (less than $8 billion annually).
3. A modest program ($4 to $5.7 billion a year) that would embrace an Earth-orbiting space station with the space shuttle as its link to Earth.

In March 1970, President Nixon opted for the shuttle-serviced space station as a long-range goal pending the development of the shuttle vehicle. In effect, the shuttle was originally simply the transportation element in a broad, multiobjective space plan. To proceed toward the ultimate plan, the shuttle would have to be built. Thus, the space shuttle became the major short-term focus of NASA, a focus that is really a means of another end.

SPACE SHUTTLE DESIGN DECISIONS

The space shuttle went through evolutionary design changes. The first design was a "fly back" concept in which two manned stages are integrated. The first stage was a big rocket-powered vehicle that would carry the smaller stage piggyback. The carrier would provide the thrust for liftoff and flight through the atmosphere. It would release the passenger (the orbiting vehicle) and return to Earth. The orbiter, with its crew and payload, would continue into space under its own rocket power. After completing its mission, it would fly back to Earth. The two-stage design carried its rocket propellants internally and had much larger flight deck and cargo bay than the later designs. The two-stage craft seemed to be an efficient means of achieving routine economical flight to space. However, the size of the craft called for huge development costs ($10 to $13 billion). Thus, it did not receive support in both Congress and Office of Management and Budget.

In 1971, NASA, now aware that low cost rather than system capability would garner support for the shuttle plan, went back to the drawing board. One proposal that emerged was to eliminate the internal tanks and carry the propellant in a single, disposable external tank. This allowed for a smaller and cheaper orbiter without significant performance loss. For the launch system, one proposal was a winged but unmanned recoverable liquid-fuel vehicle based on the successful Saturn-5 rocket from the Apollo program. Other plans considered simpler but also recoverable liquid-fuel systems, expendable solid rockets, and reusable solid rocket booster (SRB). Solid-fuel systems had been used in the past for some small unmanned spacecraft, but using them for manned flight was a technology new to the space program. However, the SRB won approval over the liquid rocket which offered lower operating costs but a higher development cost. The present space shuttle configuration was what emerged from this round of design effort. The space shuttle is a three-component system made up of

1. The orbiter
2. An expendable external fuel tank carrying liquid propellants for the orbiter's engines
3. Two recoverable SRBs

STEP Case Study: Space Shuttle Challenger

A five-orbiter space shuttle system was estimated to have a cost of $6.2 billion in 1972 to develop and test. This was about half the cost of the two-stage "fly back" design. To achieve this lower development cost, NASA had to accept higher operating costs and sacrifice full reusability. The compromise design retained recoverability and reuse of two of the three elements. The final configuration was selected in March 1972.

SHUTTLE DEVELOPMENT PROCESS

In August 1972, Rockwell International Corporation was awarded a contract for design and development of the space shuttle orbiter. Martin Marietta was assigned the development and fabrication of the external tank, Morton Thiokol (MTI) Corporation was awarded the contract for the SRBs, and Rocketdyne, a division of Rockwell, was selected to develop the orbiter main engines.

Managerial responsibility for the program was divided among three of NASA's field centers. Johnson Space Center (JSC) in Houston was assigned the management of the orbiter. Marshall space Flight Center in Huntsville, Alabama was responsible for the orbiter's main engines. Kennedy Space Center (KSC) in Florida had the job of assembling the space shuttle components, checking them out, and conducting launches.

STEP Case Discussion Questions:

1. What coordination or decision problems may arise from the three pronged managerial setup described above?
2. What could the separate contractors of the shuttle system do to assure a unified design of the overall system?

CHRONOLOGY OF DEVELOPMENTS

The shuttle development years of the 1970s faced budgetary difficulties. The envisioned five-orbiter fleet was reduced to four. These difficulties were compounded by engineering problems. System changes that pushed the frontiers of technology led to cost overruns and schedule slippages. The initial test flights were delayed by more than 2 years. The first test flights were finally conducted at Dryden Flight Research Facility in California in 1977. The orbiter Enterprise was the test craft. The Enterprise was carried on a modified Boeing 747 and released for a gliding approach and landing at the Mojave Desert test center. Five such flights were made. They served to validate the computer operations, subsonic handling, and unpowered landing performance. Extensive ground tests were conducted from 1977 to 1980. These included system vibration tests and main engine firings.

By early 1981, the space shuttle was ready for an orbital flight test program. The test flights covered over 1000 tests and data collection procedures. All flights were to be launched from Kennedy and terminate at Edwards Air Force Base. The test flight series originally called for six missions. This was later reduced to four as shown below:

1. April 12–14, 1981: STS-1 (Space Transportation System-1) Orbiter *Columbia* went on a 2-day demonstration of ability to go into orbit and return.
2. November 12–14, 1981: STS-2. *Columbia* tested the remote manipulator system and carried a payload of Earth survey instruments. The failure of a fuel cell shortened the flight by about 3 days.
3. March 22–30, 1982: STS-3. *Columbia* went on the longest of the test series. Special test of the robot arm was conducted and experiments in material processing were performed.
4. June 27–July 4, 1982: STS-4. *Columbia* carried the first department of defense payload. The conclusion of this flight ended the test flight program.

Ninety-five percent of the test objectives were accomplished. The time between flights was reduced from 7 months to 4, and then to 3. NASA then declared the space shuttle to be "operational," which meant that payload requirements in subsequent flights would be of more interest than spacecraft testing. The operational phase of the space shuttle program thus began in November 1982. The STS-sequential flight numbering was changed after STA-9 to two numbers followed by a letter, for example, 41-B. The first digit indicates the fiscal year of the scheduled launch (4 for 1984). The second digit identifies the launch site (1 for Kennedy, 2 for Vandenberg Air Force Base). The letter corresponds to the alphabetical sequence for the fiscal year. Thus, B means the second scheduled mission in the fiscal year. Due to schedule changes, some flights were not made according to the sequence of numbering. Thus, 51-D actually occurred before 51-B. Presented below is the chronology of shuttle flights after the test phase.

1. November 11–16, 1982: STS-5(*Columbia*)
2. April 4–9, 1983: STS-6(*Challenger*)
3. June 18–24, 1983: STS-7(*Challenger*)
4. August 30–September 6, 1983: STS-8(*Challenger*)
5. November 28–December 8, 1983: STS-9(*Columbia*)
6. February 3–11, 1984: 41-B(*Challenger*)
7. April 6–13, 1984: 41-C(*Challenger*)
8. August 30–September 5, 1984: 41-D(*Discovery*)
9. October 5–13, 1984: 41-G(*Challenger*)
10. November 8–16, 1984: 51-A(*Discovery*)
11. January 24–27, 1985: 51-C(*Discovery*)
12. April 12–19, 1985: 51-D(*Discovery*)
13. April 29–May 6, 1985: 51-B(*Challenger*)
14. June 17–24, 1985: 51-G(*Discovery*)
15. July 29–August 6, 1985: 51-F(*Challenger*)
16. August 27–September 3, 1985: 51-I(*Discovery*)
17. October 3–10, 1985: 51-J(*Atlantis*)
18. October 30–November 6, 1985: 61-A(*Challenger*)
19. November 26–December 3, 1985: 61-B(*Atlantis*)
20. January 12–18, 1986: 61-C(*Columbia*)
21. January 28, 1986: 51-L(*Challenger*: The accident)

STEP Case Study: Space Shuttle *Challenger* 287

The space shuttle program completed 24 successful missions (including the four test flights) over a 57-month period. *Columbia* made seven trips, *Discovery* made six, *Atlantis* two, and *Challenger* made nine trips prior to the accident. A shuttle crew may consist of up to eight people, but the limit is generally seven. The crew consists of the commander, the captain, the pilot, second in command, and two or more mission specialists. One or more payload specialists can be accommodated. A mission specialist coordinates activities of the craft and crew. A payload specialist may manage specific experiments. The commander, pilot, and mission specialists are career astronauts assigned to the mission by NASA. Payload specialists are assigned by payload sponsors in coordination with NASA.

THE *CHALLENGER* MISSION

Changes in the *Challenger* launch schedule complicated the preparations for mission 51-L. The sequence of interrelated activities involved in producing the detailed schedule and supporting logistics necessary for a successful mission requires considerable level of effort and close coordination. Flight 51-L was originally scheduled for July 1985, but schedule slippages forced the delay till January 1986. Planning for the mission began in 1984. Ten major changes in payload items caused disruption in the preparation process. Because the 12–18 months of preparation involve repetitive cycles that progressively define a flight plan in more specific detail, significant changes can require extensive time and effort to incorporate. The closer to the launch the changes occur, the more difficult and disruptive it becomes to repeat the preparation cycles.

LAUNCH DELAYS

The launching of 51-L was postponed three times and scrubbed once from the planned date of January 22, 1986. The launch first slipped from January 23 to January 25. That date was later changed to January 26, 1986, primarily because of KSC work requirements caused by the delay in the launching of mission 61-C (*Columbia*). The third postponement of the launch date occurred on the evening of January 25 due to a forecast of bad weather for January 26. The launch was rescheduled for January 27. At 9:10 a.m., on the new launch day, the countdown was halted when the ground crew reported a problem with an exterior hatch handle. By the time the problem was solved at 10:30 a.m., winds at the runway designated for a "return-to-launch-site" abort had picked up considerably and exceeded the allowable crosswinds. The launch attempt for January 27 was thus canceled at 12:35 p.m. that day. The countdown was rescheduled for January 28.

STEP Case Discussion Question:

Based on the long record of launch delays, should have NASA temporarily suspended all launches in order to conduct extensive evaluations of the technical or managerial problems? Consider arguments to support both the pros and cons of the decision to continue launching.

LAUNCH PROBLEMS

The weather for January 28 was forecast to be very cold, with overnight temperatures in the low twenties. The management team directed engineers to assess the possible effects of temperature on the launch. No critical issues were identified by the management officials. So, countdown proceeded while weather evaluation continued. Ice had accumulated in the launch pad area during the night and it caused considerable concern for the launch team. The ice inspection team was sent to the pad at 1:35 a.m., January 28, and returned to the launch control center at 3:00 a.m. After reviewing the team's report at a program meeting, the space shuttle program manager decided to continue the countdown. Another ice inspection was scheduled at launch minus 3 h. During the night, prior to fueling, a problem developed with a fire detector in the ground liquid hydrogen storage tank. Though the problem was tracked to a hardware fault and repaired, fueling was delayed by two and one-half hours. However, the launch delay was reduced to only 1 h by continuing past a planned hold at launch minus 3 h. Because of rain forecast at Casablanca, the alternate abort site, the site was scrubbed at 7:30 a.m. The change had no impact on the mission since weather at the primary transatlantic abort landing site at Dakar, Senegal, was acceptable.

At a flight crew weather briefing early in the morning, the temperature and ice on the pad were discussed. But neither then nor in earlier weather discussions was the crew told of any concern about the effect of low temperature on the shuttle system. By 8:36 a.m., all the seven crew members (Francis R. Scobee, Michael J. Smith, Ellison S. Onizuka, Judith A. Resnik, Ronald E. McNair, Christa McAuliffe, and Gregory B. Jarvis) were in their seats in the *Challenger*. At 8:44 a.m., the ice team completed its second inspection. After hearing their report, the program manager decided to allow additional time for ice to melt on the pad. He also told the ice team to perform one final ice assessment at launch minus 20 min. By the time the countdown resumed, the launch had been delayed a second hour beyond the liftoff time of 9:38 a.m. The final ice inspection was completed at 11:15 a.m. During the hold at launch minus 9 min, the mission crew and all members of the launch team gave their approval for launch. The flight began at 11:38 a.m., January 28, 1986.

STEP Case Discussion Question:

Should the crew have had a say on whether or not to launch?

THE ACCIDENT

There was no indication of problem from liftoff until the signal from the shuttle was lost. The main engines operated satisfactorily as expected. Voice communications with the crew were normal. The crew called to indicate the shuttle had begun its roll to head due east and to establish communication after launch. Fifty-seven seconds later, mission control informed the crew that the engines had successfully throttled up and all other systems were satisfactory. The commander's acknowledgment of this call was the last voice communication from the *Challenger*.

No alarms were sounded in the cockpit. The crew apparently had no indication of a problem before the rapid explosion of the shuttle which occurred 73 s after liftoff. The first evidence of the accident came from live video coverage. There were no survivable abort options during thrusting of the SRBs. There was nothing that either the crew or the ground controllers could have done to stop the catastrophe.

STEP Case Study: Space Shuttle *Challenger*

The consensus of the commission and other agencies that investigated the accident is that the accident was caused by a failure in the joint between the two lower segments of the right solid rocket motor. The specific failure was the destruction of the seals that are intended to prevent hot gases from leaking through the joint during the propellant burn of the rocket motor. There is no evidence that any other element of the shuttle contributed to the failure.

STEP Case Discussion Question:

Do you feel that a lack of coordinated hardware design contributed to the failure of the seals?

THE INVESTIGATION

Many aspects of the *Challenger* mission were investigated by the Presidential Commission. All the technical and management factors were examined in detail. For the purpose of this case study, only the management issues are addressed. The following presentations including testimonies are excerpts from the executive summary of the Presidential Commission. For a more detailed coverage, refer to the executive summary itself or the full Commission report, volumes I to V.

FLAWS IN THE DECISION PROCESS

The decision to launch the *Challenger* was flawed. Those who made that decision were unaware of the recent history of problems concerning the O-rings and the joint and were unaware of initial written recommendation of the contractor advising against the launch at temperatures below 53°F and the continuing opposition of the engineers at Thiokol after the management reversed its position. They did not have a clear understanding of Rockwell's concern that it was not safe to launch because of ice on the pad. If the decision makers had known all of the facts, it is highly unlikely that they would have decided to launch 51-L on January 28, 1986. The discussion that follows is based on excerpts from the testimony of those involved in the management judgments that led to the launch of the *Challenger* under conditions described. This testimony reveals failures in communication that resulted in a decision to launch 51-L based on incomplete and sometimes misleading information, a conflict between engineering data and management judgments, and a NASA management structure that permitted internal flight safety problems to bypass key shuttle managers.

The Shuttle Flight Readiness Review (FRR) is a step-by-step activity established by NASA to certify the readiness of all components of the space shuttle assembly. The process is focused upon the Level I FRR, held approximately 2 weeks before a launch. The Level I review, shown in Figure 11.3, is a conference chaired by the NASA associate administrator for Space Flight and supported by the NASA chief engineer, the program manager, the center directors, and project managers from Johnson, Marshall, and Kennedy, along with senior contractor representatives.

The formal portion of the process is initiated by directive from the associate administrator for Space Flight. The directive outlines the schedule for the Level I FRR and for the steps that precede it. The process begins at Level IV, shown in

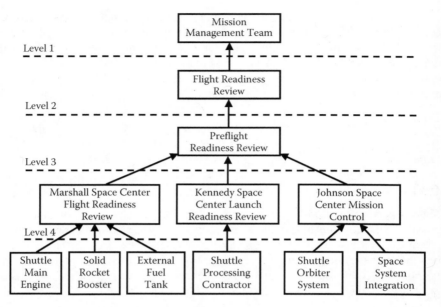

FIGURE 11.3 Flight readiness review hierarchies.

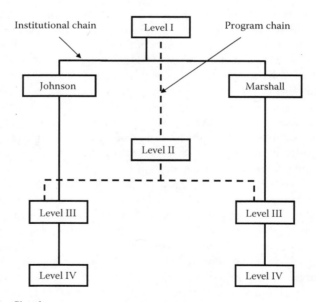

FIGURE 11.4 Shuttle program management structure.

Figure 11.4, with the contractors formally certifying (in writing) the flight readiness of the elements for which they are responsible. Certification is made to the appropriate Level III NASA project managers at Johnson and Marshall. Additionally, at Marshall the review is followed by a presentation directly to the center director. At

STEP Case Study: Space Shuttle *Challenger*

Kennedy the Level III review, chaired by the center director, verifies readiness of the launch support elements.

The next step in the process is the Certification of Flight Readiness to the Level II Program Manager at Johnson. In this review, each Space Shuttle program element endorses that it has satisfactorily completed the manufacture, assembly, test, and checkout of the pertinent element, including the contractors' certification that design and performance are up to standard. The FRR process culminates in the Level I review. In the initial notice of the review, the Level I directive establishes a Mission Management Team for the particular mission. The team assumes responsibility for each Shuttle's readiness for a period commencing 48 h before launch and continuing through postlanding activities. On call throughout the entire period, the Mission Management Team supports the associate administrator for Space Flight and the program manager.

A structured Mission Management Team meeting, called L-1, is held 24 h prior to each scheduled launch. Its agenda includes closeout of any open work, a closeout of any FRR action items, a discussion of new or continuing anomalies, and an updated briefing on anticipated weather conditions at the launch site and at the abort landing sites in different parts of the world. It is standard practice of Level I and II officials to encourage the reporting of new problems or concerns that might develop in the interval between the FRR and the L-1 meeting, and between the L-1 launch. In a procedural sense, the process described was followed in the case of Flight 51-L. However, in the launch preparation for 51-L, the relevant concerns of Level III NASA personnel and element contractors were not adequately communicated to the NASA Level I and II management responsible for the launch.

STEP Case Discussion Question:

It seems NASA has a well established and potentially effective managerial system. However, the implementation of the system might have been flawed. Why did that standard system not work for Challenger? Many things that could go wrong did just that. Is Murphy's Law manifesting itself here?

Two of the specific concerns that were not properly addressed in the *Challenger* preparations are

1. The objections to launch voiced by MTI engineers about the detrimental effect of cold temperatures on the performance of the Solid Rocket Motor joint seal
2. The degree of concern of Thiokol and Marshall about the erosion of the joint seals in prior Shuttle flights, notably 51-C (January 1985) and 51-B (April 1985)

On January 23, Jesse Moore, associate administrator for space flight, issued a directive stating that the FRR had been conducted on January 15 and that 51-L was ready to fly pending closeout of open work, satisfactory countdown, and completion of remaining FRR action items, which were to be closed out during the L-1 meeting. No problems with the SRB were identified.

Since December 1982, the O-rings had been designated a "Criticality 1" feature of the SRB design, a term denoting a failure point (without back-up) that could cause a loss of life or vehicle if the component fails. In July 1985, after a nozzle joint on STS 51-B showed erosion of a secondary O-ring, indicating that the primary seal failed, a launch constraint was placed on flight 51-F and subsequent launches. These constraints had been imposed and regularly waived by the SRB project manager at Marshall, Lawrence B. Mulloy. Neither the launch constraint, the reason for it, or the six consecutive waivers prior to 51-L were known to Moore (Level I) or Aldrich (Level II) or Thomas at the time of the FRR process for 51-L.

There were other paths of system reporting that were designed to bring forward information about the SRB joint anomalies. One path was the task force of Thiokol engineers and Marshall engineers who had been conducting subscale pressure tests during 1985, a source of documented rising concern and frustration on the part of some of the Thiokol participants and a few of the Marshall participants. But Level II was not in the line of reporting for this activity. Another path was the examination at each FRR of evidence of earlier flight anomalies. For 51-L, the data presented in this latter path, while it reached Levels I and II, never referred to either test anomalies or flight anomalies with O-rings.

No mention of the O-ring problems in the SRB joint appeared in the Certification of Flight Readiness, signed for Thiokol on January 9, 1986, by Joseph Kilminster, for the SRB set designated BI026. Similarly, no mention appeared in the certification endorsement, signed on January 15, 1986, by Kilminster and by Mulloy. In the 51-L readiness reviews, it appears that neither Thiokol management nor the Marshall Level III project managers believed that the O-ring blow-by and erosion risk was critical. The testimony and contemporary correspondence show that Level III believed there was ample margin to fly with O-ring erosion, provided the leak check was performed at 200 lb per square inch. Following the January 15 FRR each element of the Shuttle was certified as flight-ready. The Level I Mission Management Team meeting took place as scheduled at 11:00 a.m., on January 25. No technical issues appeared at this meeting or in the documentation, and all FRR actions were reported closed out.

STEP Case Discussion Questions:

1. *Do the several layers of managerial reporting impede critical communication requirements? If so, what is a possible remedy?*
2. *How could the seriousness of the ring problem have been better impressed upon the Thiokol management and the Marshall project managers?*

TESTIMONIES

Mr. Mulloy testified as follows regarding the FRR record about O-ring concerns:

Chairman Rogers: ... Why wasn't that a cause for concern on the part of the whole NASA organization?
Mr. Mulloy: It was cause for concern, sir.
Chairman Rogers: Who did you tell about this?
Mr. Mulloy: Everyone, sir.

Chairman Rogers: And they all knew about it at the time of 51-L?
Mr. Mulloy: Yes, sir. You will find in the FRR record that went all the way to the L-1 review.

But contrary to the testimony of the SRB project manager, the seriousness of concern was not conveyed in FRR to the Level I personnel. The only remaining issue facing the Mission Management Team at the L-1 review was the approaching cold front, with forecasts of rain showers and temperatures in the mid-1960s. There had also been heavy rain since 51-L had been rolled out to the launch pad, approximately 7 in. compared with the 2.5 in. that would have been normal for that season. At 12:36 p.m., on the 27th, the Mission Management Team scrubbed the launch for that day due to high cross winds at the launch site. In the accompanying discussion that ran for about half an hour, all appropriate personnel were polled as to the feasibility of a launch within 24 h. Participants were requested to identify any constraints. This meeting, aimed at launch at 9:38 a.m. on January 28, pronounced no constraints or concerns about the performance of the SRBs.

At 2:00 p.m., on the 27th, the Mission Management Team met again. At that time, the weather was expected to clear, but it appeared that temperatures would be in the low twenties for about 11 h. Issues were raised with regard to the cold weather effects on the launch facility, including the water drains, the eye wash and shower water, fire suppression system, and overpressure water trays. It was decided to activate heaters in the Orbiter, but no concerns were expressed about the O-rings in the SRBs. The decision was to proceed with the countdown and with fueling, but all members of the team were asked to review the situation and call if any problem arose.

At about 2:30 p.m., at Thiokol's Wasatch plant, Robert Ebeling, after learning of the predicted low temperature for launch, convened a meeting with Roger Boisjoly and with other Thiokol engineers. Ebeling was concerned about predicted cold temperatures at KSC. In a postaccident interview, Ebeling recalled the substance of the meeting:

> The meeting lasted 1 h, but the conclusion of that meeting was engineering—especially Arnie, Roger Boisjoly, Brian Russell, myself, Jerry Burns, they come to mind—were very adamant about their concerns on this lower temperature, because we were way below our database and we were way below what we qualified for.

Later in the afternoon on the same day, Allan McDonald—Thiokol's liaison for the SRB project at KSC—received a telephone call from Ebeling, expressing concern about the performance of the SRB field joints at low temperatures. During testimony before the Commission on February 27, McDonald recounted that conversation:

Mr. McDonald: Well, I had first become aware of the concern of the low temperatures that were projected for the Cape, it was late in the afternoon on the 27th. I was at Carver Kennedy's house. He is a vice president of, as I mentioned, our space operations center at the Cape, and supports the stacking of the solid rocket motors (SRMs). And I had a call from Bob Ebeling. He is the manager of our ignition system and final assembly, and he worked for me as program manager at Thiokol in Utah. And he called me and said that they had just received some word earlier that the weatherman was projecting temperatures as low as 18°F some time in the early morning hours of the 28th, and that they had some meetings with some of the engineering people and had some concerns about the O-rings getting to those kinds of temperatures.

And he wanted to make me aware of that and also wanted to get some more updated and better information on what the actual temperature was going to be depicted, so that they could make some calculations on what they expected the real temperature the O-rings may see...

I told him that I would get that temperature data for him and call him back. Carver Kennedy then, when I hung up, called the launch operations center to get the predicted temperatures from pad B, as well as what the temperature history had been during the day up until that time.

... He obtained those temperatures from the launch operations center, and they basically said that they felt it was going to get near freezing or freezing before midnight. It would get as low as 22° as a minimum in the early morning hours, probably around 6:00 o'clock, and that they were predicting a temperature of about 26° at the intended time, about 9:38 the next morning.

I took that data and called back to the plant and sent it to Bob Ebeling and relayed that to him, and told him he ought to use this temperature data for his predictions, but I thought this was very serious and to make sure that he had the vice president, engineering, involved in this and all of his people, that I wanted them to put together some calculations and a presentation of material.

Chairman Rogers: Who's the vice president, engineering?

Mr. McDonald: Mr. Bob Lund is our vice president, engineering, at our MTI facility in Utah.

To make sure he was involved in this, and that this decision should be an engineering decision, not a program management decision. And I told him that I would like him to make sure they prepared some charts and were in a position to recommend the launch temperature and to have the rationale for supporting that launch temperature.

I then hung up and I called Mr. Mulloy. He was staying at the Holiday Inn in Merritt Island and they couldn't reach him, and so I called Cecil Houston—Cecil Houston is the resident manager for the Marshall Space Flight Center office at KSC—and told him about our concerns with the low temperatures and the potential problem with the O-rings.

And he said that he would set up a teleconference. He had a four-wire system next to his office. His office is right across from the Vehicle Assembly Building (VAB) in the trailer complex C over there. And he would set up a four-wire teleconference involving the engineering people at Marshall Space Flight Center at Huntsville, our people back at Thiokol in Utah; and that I should come down to his office and participate at Kennedy from there, and that he would get back with me and let me know when that time would be.

Soon thereafter Cecil Houston called Dr. Judson Lovingood, Deputy Shuttle Project Manager at Marshall Space Flight Center, to inform him of the concerns about the O-rings and asked Lovingood to set up a teleconference with senior project management personnel, with George Hardy, Marshall's deputy director of science and engineering, and with MTI personnel. Lovingood called Stanley Reinartz, shuttle project manager, a few minutes later and informed him of the planned teleconference. The first phase of the teleconference began at 5:45 p.m., Eastern Standard Time (EST); participants included Reinartz, Lovingood, Hardy, and numerous people at Kennedy, Marshall, and Thiokol-Wasatch (Allan McDonald missed this phase; he did not arrive at Kennedy until after 8:00 p.m.). Concerns for the effect of low temperature on the O-rings and the joint seal were presented by Morton Thiokol, along with an opinion that launch should be delayed. A recommendation was also made that Aldrich, program manager at Johnson (Level II), be informed of these concerns. The following are excerpts from testimony before the Presidential Commission relating to the teleconference:

Dr. Keel: You just indicated earlier that, based upon that teleconference, you thought there was a good possibility of delay. Is that what Thiokol was recommending then, delay?

Dr. Lovingood: That is the way I heard it, and they were talking about the 51-C experience and the fact that they had experienced the worst case blow-by as far as the arc and the soot and so forth. And also, they talked about the resiliency data that they had. So it appeared to me—and we didn't have all of the proper people there. That was another aspect of this. It appeared to me that we had better sit down and get the data so that we could understand exactly what they were talking about and assess that data. And that is why I suggested that we go ahead and have a telecon within the center, so that we could review that.

Dr. Keel: So as early as after that first afternoon conference at 5:45 p.m., it appeared that Thiokol was basically saying delay. Is that right?

Dr. Lovingood: That is the way it came across to me. I don't know how other people perceived it, but that's the way it came across to me.

Dr. Keel: Mr. Reinartz, how did you perceive it?

Mr. Reinartz: I did not perceive it that way. I perceived that they were raising some questions and issues which required looking into by all the right parties, but I did not perceive it as a recommendation delay.

Dr. Keel: Some prospects for delay?

Mr. Reinartz: Yes, sir, that possibility is always there.

Dr. Keel: Did you convey that to Mr. Mulloy and Mr. Hardy before the 8:15 conference?

Mr. Reinartz: Yes, I did. And as a matter of fact, we had a discussion. Mr. Mulloy was just out of communication for about an hour, and then after that I got in contact with him, and we both had a short discussion relating to the general nature of the concerns with Dr. Lucas and Mr. Kingsbury at the motel before we both departed for the telecon that we had set up out at the Cape.

Dr. Keel: But based upon that, Mr. Lovingood, that impression, you thought it was a significant enough possibility that Mr. Aldrich should have been contacted?

Dr. Lovingood: Yes.

Dr. Keel: In addition, did you recommend that Mr. Lucas, who is director of Marshall, of course, and Mr. Kingsbury who is Mr. Hardy's boss, participate in the 8:15 conference?

Dr. Lovingood: Yes, I did.

Dr. Keel: And you recommended that to whom?

Dr. Lovingood: I believe I said that over the net. I said that I thought we ought to have an inter-center meeting involving Dr. Lucas and Mr. Kingsbury, and then plan to go on up the line to Level II and Level I.

And then it was after we broke off that first telecon I called Stan to the motel and told him that he ought to go ahead and alert Arnie to that possibility.

Dr. Keel: And Mr. Reinartz, you then visited the motel room of Mr. Lucas with Mr. Kingsbury, and also was Mr. Mulloy with you then?

Mr. Reinartz: Yes, sir, he was. In the first couple of minutes I believe I was there by myself, and then Mr. Mulloy joined us.

Dr. Keel: And did you discuss with them Mr. Lovingood's recommendation that the two of them, Lucas and Kingsbury, participate?

Mr. Reinartz: No, sir. I don't recall discussing Mr. Lovingood's recommendations. I discussed with them the nature of the telecon, the nature of the concerns raised by Thiokol, and the plans to gather the proper technical support people at Marshall for examination of the data. And I believe that was the essence of the discussion.

Chairman Rogers: But you didn't recommend that the information be given to Level II or Level I?

Mr. Reinartz: I don't recall that I raised that issue with Dr. Lucas. I told him what the plans were for proceeding. I don't recall, Mr. Chairman, making any statement regarding that.

Mr. Hotz: Mr. Reinartz, are you telling us that you in fact are the person who made the decision not to escalate this to a Level II item?

Mr. Reinartz: This is correct, sir.

STEP Case Discussion Question:

Is Mr. Reinartz's judgement appropriate? What should he have done?

At approximately 8:45 p.m., Phase 2 of the teleconference commenced, the Thiokol charts and written data having arrived at KSC by telefax. (A table of teleconference participants is included with chronology of events.) The charts presented a history of the O-ring erosion and blow-by in the SRB joints of previous flights presented the results of subscale testing at Thiokol and the results of static tests of SRMs. In the following testimony, Roger Boisjoly, Allan McDonald, and Larry Mulloy expressed their recollections of this teleconference up to the point when an off-net caucus was requested:

STEP Case Study: Space Shuttle *Challenger*

Mr. Boisjoly: I expressed deep concern about launching at low temperature. I presented Chart 2–1 with emphasis—now, 2–1, if you want to see it, I have it, but basically that was the chart that summarized the primary concerns, and that was the chart that I pulled right out of the Washington presentation without changing one word of it because it was still applicable, and it addresses the highest concern of the field joint in both the ignition transient condition and the steady-state condition, and it really sets down the rationale for why we were continuing to fly. Basically, if erosion penetrates the primary O-ring seal, there is a higher probability of no secondary seal capability in the steady-state condition. And I had two subbullets under that which stated bench testing showed O-ring not capable of maintaining contact with metal parts, gap, opening rate to maximum operating pressure. I had another bullet which stated bench testing showed capability to maintain O-ring contact during initial phase (O-170 ms of transient). That was my comfort basis of continuing to fly under normal circumstances, normal being within the database we had.

I emphasized, when I presented that chart about the changing of the timing function of the O-ring as it attempted to seal. I was concerned that we may go from that first beginning region into that intermediate region, from 0 to 170 being the first region, and 170 to 330 being the intermediate region where we didn't have a high probability of sealing or seating.

I then presented Chart 2–2 with added concerns related to the timing function. And basically on that chart, I started off talking about a lower temperature than current database results in changing the primary O-ring sealing timing function, and I discussed the SRM-15(Flight 51-C, January, 1985) observations, namely, the 15A (Left SRM, Flight 51-C) motor had 80° arc black grease between the O-rings, and make no mistake about it, when I say black, I mean black just like coal. It was jet black. And SRM-15B(Right SRM, Flight 51-C) had a 110° arc of black grease between the O-rings. We would have low O-ring squeeze due to low temperature which I calculated earlier in the day. We should have higher O-ring Shore hardness…

Now, that would be harder. And what that material really is, it would be likened to trying to shove a brick into a crack versus a sponge. That is good analogy for purposes of this discussion. I also mentioned that thicker grease, as a result of lower temperatures, would have a higher viscosity. It wouldn't be as slick and slippery as it would be at room temperature. And so it would be a little bit more difficult to move across it.

We would have higher O-ring pressure actuation time, in my opinion, and that is what I presented.… These are the sum and substance of what I just presented. If action time increases, then the threshold of secondary seal pressurization capability is approached. That was my fear. If the threshold is reached, then secondary seal may not be capable of being pressurized, and that was the bottom line of everything that had been presented up to that point.

Chairman Rogers: Did anybody take issue with you?

Mr. Boisjoly: Well, I am coming to that. I also showed a chart of the joint with an exaggerated cross section to show the seal lifted off, which has been shown to everybody. I was asked, yes, at that point in time I was asked to quantify my concerns, and I said I couldn't. I couldn't quantify it. I had no data to quantify it, but I did say I knew that it was away from goodness in the current database. Someone on the net commented that we had soot blow-by on SRM-22 (Flight 61-A, October, 1985) which was launched at 75°. I don't remember who made the comment, but that is where the first comment came in about the disparity between my conclusion and the observed data because SRM-22 (Flight 61-A, October, 1985) had blow-by at essentially a room temperature launch.

I then said that SRM-15 (Flight 51-C, January, 1985) had much more blow-by indication and that it was indeed telling us that lower temperature was a factor. This was supported by inspection of flown hardware by myself. I was asked again for data to support my claim, and I said I have none other than what is being presented, and I had been trying to get resilience data, Arnie and I both, since last October, and that statement was mentioned on the net.

Others in the room presented their charts, and the main telecon session concluded with Bob Lund, who is our vice president of engineering, presenting his conclusions and recommendations charts which were based on our data input up to that point. Listeners on the telecon were not pleased with the conclusions and the recommendations.

Chairman Rogers: What was the conclusion?

Mr. Boisjoly: The conclusion was we should not fly outside of our database, which was 53°. Those were the conclusions. And we were quite pleased because we knew in advance, having participated in the preparation, what the conclusions were, and we felt very comfortable with that.

Mr. Acheson: Who presented that conclusion?

Mr. Boisjoly: Mr. Bob Lund. He had prepared those charts. He had input from other people. He had actually physically prepared the charts. It was about that time that Mr. Hardy from Marshall was asked what he thought about the MTI recommendation, and he said he was appalled at the MTI decision. Mr. Hardy was also asked about launching, and he said no, not if the contractor recommended not launching, he would not go against the contractor and launch.

There was a short discussion that ensued about temperature not being a discriminator between SRM-15 (Flight 51-C) and SRM-22 (Flight 61-A), and shortly after, I believe it was Mr. Kilminster who asked—excuse me, I'm getting confused here. Mr. Kilminster was asked by NASA if he would launch, and he said no because the engineering recommendation was not to launch.

Then MTI management then asked for a 5 min caucus. I'm not sure exactly who asked for that, but it was asked in such a manner that I remember it was asked for, a 5 min caucus, which we put on—(we put) the line on mute and went off-line with the rest of the net.

STEP Case Study: Space Shuttle *Challenger* 299

Chairman Rogers: Mr. Boisjoly, at the time that you made the–that Thiokol made the recommendation not to launch, was that the unanimous recommendation as far as you knew?
Mr. Boisjoly: Yes. I have to make something clear. I have been distressed by the things that have been appearing in the paper and things that have been said in general, and there was never one positive, pro-launch statement ever made by anybody. There have been some feelings since then that folks have expressed that they would support the decision, but there was not one positive statement for launch ever made in that room.

McDonald's Testimony

Mr. McDonald: I arrived at the KSC at about 8:15 p.m., and when I arrived there others who had already arrived were Larry Mulloy, who was there—he is the manager, the project manager for the SRB for Marshall. Stan Reinartz was there and he is the manager of the Shuttle Project Office. He's Larry Mulloy's boss. Cecil Houston was there, the resident manager for Marshall. And Jack Buchanan was there. He happens to be our manager, Morton Thiokol's manager of our launch support services office at Kennedy. The telecon hadn't started yet. It came on the network shortly after I got there....
Chairman Rogers: Was it essentially a telephone conference or was there actually a network of pictures?
Mr. McDonald: It was a telephone conference....

But I will relay...what I heard at the conference as best I can. The teleconference started I guess close to 9:00 o'clock and, even though all the charts weren't there, we were told to begin and that MTI should take the lead and go through the charts that they had sent to both centers.

The charts were presented by the engineering people from Thiokol, in fact by the people who had made those particular charts. Some of them were typed, some of them were handwritten. And they discussed their concerns with the low temperatures relative to the possible effects on the O-rings, primarily the timing function to seal the O-rings.

They presented a history of some of the data that we had accumulated both in static test and in flight tests relative to temperatures and the performance of the O-rings, and reviewed the history of all of our erosion studies of the O-rings, in the field joints, any blow-by of the primary O-ring with soot or products of combustion or decomposition that we had noted, and the performance of the secondary O-rings.

And there was an exchange amongst the technical people on that data as to what it meant...But the real exchange never really came until the conclusions and recommendations came in.

At that point in time, our vice president, Mr. Bob Lund, presented those charts and he presented the charts on the conclusions and recommendations. And the bottom line was that the engineering people would not recommend a launch below 53°F. The basis for the recommendation was primarily our concern with the launch that had occurred about a year earlier, in January of 1985, I believe it was 51-C.

Mulloy's Testimony

Mr. Mulloy: That telecon was a little late starting. It was intended to be set up at 8:15... and the teleconference was begun at 8:45.

And Thiokol will then present to you today that data that they presented to us in that telecon. I will not do that. The bottom line of that, though, initially was that Thiokol engineering, Bob Lund, who is the vice president and director of engineering, who is here today, recommended that 51-L not be launched if the O-ring temperatures predicted at launch time would be lower than any previous launch and that was 53°.

Dr. Walker: May I ask a question? I wish you would distinguish between the predicted bulk temperatures and the O-ring temperatures. In fact, as I understand it, you really don't have any official O-ring temperature prediction in your models, and it seems that the assumption has been that the O-ring temperature is the same as the bulk temperature, which we know is not the case.

Mr. Mulloy: You will see, sir, in the Thiokol presentation today that that is not the case. This was a specific calculation of what the O-ring temperature was on the day of the January 1985 launch. It is not the bulk temperature of the propellant, nor is it the ambient temperature of the air.

It was Thiokol's calculation of what the lowest temperature an O-ring had seen in previous flights, and the engineering recommendation was that we should not move outside of that experience base.

I asked Joe Kilminster, who is the program manager for the booster program at Thiokol, what his recommendation was, because he is the gentleman that I get my recommendations from in the program office. He stated that, based on that engineering recommendation, that he could not recommend launch.

At that point I restated, as I have testified to, the rationale that was essentially documented in the 1982 Critical Items List that stated that the rationale had been that we were flying with a simplex joint seal. And you will see in the Thiokol presentation that the context of their presentation is that the primary ring, with the reduced temperatures and reduced resiliency, may not function as a primary seal and we would be relying on secondary.

And without getting into their rationale and getting ahead, the point, the bottom line, is that we were continuing—the assessment was, my assessment at that time was, that we would have an effective simplex seal, based upon the engineering data that Thiokol had presented, and that none of those engineering data seemed to change that basic rationale.

Stan Reinartz then asked George Hardy, the deputy director of science and engineering at Marshall, what his opinion was. George stated that he agreed that the engineering data did not seem to change this basic rationale, but also stated on the telecon that he certainly would not recommend launching if Thiokol did not.

At that time Joe Kilminster requested a 5 min off-net caucus, and that caucus lasted approximately 30 min.

STEP Case Study: Space Shuttle *Challenger*

The teleconference was recessed at approximately 10:30 p.m. The off-net caucus of Thiokol personnel started and continued for about 30 min at the Wasatch office. The major issues, according to the testimony of Jerry Mason, senior vice president for Wasatch Operations, were the effect of temperature upon the O-rings and the history of erosion of the O-rings:

Mr. Mason: Now, in the caucus we revisited all of our previous discussions, and the important things that came out of that was that, as we had recognized, we did have the possibility that the primary O-ring might be slower to move into the seating position and that was our concern, and that is what we had focused on originally.

The fact that we couldn't show direct correlation with the O-ring temperature was discussed, but we still felt that there was some concern about it being colder.

We then recognized that, if the primary did move more slowly, that we could get some blow-by and erosion on the primary. But we had pointed out to us in that caucus a point that had not come across clearly in our earlier discussions, and that is that we had run tests where we deliberately cut large pieces out of the O-rings to see what the threshold of sealing was, and we found we could go to 125 thousandths of a cut out of the O-ring and it would still seal.

About 10 engineers participated in the caucus, along with Mason, Kilminster, C. G. Wiggins (Vice President, Space Division), and Lund. Arnold Thompson and Boisjoly voiced very strong objections to launch, and the suggestion in their testimony was that Lund was also reluctant to launch:

Mr. Boisjoly: Okay, the caucus started by Mr. Mason stating a management decision was necessary. Those of us who opposed the launch continued to speak out, and I am specifically speaking of Mr. Thompson and myself because in my recollection he and I were the only ones that vigorously continued to oppose the launch. And we were attempting to go back and rereview and try to make clear what we were trying to get across, and we couldn't understand why it was going to be reversed. So we spoke out and tried to explain once again the effects of low temperature. Arnie actually got up from his position which was down the table, and walked up the table and put a quarter pad down in front of the table, in front of the management folks, and tried to sketch out once again what his concern was with the joint, and when he realized he wasn't getting through, he just stopped.

I tried one more time with the photos. I grabbed the photos, and I went up and discussed the photos once again and tried to make the point that it was my opinion from actual observations that temperature was indeed a discriminator and we should not ignore the physical evidence that we had observed. And again, I brought up the point that SRM-15 (Flight 51-C, January, 1985) had a 110° arc of black grease while SRM-22 (Flight 61-A, October, 1985) had a relatively different amount, which was less and wasn't quite as black. I also stopped when it was apparent that I couldn't get anybody to listen.

Dr. Walker: At this point did anyone else speak up in favor of the launch?

Mr. Boisjoly: No, sir. No one said anything, in my recollection, nobody said a word. It was then being discussed amongst the management folks. After Arnie and I had our last say, Mr. Mason said we have to make a management decision. He turned to Bob Lund and asked him to take off his engineering hat and put on his management hat. From this point on, management formulated the points to base their decision on. There was never one comment in favor, as I have said, of launching by any engineer or other nonmanagement person in the room before or after the caucus. I was not even asked to participate in giving any input to the final decision charts.

I went back on the net with the final charts or final chart, which was the rationale for launching, and that was presented by Mr. Kilminster. It was hand written on a notepad, and he read from that notepad. I did not agree with some of the statements that were being made to support the decision. I was never asked nor polled, and it was clearly a management decision from that point.

I must emphasize, I had my say, and I never (would) take (away) any management right to take the input of an engineer and then make a decision based upon that input, and I truly believe that. I have worked at a lot of companies, and that has been done from time to time, and I truly believe that, and so there was no point in me doing anything any further than I had already attempted to do. I did not see the final version of the chart until the next day. I just heard it read. I left the room feeling badly defeated, but I felt I really did all I could to stop the launch. I felt personally that management was under a lot of pressure to launch and that they made a very tough decision, but I didn't agree with it.

One of my colleagues who was in the meeting summed it up best. This was a meeting where the determination was to launch, and it was up to us to prove beyond a shadow of a doubt that it was not safe to do so. This is in total reverse to what the position usually is in a preflight conversation or a flight readiness review. It is usually exactly opposite that.

Dr. Walker: Do you know the source of the pressure on management that you alluded to?

Mr. Boisjoly: Well, the comments made over the net is what I felt, I can't speak for them, but I felt it—I felt the tone of the meeting exactly as I summed up, that we were being put in a position to prove that we should not launch rather than being put in the position and prove that we had enough data to launch. And I felt that very real.

Dr. Walker: These were the comments from the NASA people at Marshall and at KSC?

Mr. Boisjoly: Yes.

Dr. Feynman: I take it you were trying to find proof that the seal would fail?

Mr. Boisjoly: Yes.

Dr. Feynman: And of course, you didn't, you couldn't because five of them didn't, and if you had proved that they would have all failed, you would have found yourself incorrect because five of them didn't fail.

Mr. Boisjoly: That is right. I was very concerned that the cold temperatures would change that timing and put us in another regime, and that was the whole basis of my fighting that night.

After the discussion between MTI management and the engineers, a final management review was conducted by Mason, Lund, Kilminster, and Wiggins. Lund and Mason recall this review as an unemotional, rational discussion of the engineering facts as they knew them at that time.

Mr. Lund: We tried to have the telecon, as I remember it was about 6:00 o'clock (MST), but we didn't quite get things in order, and we started transmitting charts down to Marshall around 6:00 or 6:30 (MST), something like that, and we were making charts in real time and seeing the data, and we were discussing them with the Marshall folks who went along.

We finally got the—all the charts in, and when we got all the charts in I stood at the board and tried to draw the conclusions that we had out of the charts that had been presented, and we came up with a conclusions chart and said that we didn't feel like it was a wise thing to fly.

Question: What were some of the conclusions?

Mr. Lund: I had better look at the chart. Well, we were concerned the temperature was going to be lower than the 50 or the 53 that had flown the previous January, and we had experienced some blow-by, and so we were concerned about that, and although the erosion on the O-rings, and it wasn't critical, that, you know, there had obviously been some little puff go through. It had been caught. There was no real extensive erosion of that O-ring, so it wasn't a major concern, but we said, gee, you know, we just don't know how much further we can go below the 51° or 53° or whatever it was. So we were concerned with the unknown. And we presented that to Marshall, and that rationale was rejected. They said that they didn't accept that rationale, and they would like us to consider some other thoughts that they had had.

... Mr. Mulloy said he did not accept that, and Mr. Hardy said he was appalled that we would make such a recommendation. And that made me ponder of what I'd missed, and so we said, what did we miss, and Mr. Mulloy said, well, I would like you to consider these other thoughts that we have had down here. And he presented a very strong and forthright rationale of what they thought was going on in that joint and how they thought that the thing was happening, and they said, we'd like you to consider that when they had some thoughts that we had not considered.... So after the discussion with Mr. Mulloy, and he presented that, we said, well, let's ponder that a little bit, so we went off-line to talk about what we—

Question: Who requested to go off-line?

Mr. Lund: I guess it was Joe Kilminster.... And so we went off-line on the telecon.... So we could have a roundtable discussion here.

Question: Who were the management people that were there?

Mr. Lund: Jerry Mason, Cal. Wiggins, Joe, I, manager of engineering design, the manager of applied mechanics. On the chart.

Before the Commission on February 25, 1986, Mr. Lund testified as follows regarding why he changed his position on launching *Challenger* during the management caucus when he was asked by Mr. Mason "To take off his engineering hat and put on his management hat:"

Chairman Rogers: How do you explain the fact that you seemed to change your mind when you changed your hat?

Mr. Lund: I guess we have got to go back a little further in the conversation than that. We have dealt with Marshall for a long time and have always been in the position of defending our position to make sure that we were ready to fly, and I guess I didn't realize until after that meeting and after several days that we had absolutely changed our position from what we had been before. But that evening I guess I had never had those kinds of things come from the people at Marshall. We had to prove to them that we weren't ready, and so we got ourselves in the thought process that we were trying to find some way to prove to them it wouldn't work, and we were unable to do that. We couldn't prove absolutely that that motor wouldn't work.

Chairman Rogers: In other words, you honestly believed that you had a duty to prove that it would not work?

Mr. Lund: Well, that is kind of the mode we got ourselves into that evening. It seems like we have always been in the opposite mode. I should have detected that, but I did not, but the roles kind of switched...

Supplemental testimony of Mr. Mason obtained in a Commission staff interview is as follows:

Question: Do you recall Mr. Hardy and Mr. Mulloy's comments after—I think after Mr. Kilminster had got done, or Mr. Lund got done presenting the charts? They had some comments. Do you recall—

Mr. Mason: Oh, yes, it was over and over. Hardy said that, "I'm appalled at your recommendation."...

Question: Well, did Mr. Hardy's "appalled" remark and Mr. Mulloy's "can't launch, we won't be able to launch until April" remark, how did that affect your thinking and affect your decision?

Mr. Mason: My personal thinking, I just, you know, it didn't make that much difference.... And the comments that they made, in my view, probably had got more reaction from the engineer(s) at the lower level than they would from the manager(s), because we deal with people, and managers all the time....

Mr. McDonald indicated that during the period of the internal MTI caucus he continued to argue for delay with Mulloy, challenging, among other things, the rationale that the rocket motor was qualified down to 40°F. Present were Reinartz, Jack Buchanan, the manager of MTI Launch Support Services at Kennedy and Cecil Houston. McDonald's testimony described that conversation:

STEP Case Discussion Question:

At this point, what differences of opinion would you expect between managers and engineers about the decision to launch? Which functional area would normally be more in favor of launching? Why?

Mr. McDonald: ... while they were off-line, reevaluating or reassessing this data... I got into a dialogue with the NASA people about such things as qualification and launch commit criteria (LCC). The comment I made was it is my understanding that the motor was supposedly qualified to 40° to 90°. I've only been on the program less than 3 years, but I don't believe it was. I don't believe that all of those systems, elements, and subsystems were qualified to that temperature.

And Mr. Mulloy said well, 40° is propellant mean bulk temperature, and we're well within that. This is a requirement. We're at 55° for that, and that the other elements can be below that... that, as long as we don't fall out of the propellant mean bulk temperature. I told him I thought that was asinine because you could expose that large SRM to extremely low temperatures—I don't care if it's 100 below zero for several hours—with that massive amount of propellant, which is a great insulator, and not change that propellant mean bulk temperature but only a few degrees, and I don't think the spec really meant that.

But that was my interpretation because I had been working quite a bit on the filament wound case SRM. It was my impression that the qualification temperature was 40°–90°, and I knew everything wasn't qualified to that temperature, in my opinion. But we were trying to qualify that case itself at 40°–90° for the filament wound case. I then said I may be naive about what generates launch commit criteria, but it was my impression that launch commit criteria was based upon whatever the lowest temperature, or whatever loads, or whatever environment was imposed on any element or subsystem of the Shuttle. And if you are operating outside of those, no matter which one it was, then you had violated some launch commit criteria.

This was my impression of what that was. And I still didn't understand how NASA could accept a recommendation of fly below 40°. I could see why they took issue with the 53, but I could never see why they would have accepted a recommendation below 40°, even though I didn't agree that the motor was fully qualified to 40. I made the statement that if we're wrong and something goes wrong on this flight, I wouldn't want to have to be the person to stand up in front of board of inquiry and say that I went ahead and told them to go ahead and fly this thing outside what the motor was qualified to. I made that very statement.

Mr. Mulloy's recollections of these discussions are as follows:

Mr. Mulloy: Mr. Kilminster then requested an off-net caucus. It has been suggested, implied, or stated that we directed Thiokol to go reconsider these data. That is not true. Thiokol asked for a caucus so that they could consider the discussions that had ensued and the comments that Mr. Hardy and I and others had made. This caucus, as has been stated, was going to start at that point, and Mr. McDonald interjected into the teleconference. At that point, he made the first comment that he had made during this entire teleconference.

Mr. McDonald testified for quite a while yesterday about his thoughts on this, but he did not say any of them until this point. At that point, he stated that he thought what George Hardy said was a very important consideration, and that consideration was, and he asked Mr. Kilminster to be sure and consider the comment made by George Hardy during the course of the discussions, that the concerns expressed were for primary O-ring blow-by and that the secondary O-ring was in a position to seal during the time of blow-by and would do so before significant joint rotation had occurred. They then went into their caucus, having asked for 5 min—

Mr. Hotz: ... It figures quite prominently in the discussion that you were quoted as saying, do you expect us to wait till April to launch?

Mr. Mulloy: Yes, sir.

Dr. Walker: Is that an accurate statement or not?

Mr. Mulloy: It is certainly a statement that is out of context, and the way I read the quote, sir—and I have seen it many times, too many times—the quote I read was: My God, Thiokol, when do you want me to launch, next April? Mr. McDonald testified to another quote that says: You guys are generating new Launch Commit Criteria. Now, both of those I think kind of go together, and that is what I was saying. I don't know whether that occurred during the caucus or subsequent to. I just simply can't remember that.

Mr. Hotz: Well, never mind the timing.

Mr. Mulloy: Well, yes, sir. I'm going to answer your question now. I think those quotes derive from a single thought that may have been expressed by me using some of those words. I have not yet encountered anyone other than those at KSC who heard those words, so I don't believe they were transmitted over the net. The total context I think in which those words may have been used is, there are currently no Launch Commit Criteria (LCC) for joint temperature. What you are proposing to do is to generate a new LCC on the eve of launch, after we have successfully flown with the existing LCC 24 previous times. With this LCC, i.e., do not launch with a temperature greater [sic] than 53°, we may not be able to launch until next April. We need to consider this carefully before we jump to any conclusions. It is all in the context, again, with challenging your interpretation of the data, what does it mean and is it logical, is it truly logical that we really have a system that has to be 53° to fly?

At approximately 11 p.m., the Thiokol/NASA teleconference resumed, the Thiokol management stating that they had reassessed the problem, that the temperature effects were a concern, but that the data were admittedly inconclusive. Kilminster read the rationale recommending launch and stated that that was Morton Thiokol's recommendation. Hardy requested that it be sent in writing by telefax both to Kennedy and to Marshall, and it was seat. The testimony of Mulloy and Hardy regarding the remainder of the teleconference and their rationale for recommending launch follows:

Mr. Mulloy: Okay, sir. At the completion of the caucus, of course, Mr. Kilminster came back on the loop and stated they had assessed all the data and considered the discussions that had ensued for the past couple of hours and the discussions that occurred during their caucus.

Chairman Rogers: Was it a couple of hours?

Mr. Mulloy: Yes, sir. We started at 8:45 and I believe it was probably 11:00 o'clock before he came back on the loop. It was a long discussion. And I must emphasize that I had no knowledge of what interchange occurred during the caucus at Thiokol, because all sites were on mute. We were on mute at KSC. No communications occurred between myself and Mr. Hardy at Huntsville, nor did any communication occur between KSC and Thiokol during that caucus. After Mr. Kilminster made that recommendation, Mr. Reinartz then asked if there were any further comments, and to my recollection there were none. There were no further comments made.

I then asked Mr. Kilminster to send me a copy of his flight readiness rationale and recommendation. The conference was then terminated at approximately 11:15. I have no knowledge of, as has been testified, of Mr. McDonald being asked to sign that documentation. That would have been unusual, because Mr. Kilminster signs all flight readiness documentation.

Now, after the teleconference was complete, Mr. McDonald informed Mr. Reinartz and me that if the Thiokol engineering concern for the effect of cold was not sufficient cause to recommend not launching, there were two other considerations, launch pad ice and recovery area weather. I stated that launch pad ice had been considered by the Mission Management Team—

Chairman Rogers: Excuse me. Could you identify that discussion where that took place?

Mr. Mulloy: That was after the teleconference was completed, after Mr. Kilminster made his recommendation, after Mr. Reinartz asked "are there any other comments?" There were no other comments on the telecon from anyone.... I stated that launch pad ice had been considered by the Mission Management Team before deciding to proceed and that a further periodic monitoring of that condition was planned. I further stated that I had been made aware of the recovery area weather previously and planned to place a call to Mr. Aldrich and advise him that the weather in the recovery area exceeded the LCC. As I stated earlier, when you asked what were the LCC, one of them was that the recovery area weather has limitations on it. The report we had, that Mr. McDonald confirmed, was that we were outside of those limits. Now, I must point out that that is not a hard LCC. That is an advisory call, and the LCC so states that. It does require that we discuss the condition.

So at about 11:30 p.m., Mr. Cecil Houston established a teleconference with Mr. Aldrich and Mr. Sestile at KSC. I informed Mr. Aldrich that the weather in the recovery area could preclude immediate recovery of the SRBs, since the ships were in a survival mode and they were moving back toward Cape Kennedy at about three knots, and the estimate provided to us by Mr. Sestile was that they would be probably 40 miles from the SRB impact area at the time of launch, at 9:38 and then, continuing at three knots, it was going to be some period of time before they could get back and locate the boosters.

The concern I had for that was not loss of the total booster, but loss of the main parachutes for the booster, which are separated at water impact, and loss of the frustum of the boosters, which has the drogue parachute on it, which comes down separately, because with the 50 knot winds we had out there and with the kind of sea states we had, by the time the recovery ships got back out there, there was little probability of being able to recover those. I informed Mr. Aldrich of that, and he decided to proceed with the launch after that information. I did not discuss with Mr. Aldrich the conversations that we had just completed with MTI.

Chairman Rogers: Could you explain why?

Mr. Mulloy: Yes, sir. At the time, and I still consider today, that was a level III issue, Level III being an SRB element or an external tank element or space shuttle main engine element or an Orbiter. There was no violation of LCC. There was no waiver required in my judgment at that time and still today. And we work many problems at the Orbiter and the SRB and the External Tank level that never get communicated to Mr. Aldrich or Mr. Moore. It was clearly a Level III issue that had been resolved.

... There were 27 full-scale seal tests with an O-ring groove damage tolerances, damage in the grooves, and damage tolerance on O-rings. And then there were two cold gas tests. And these data were presented on the night of the 27th. All of that was at ambient temperature. And then we did discuss what is a development qualification motor experience range, and that is shown on the chart. We had experience everywhere from 40° to 85°. There then were data presented on two cold gas tests at 30°, where the O-ring was pressurized at the motor pressurization rate at 30°, which would indicate that an O-ring would operate before joint rotation at 30°.

Dr. Ride: Was that actually in a joint?

Mr. Mulloy: No, it is not. It is a full-scale O-ring, full-scale groove, in a scaled test device, where the pressurize rate on that O-ring is zero to 900 psi (pounds per square inch) in 600 ms at a temperature of 30°.

Dr. Walker: You would say, then, the O-ring was qualified to a temperature of 30°? Would that be an accurate statement?

Mr. Mulloy: The day that we were looking at it, on the 27th, these two tests that we did indicated that it would perform at 30° under the motor pressurization rate before the joint rotated.

Dr. Walker: What about, let's consider the putty and the O-ring, because that is really the system that responds to the pressure surge. What temperature was the putty/O-ring system qualified to?

Mr. Mulloy: The lowest that I'm aware of—and we're still flushing this out, because this is kind of what we talked about on the 27th, but the lowest that I'm aware of is the 40° test on one of the development motors.

Dr. Walker: And, of course, during those tests the putty was modified before the test. The putty was not just laid up and then the seal made. The putty was then smoothed out or some attempt was made to remove the volcanoes, I think.

Mr. Mulloy: Because the horizontal assembly caused that.

Now, there's one other significant point on this chart that we did discuss, that we didn't have the quantities on on the 27th, and I mentioned this earlier. We have 150 case segment proof tests, with a large number of joints with a simulation of a cold O-ring. That is the 90 durometer with a .275, and that was at about 35°.

So those are the certification data that we kind of discussed, all of which we didn't discuss. The two cold gas tests we did, the segment proof test we did, the development and qualification motor test we did, as a basis for understanding what we could expect to happen at colder temperatures on the joints.

Mr. Hardy testified as follows:

Mr. Hardy: At the teleconference on the evening of January 27, 1986, Thiokol engineering personnel in Utah reviewed charts that had been datafaxed to Huntsville and KSC participants just prior to the beginning of the conference. Now, I am not going to repeat a lot of what you have already heard, but I will give you some of my views on the whole matter.

The presentations were professional in nature. There were numerous questions and answers. There was a discussion on various data and points raised by individuals at Thiokol or at Marshall or at Kennedy. I think it was a rather full discussion. There were some 14 charts presented, and as has been mentioned earlier, we spent about two, two and a half hours reviewing this. To my knowledge, anyone who desired to make a point, as a question or express a view was in no way restrained from doing so.

As others have mentioned, I have heard this particular teleconference characterized as a heated discussion. I acknowledge that there were penetrating questions that were asked, I think, from both, from all people involved. There were various points of view and an interpretation of the data that was exchanged. The discussion was not, in my view, uncharacteristic of discussions on many flight readiness issues on many previous occasions. Thiokol engineering concluded their presentation with recommendation that the launch time be determined consistent with flight experience to date, and that is the launch with the O-ring temperatures at or greater than 53° F.

Mr. Kilminster at Thiokol stated ... to the best of my recollection, that with that engineering assessment, he recommended we not launch on Tuesday morning as scheduled. After some short discussion, Mr. Mulloy at KSC summarized his assessment of the data and his rationale with that data, and I think he has testified to that.

Mr. Reinartz, who was at KSC, asked me for comment, and I stated I was somewhat appalled, and that was referring specifically to some of the data or the interpretation of some of the data that Thiokol had presented with respect to its influence on the joint seal performance relative to the issue under discussion, which specifically was the possibility that the primary seal may take longer to actuate and therefore to blow by the primary seal. The blow-by of the primary seal may be longer, and I am going to elaborate on that a little further in this statement.

Then I went on to say that I supported the assessment of data presented essentially as summarized by Mr. Mulloy, but I would not recommend launch over Thiokol's objections. Somewhere about this time, Mr. Kilminster at Utah stated that he wanted to go off the loop to caucus for about 5 min. I believe at this point Mr. McDonald, the senior Thiokol representative at KSC for this launch suggested to Mr. Kilminster that he consider a point that I think I had made earlier, that the secondary O-ring is in the proper position to seal if blow-by of the primary O-ring occurred.

I clearly interpreted this as a somewhat positive statement of supporting rationale for launch ... The status of the caucus by Thiokol lasted some 30, 35 min. At Huntsville during this Thiokol caucus, we continued to discuss the data presented. We were off the loop, we were on mute. We were around a table in small groups. It was not an organized type discussion. But I did take that opportunity to discuss my assessment and understanding of the data with several of my key advisors, and none of us had any disagreement or differences in our interpretation of what we believe the data were telling us with regard to the primary issue at hand.

When Thiokol came back on line, Mr. Kilminster reviewed rationale that supported proceeding with the launch and so recommended Mr. Reinartz asked if anyone in the loop had a different position or disagreed or something to that effect, with the Thiokol recommendation as presented by Mr. Kilminster. There were no dissenting responses. The telecon was terminated shortly after, and I have no knowledge of any subsequent events or discussions between personnel at KSC or at Thiokol on this matter.

At about 5:00 a.m., on January 28, a discussion took place among Messrs. Mulloy, Lucas, and Reinartz in which Mulloy reported to Lucas only that there had been a discussion with Thiokol over their concerns about temperature effects on the O-rings, and that it had been resolved in favor of launch. The following testimony of Mr. Mulloy and Dr. Lucas recount that discussion:

General Kutyna: ... Larry, let me follow through on that, I am kind of aware of the launch decision process, and you said you made the decision at your level on this thing. If this were an airplane, an airliner, and I just had a 2 h argument with Boeing on whether the wing was going to fall off or not, I think I would tell the pilot, at least mention it. Why didn't we escalate a decision of this importance?
Mr. Mulloy: I did, sir.
General Kutyna: You did?
Mr. Mulloy: Yes, sir.
General Kutyna: Tell me what levels above you.
Mr. Mulloy: As I stated earlier, Mr. Reinartz, who is my manager, was at the meeting, and on the morning, about 5:00 o'clock in the operations support room where we all were, I informed Dr. Lucas of the content of the discussion.
General Kutyna: But this is not in the launch decision chain.
Mr. Mulloy: No, sir. Mr. Reinartz is in the launch decision chain, though.
General Kutyna: And is he the highest level in that chain?

Mr. Mulloy: No. Normally it would go from me to Mr. Reinartz to Mr. Aldrich to Mr. Moore.

STEP Case Discussion Question:

Obviously, a few people are attempting to get the right information to the right people, but there are some trying to obstruct the flow of information. What implication does this have on project integration?

Dr. Lucas' testimony is as follows:

Chairman Rogers: Would you please tell the Commission when you first heard about the problem of the O-rings and the seals insofar as it involves launch 51-L? And I don't want you to go way back, but go back to when you first heard. I guess it was on January 27th, was it?

Dr. Lucas: Yes, sir. It was on the early evening of the 27th, I think about 7:00 p.m., when I was in my motel room along with Mr. Kingsbury. And about that time, Mr. Reinartz and Mr. Mulloy came to my room and told me that they had heard that some members of Thiokol had raised a concern about the performance of the SRBs in the low temperature that was anticipated for the next day, specifically on the seals, and that they were going out to the KSC to engage in a telecon with the appropriate engineers back at Marshall Space Flight Center in Huntsville and with corresponding people back at the Wasatch division of Thiokol in Utah. And we discussed it a few moments and I said, fine, keep me informed, let me know what happens.

Chairman Rogers: And when was the next time you heard something about that?

Dr. Lucas: The next time was about 5:00 a.m., on the following morning, when I went to the KSC and went to the launch control center. I immediately saw Mr. Reinartz and Mr. Mulloy and asked them how the matter of the previous evening was dispositioned.

Chairman Rogers: You had heard nothing at all in between?

Dr. Lucas: No, sir.

Chairman Rogers: So from 8:00 o'clock that evening until 5:00 o'clock in the morning, you had not heard a thing?

Dr. Lucas: It was about 7:00, I believe, sir. But for that period of time, I heard nothing in the interim...

Chairman Rogers: ... And you heard Mr. Reinartz say he didn't think he had to notify you, or did he notify you?

Dr. Lucas: He told me, as I testified, when I went into the control room, that an issue had been resolved, that there were some people at Thiokol who had a concern about the weather, that that had been discussed very thoroughly by the Thiokol people and by the Marshall Space Flight Center people, and it had been concluded agreeably that there was no problem, that he had a recommendation by Thiokol to launch and our most knowledgeable people and engineering talent agreed with that. So from my perspective, I didn't have—I didn't see that as an issue.

Chairman Rogers: And if you had known that Thiokol engineers almost to a man opposed the flight, would that have changed your view?

Dr. Lucas: I'm certain that it would.

Chairman Rogers: So your testimony is the same as Mr. Hardy's. Had he known, he would not have recommended the flight be launched on that day.

Dr. Lucas: I didn't make a recommendation one way or the other. But had I known that, I would have then interposed an objection, yes.

Chairman Rogers: I gather you didn't tell Mr. Aldrich or Mr. Moore what Mr. Reinartz had told you?

Dr. Lucas: No, sir. That is not the reporting channel. Mr. Reinartz reports directly to Mr. Aldrich. In a sense, Mr. Reinartz informs me as the institutional manager of the progress that he is making in implementing his program, but that I have never on any occasion reported to Mr. Aldrich.

Chairman Rogers: And you had subsequent conversations with Mr. Moore and Mr. Aldrich prior to the flight and you never mentioned what Mr. Reinartz had told you?

Dr. Lucas: I did not mention what Mr. Reinartz told me, because Mr. Reinartz had indicated to me there was not an issue, that we had a unanimous position between Thiokol and the Marshall Space Flight Center, and there was no issue in his judgment, nor in mine as he explained it to me.

Chairman Rogers: But had you known, your attitude would have been totally different?

Dr. Lucas: Had I had the advantages at that time of the testimony that I have heard here this week, I would have had a different attitude, certainly.

Chairman Rogers: In view of the fact that you were running tests to improve the joint, didn't the fact that the weather was so bad and Reinartz had told you about the questions that had been raised by Thiokol, at least, didn't that cause you serious concern?

Dr. Lucas: I would have been concerned if Thiokol had come in and said, we don't think you should launch because we've got bad weather.

Chairman Rogers: Well, that's what they did, of course, first. That is exactly what they did. You didn't know that?

Dr. Lucas: I knew only that Thiokol had raised a concern.

Chairman Rogers: "Did you know they came and recommended against the launch," is the question.

Dr. Lucas: I knew that I was told on the morning of the launch that the initial position of some members of Thiokol—and I don't know who it was—had recommended that one not launch with the temperature less than 53°F.

Chairman Rogers: And that didn't cause you enough concern so you passed that information on to either Mr. Moore or Mr. Aldrich?

Dr. Lucas: No, sir, because I was shown a document signed by Mr. Kilminster that indicated that that would not be significant, that the temperature would not be—that it would be that much lower, as I recall it.

It is clear that crucial information about the O-ring damage in prior flights and about the Thiokol engineers' argument with the NASA telecon participants never reached Jesse Moore or Arnold Aldrich, the Level I and II program officials, or J. A. (Gene)

STEP Case Study: Space Shuttle *Challenger*

Thomas, the launch director for 51-L. The testimony of Aldrich describes this failure of the communication system very aptly:

Dr. Feynman: ... have you collected your thoughts yet on what you think is the cause—I wouldn't call it of the accident but the lack of communication which we have seen and which everybody is worried about from one level to another? ...

Mr. Aldrich: Well, there were two specific breakdowns at least, in my impression, about that situation. One is the situation that occurred the night before the launch and leading up to the launch where there was a significant review that has been characterized in a number of ways before the Commission and the Commission's Subpanels and the fact that that was not passed forward.

And I can only conclude what has been reported, and that is, that the people responsible for that work in the SRB project at Marshall believed that the concern was not of a significance that would be required to be brought forward because clearly the program requirements specify that critical problems should be brought forward to Level II and not only to Level II but through myself to Level I.

The second breakdown in communications, however, and one that I personally am concerned about is the situation of the variety of reviews that were conducted last summer between the NASA Headquarters Organization and the Marshall Organization on the same technical area and the fact that that was not brought through my office in either direction—that is, it was not worked through—by the NASA Headquarters Organization nor when the Marshall Organization brought these concerns to be reported were we involved.

And I believe that is a critical breakdown in process and I think it is also against the documented reporting channels that the program is supposed to operate to. Now, it in fact did occur in that matter. In fact, there is a third area of concern to me in the way the program has operated. There is yet one other way that could have come to me, given a different program structure. I'm sure you've had it reported to you as it has been reported to me that in August or I think or at least at some time late in the summer or early fall the Marshall SRB project went forward to procure some additional SRM casings to be machined and new configurations for testing of the joints.

Now it turns out that the budget for that kind of work does not come through my Level II office. It is worked directly between the Marshall Center in NASA headquarters and there again had I been responsible for the budget for that sort of work, it would have to come through me, and it would have been clear that something was going on here that I ought to know about. And so there are three areas of breakdown, and I haven't exactly answered your question. But I have explained it in the way that I best know it and—well, I can say a fourth thing.

There was some discussion earlier about the amount of material that was or was not reported on O-ring erosion in the FRRs and I researched the FRR back reports and also the flight anomaly reports that were forwarded to my center, to my office, by the SRB project and as was indicated, there is a treatment of the SRM O-ring erosion, I believe, for the STS 41-C FRR, which quantifies it and indicates some limited amount of concern.

The next time that is mentioned, I believe it is the STS 51-E, FRR in January 1985 or early in February, and that indicates, again, a reference to it but refers back to the 41-C as the only technical data. And then from there forward the comment on O-ring erosion only is that there was another instance and it is not of concern. Clearly the amount of reporting in the FRR is of concern to me, but in parallel with that, each of the flight anomalies in the STS program are required to be logged and reviewed by each of the projects and then submitted through the Level II system for formal close-out.

And in looking back and reviewing the anomaly close-outs that were submitted to Level II from the SRB project, you find that O-ring erosion was not considered to be an anomaly and, therefore, it was not logged and, therefore, there are not anomaly reports that progress from one flight to the other. Yes, that is another way that that information could have flagged the system, and the system is set up to use that technique for flagging. But if the erosion is classified as not an anomaly, it then is in some other category and the system did not force it in that direction. None of those are very focused answers, but they were all factors.

Chairman Rogers asked four key officials about their knowledge of the Thiokol objections to launch:

Chairman Rogers: ... By way of a question, could I ask, did any of your gentlemen prior to launch know about the objections of Thiokol to the launch?
Mr. Smith (KSC Director): I did not.
Mr. Thomas (Launch Director): No, sir.
Mr. Aldrich (Shuttle Program Director): I did not.
Mr. Moore (Associate Administrator for Space Flight): I did not.

Additionally, in further testimony J. A. (Gene) Thomas commented on the launch.

Mr. Hotz: ... Mr. Thomas, you are familiar with the testimony that this Commission has taken in the last several days on the relationship of temperature to the seals in the SRB?
Mr. Thomas: Yes, sir, I have been here all week.
Mr. Hotz: Is this the type of information that you feel that you should have as launch director to make a launch decision?
Mr. Thomas: If you refer to the fact that the temperature according to the Launch Commit Criteria should have been 53°, as has been testified, rather than 31, yes, I expect that to be in the LCC. That is a controlling document that we use in most cases to make a decision for launch.
Mr. Hotz: But you are not really very happy about not having had this information before the launch?
Mr. Thomas: No, sir. I can assure you that if we had had that information, we wouldn't have launched if it hadn't been 53°.

STEP Case Study: Space Shuttle *Challenger* 315

PRESIDENTIAL COMMISSION FINDINGS

The following are the findings or conclusions of the Presidential Commission based on the testimony heard.

1. There was a serious flaw in the decision-making process leading up to the launch of flight 51-L. A well-structured and managed system emphasizing safety would have flagged the rising doubts about the SRB joint seal. Had these matters been clearly stated and emphasized in the flight readiness process in terms reflecting the views of most of the Thiokol engineers and at least some of the Marshall engineers, it seems likely that the launch of 51-L might not have occurred when it did.
2. The waiving of launch constraints appears to have been at the expense of flight safety. There was no system which made it imperative that launch constraints and waivers of launch constraints be considered by all levels of management.
3. There was a propensity of management at Marshall to contain potentially serious problems and to attempt to resolve them internally rather than communicate them forward. This tendency is altogether at odds with the need for Marshall to function as part of a system working toward successful flight missions, interfacing and communicating with the other parts of the system that work to the same end.
4. The Presidential Commission concluded that the Thiokol Management reversed its position and recommended the launch of 51-L, at the urging of Marshall and contrary to the views of its engineers in order to accommodate a major customer.

STEP Case Discussion Questions:

Evaluate your own operating environment and see if you can find situations that closely identify with the four problem areas listed above. How could each problem be resolved? Should a communication matrix have been developed, monitored, and enforced by NASA for the shuttle mission?

MANAGEMENT DECISION AMBIGUITIES

Another source of problem for the *Challenger* was the unusually cold weather on the night before the launch. Reaction control system heaters on the Orbiter were activated and the SRB recovery batteries were checked and found to be functioning within specifications. There were no serious concerns regarding the External Tank. The freeze protection plan for the launch pad was implemented, but the results were not what had been anticipated. The freeze protection plan usually involves completely draining the water system. However, this was not possible because of the imminent launch of 51-L. In order to prevent pipes from freezing, a decision was made to allow water to run slowly from the system. This had never been done before, and the combination of freezing temperatures and stiff winds caused large amounts of ice to form below the

240 foot level of the fixed service structure including the access to the crew emergency egress slide wire baskets. Ice also was forming in the water trays beneath the vehicle.

These conditions were first identified by the Ice Team at approximately 2:00 a.m., on January 28 and were assessed by management and engineering throughout the night, culminating with a Mission Management Team meeting at 9:00 a.m. At this meeting, representatives for the Orbiter prime contractor, Rockwell International, expressed their concern about what effects the ice might have on the Orbiter during launch. Rockwell had been alerted about the icing conditions during the early morning and was working on the problem at its Downey, California, facility.

During Commission hearings, the president of Rockwell's Space Transportation Systems Division, Dr. Rocco Petrone, and two of his vice presidents, Robert Glaysher and Martin Cioffoletti, all described the work done regarding the ice conditions and the Rockwell position at the 9:00 a.m., meeting with regard to launch. Dr. Petrone had arrived at Kennedy on Friday, January 24. On Monday the 27th he left to return to Rockwell's facility in California, but Glaysher and Cioffoletti remained at Kennedy. Dr. Petrone testified that he first heard about the ice at 4:00 a.m., Pacific Standard Time (PST). He explained what followed:

I had gotten up and went to the support room to support this launch. We have people monitoring consoles, and I checked in, and they told me there was a concern, and when I arrived at about 4:30, I was informed we were working the problem with our aerodynamicist and debris people, but very importantly, we would have to make an input to Kennedy for a meeting scheduled at 6:00 o'clock our time and 9:00 o'clock Florida time.

We had approximately an hour of work to bring together. The work had been underway when I arrived and was continuing.

At that time I got on the phone with my Orbiter program managers just to discuss the background of where we were, how things stood, and what their concerns were locally. They described what they knew in Florida, and we also in Downey did television input, and we could see some of the ice scenes that were shown here this morning.

We arrived through a series of meetings to a top level discussion at approximately 5:30 PST, from which we drew the following conclusions: Ice on the mobile launcher itself, it could be debris. We were very concerned with debris of any kind at the time of launch. With this particular ice, one, could it hit the Orbiter? There was wind blowing from the west. That appeared not to be so, that it wouldn't hit the Orbiter but would land on the mobile launcher. The second concern was what happens to that ice at the time you light your liquid fuel engines, the SSMEs, and would it throw it around and ricochet and potentially hit the Orbiter.

The third aspect is the one that has been discussed here of aspiration, what would happen when the large SRM ignite and in effect suck in air, referred to as aspiration, and ice additionally would come down, how much unknown.

The prime thing we were concerned about was the unknown base line. We had not launched in conditions of that nature, and we just felt we had an unknown.

I then called my program managers over in Florida at 5:45 (PST) and said we could not recommend launching from here, from what we see. We think the tiles would be endangered, and we had a very short conversation. We had a meeting to go through, and I said let's make sure that NASA understands that Rockwell feels it is not safe to launch, and that was the end of my conversation.

STEP Case Study: Space Shuttle *Challenger*

Mr. Glaysher, who was at Kennedy, came to the center at approximately 7:45 a.m., EST. He conferred with Rockwell's chief engineer as well as the vice president of engineering, Dr. John Peller, at Rockwell's Downey Plant. At 9:00 a.m., after the ice debris team had reported back from the pad inspection, Glaysher was asked for Rockwell's position on launch. He discussed aspiration effects, the possible ricochet of ice from the fixed service structure, and what the ice resting on the mobile launch platform would do at ignition. Glaysher said he told the Mission Management Team when it met at 9:00 a.m., that the ice was an unknown condition, and Rockwell was unable to predict where the ice would go or the degree of potential damage to the Orbiter thermal protection system (TPS) if it were struck by the ice. He testified that his recommendation to NASA was

> My exact quote—and it comes in two parts. The first one was, Rockwell could not 100% assure that it is safe to fly which I quickly changed to Rockwell cannot assure that it is safety to fly...

Rockwell's other vice president at Kennedy, Martin Cioffoletti, described the concern about ice in a slightly different manner:

Mr. Cioffoletti: Similarly, I was called in and told about the problem and came into the 6:00 o'clock meeting which you heard about a few minutes ago, and at the conclusion of that meeting I spoke with Mr. Dick Kohrs, the deputy program manager from Johnson Space Flight Center, and he asked if we could get the Downey folks to look at the falling ice and how I might reverse toward the vehicle, and also, did we have any information on aspiration effects.

So I did call back to Downey and got the John Peller folks working on that problem, and they did, as you saw from Charlie Stevenson's sketches, predict that the ice would travel only about halfway to the vehicle, free-falling ice carried by the winds. So we felt that ice was not a problem. However, it would land on the mobile launch platform that we considered a problem. We also investigated the aspiration database we had, and we had seen the aspiration effect on previous launches where things were pulled into the SRB hole after ignition, but we had never seen anything out as far as the fixed surface tower. So we felt in fact it was an unknown. We did not have the database to operate from an aspiration effect.

At the 9:00 o'clock meeting, I was asked by Arnie Aldrich, the program manger, to give him the results of our analysis, and I essentially told him what I just told you and felt that we did not have a sufficient database to absolutely assure that nothing would strike the vehicle, and so we could not lend our 100% credence, if you will, to the fact that it was safe to fly ... I said I could not predict the trajectory that the ice on the mobile launch platform would take at SRB ignition.

Chairman Rogers: But I think NASA's position probably would be that they thought that you were satisfied with the launch. Did you convey to them in a way that they were able to understand that you were not approving the launch from your standpoint?

Mr. Cioffoletti: I felt that by telling them we did not have a sufficient database and could not analyze the trajectory of the ice, I felt he understood that Rockwell was not giving a positive indication that we were for the launch.

After Cioffoletti's testimony at the Commission hearings, Dr. Petrone was pressed for a more detailed description of Rockwell's launch recommendation:

General Kutyna: Dr. Petrone, you've got a lot more experience than I have in this business, but the few launch conferences that I have been on the question is very simple. Are you go or are you no-go for launch and "maybe" isn't an answer. I hear all kinds of qualifications and cautions and considerations here.
     Did someone ask you are you go or no-go? Was that not asked?
Dr. Petrone: At this particular meeting, as far as—and I was not in Florida, and so I cannot answer that. It had been done at earlier meetings. This was a technical evaluation of a series of problems, and we talked about debris hitting the TPS and the tiles, and the long series of reviews that we had done that morning and all led us to a conclusion that they were not safe to fly. And we transmitted that to program managers along with the technical evaluation quickly of why we had arrived at that. So much of it is how the question gets raised because earlier we had aspiration work, ricochet work, a number of things which we did, and then we came up with our recommendation.
Chairman Rogers: And your recommendation now you say it was, it was unsafe to fly?
Dr. Petrone: Correct, sir.

Two things are apparent from the Rockwell testimony. First, Rockwell did not feel it had sufficient time to research and resolve the ice on the pad problem. Second, even though there was considerable discussion about ice, Rockwell's position on launch described above was not clearly communicated to NASA officials in the launch decision chain during the hours preceding 51-L's launch.

At a meeting with Commission investigators on March 4, 1986, at Kennedy, HoraceLamberth, NASA director of Shuttle Engineering, said he did not interpret Rockwell's position at the 9:00 a.m. Mission Management Team meeting on January 28 as being "no-go." Lamberth said the language used by Rockwell was "we can't give you 100% assurance" but there was no feeling in his mind that Rockwell was voicing a no-go recommendation. "It just didn't come across as the normal Rockwell no-go safety of flight issues come across." This conclusion is confirmed in part by an interview of Dr. John Peller, Rockwell's vice president of engineering, who was assigned the ice problem early Tuesday morning. Dr. Peller, in describing a telephone conversation with the Johnson Director of Engineering, Tom Moser, stated

Dr. Peller: That was a call from Tom Moser to me, in which he asked again to understand my concerns. And I just repeated the same concerns. And he asked, "Did I think that it was likely that the vehicle would take safety critical damage?"
     And I said, "From the possibility that the vehicle would take safety critical damage," I said, "there's a probability in a sense that it was probably an unlikely event, but I could not prove that it wouldn't happen..."

STEP Case Study: Space Shuttle *Challenger*   319

> ... I never used the words "no-go" for launch. I did use the words that we cannot prove it is safe. And normally that's what we were asked to do. We were unable to do that in this particular case, although it was a strange case, that we normally don't get involved in.

Arnold Aldrich, NASA Mission Management team leader, described NASA's view of the ice situation and his recollection of Rockwell's position. He said that on Tuesday morning the mission management team did a detailed analysis of the ice on the fixed service structure. Representatives from the ice team, Rockwell, and the directors of engineering (Horace Lamberth) and the Orbiter project (Richard Colonna) all considered the problem. Aldrich reported the discussion as follows:

> Following the discussion of the acceptability of the ice threat to the Orbiter, based upon the conditions described in detail of the fixed service structure—and more of that you've seen here portrayed well this morning—I asked the NASA managers involved for their position on what they felt about the threat of that to the Orbiter.
>
> Mr. Lamberth reported that KSC engineering had calculated the trajectories, as you've heard, of the falling ice from the fixed service structure east side, with current 10-knot winds at 300°, and predicted that none of this ice would contact the Orbiter during its ignition or launch sequence and that their calculations even showed that if the winds would increase to 15 knots, we still would not have contact with the Orbiter.
>
> Mr. Colonna, Orbiter project manager, reported that similar calculations had been performed in Houston by the mission evaluation team there. They concurred in this assessment. And further, Mr. Colonna stated that, even if these calculations were significantly in error, that it was their belief that falling ice from the fixed service structure, if it were in fact to make its way to the Orbiter, it would only be the most lightweight ice that was in that falling stream, and it would impact the Orbiter at a very oblique angle.
>
> Impacts of this type would have very low probability of causing any serious damage to the Orbiter, and at most would result in postflight turnaround repairs.
>
> At this point I placed a phone call to Mr. Moser that I had previously mentioned, director of Engineering at the JSC, who was in the mission evaluation room, and he confirmed the detailed agreement with Mr. Lamberth's and Mr. Colonna's position....
>
> And both Mr. Lamberth and Mr. Colonna reported that their assessment was that the time it took for the ice to fall, to hit the Orbiter and to rebound, and the location of the fixed service structure on the MLP (mobile launch platform) would not cause that ice in their view to be a concern to rebound and come up and impact the rear end of the Orbiter.
>
> Following these discussions, I asked for a position regarding proceeding with the launch. Mr. Colonna, Mr. Lamberth, and Mr. Moser all recommended that we proceed.
>
> At that time, I also polled Mr. Robert Glaysher, the vice president, Orbiter project manager, Rockwell International STS Division, and Mr. Marty Cioffoletti, Shuttle Integration Project Manager, Rockwell International STS Division. Mr. Glaysher stated–and he had been listening to this entire discussion and had not been directly involved with it, but had been party to this the whole time.
>
> His statement to me as best I can reconstruct it to report to you at this time was that, while he did not disagree with the analysis that JSC and KSC has reported, that they would not give an unqualified go for launch as ice on the launch complex was a condition which had not previously been experienced, and thus this posed a small additional, but unquantifiable, risk. Mr. Glaysher did not ask or insist that we not launch, however.

At the conclusion of the above review, I felt reasonably confident that the launch should proceed.

In addition to Rockwell's input, Mr. Aldrich also had reports from other contractors and the ice, frost, and debris team at the 9:00 session. Ice on the vehicle assembly appeared to be of no concern; sheet ice in the noise suppression trays had been broken up and removed; as previously noted the ice team reported that there was ice on the fixed service structure between 95 ft above ground and 215 ft; no ice above 255 ft. The north and west sides had large amounts of ice and icicles. The final assessment was made that the ice on the fixed service structure would not strike or damage the Orbiter tiles or the vehicle assembly during ignition or ascent, owing to the considerable horizontal distance between the service structure and the vehicle assembly. The decision was made to launch pending a final ice team review of the launch complex in order to assess any changes in the situation. This inspection was completed following the Mission Management Team meeting and the ice team report indicated no significant change.

ADDITIONAL COMMISSION FINDINGS

The testimonies on the ice problem were reviewed by the commission and their findings are as presented below:

1. An analysis of all of the testimony and interviews establishes that Rockwell's recommendation on launch was ambiguous. The commission finds it difficult, as did Mr. Aldrich, to conclude that there was a no-launch recommendation. Moreover, all parties were asked specifically to contact Aldrich or Moore about launch objectives due to weather. Rockwell made no phone calls or further objections to Aldrich or other NASA officials after the 9:00 Mission Management Team meeting and subsequent to the resumption of the countdown.
2. The commission is also concerned about the NASA response to the Rockwell position at the 9:00 a.m., meeting. While it is understood that decisions have to be made in launching a Shuttle, the commission is not convinced Levels I and II appropriately considered Rockwell's concern about the ice. However, ambiguous Rockwell's position was, it is clear that they did tell NASA that the ice was an unknown condition. Given the extent of the ice on the pad, the admitted unknown effect of the SRM and Space Shuttle Main Engines ignition on the ice, as well as the fact that debris striking the Orbiter was a potential flight safety hazard, the commission finds the decision to launch questionable under those circumstances. In this situation, NASA appeared to be requiring a contractor to prove that it was not safe to launch, rather than proving it was safe. Nevertheless, the commission has determined that the ice was not a cause of the 51-L accident and does not conclude that NASA's decision to launch specifically overrode a no-launch recommendation by an element contractor.
3. The Commission concluded that the freeze protection plan for launch pad 39B was inadequate. The Commission believes that the severe cold and presence of so much ice on the fixed service structure made it inadvisable

to launch on the morning of January 28, and that margins of safety were whittled down too far.

Additionally, access to the crew emergency slide wire baskets was hazardous due to ice conditions. Had the crew been required to evacuate the orbiter on the launch pad, they would have been running on an icy surface. The Commission believes the crew should have been made aware of the situation and based on the seriousness of the condition; greater consideration should have been given to delaying the launch.

STEP Case Discussion Questions:

1. As mentioned above, it seems launch officials wanted a contractor to furnish a proof for "no-go" rather than a proof for "go." Does this qualify as "management by crisis?"
2. The Triple C principle (Chapter 5) recommends an integrated approach to project Communication, Cooperation, and Coordination. How could the Triple C be instituted for a complex project in order to avoid the problems presented in this case study?

POSTINVESTIGATION DEVELOPMENTS

Some of the developments following the *Challenger* accident and investigations include

1. The NASA official, who gave final approval to launch *Challenger*, Jesse W. Moore, resigned from NASA.
2. It was announced in June 1986 that a redesign of the booster rocket joints would incorporate full redundancy so that a single failure would not jeopardize future space flights. The new joint would be rated as a "criticality-1R" item. The "R" in the rating implies that there is a redundant capability.
3. The Reagan administration, in August 1986, ordered NASA to get out of the commercial launching business. This marked the termination of a two-decade struggle by NASA to run a space shuttle service that would pay for itself. The order was based on the belief that space services could be provided better and cheaper by the private sector.
4. Critics called for a new supplier of shuttle booster rockets. It was revealed in August 1986 that MTI Corporation skipped three of seven mandatory government safety inspections of O-ring seals in *Challenger's* right-hand solid booster rocket.
5. In October 1986, an independent panel of scientific and engineering experts endorsed NASA's decision to continue testing its redesigned shuttle booster rockets in the same horizontal position as before the *Challenger* accident. The panel was organized by the National Research Council at the recommendation of the Presidential Commission. The Commission had urged NASA to consider vertical tests in order to duplicate actual launch configuration as closely as possible.

6. Robert Crippen, the astronaut assigned the job of revamping shuttle management, recommended that NASA should learn to analyze the full history and trend of spaceship problems. He said the short-term troubleshooting by NASA "failed to put together the whole story" of the booster rocket.
7. Several of the families of the astronauts who died in the *Challenger* accident filed lawsuits against NASA and the shuttle contractors. In particular, the family of Michael J. Smith (pilot) filed a $15 million negligence claim on July 2, 1986. The families of Judith A. Resnik (mission specialist) and Ronald E. McNair (mission specialist) also filed independent lawsuits.
8. In December 1986, The Justice Department announced that it has reached confidential out-of-court settlements with the families of four of the seven astronauts who died in the accident. The families were those of Christa McAuliffe (school teacher), Francis R. Scobee (mission commander), Ellison S. Onizuka (mission specialist), and Gregory B. Jarvis (payload specialist).
9. By January 1987, about $1.2 million had been donated to a fund established for the education, health, and support of the 11 children of the *Challenger* crew.
10. It was revealed for the first time in August, 1986 that internal NASA memoranda had warned that the O-ring seals could leak and cause a catastrophe. This information had not been provided to the Presidential Commission that investigated the *Challenger* accident.
11. In October 1986, a congressional committee says NASA may not have the technical and scientific expertise to conduct the space shuttle program. The House Science and Technology Committee promised to conduct an in-depth review of NASA's technical ability during the 1987 fiscal year.
12. NASA officials announced in January 1987 that the agency is creating a nap plan for launch managers so that they could get adequate rest during the hectic pace of shuttle flight preparations.
13. In May 1987, retired astronaut Walter Schirra recommended an agenda for the space program. He called for a space program manned by high-tech astronauts dedicated to no-frills missions. He urged NASA to delay and totally redesign the next space shuttle.
14. On January 10, 1987, NASA announced the selection of the next *Challenger* crew consisting of previous space fliers. It was the first time that a space shuttle crew would be comprised entirely of astronauts who had flown in space before.

SPACE SHUTTLE *COLUMBIA* DISASTER

Another space shuttle accident occurred on February 1, 2003. Space Shuttle *Columbia*, on STS-107 mission, disintegrated over Texas during re-entry into the Earth's atmosphere. All seven crew members perished. The shuttle was on its 28th mission. The explosive accident occurred shortly before the Shuttle was to land.

The loss of *Columbia* was a result of damage sustained during launch when a piece of foam insulation, the size of a small briefcase, broke off from the space

STEP Case Study: Space Shuttle *Challenger* 323

shuttle external tank, which is the main propellant tank. The break-off was due to the high aerodynamic forces of the launch process. The debris struck the leading edge of the left wing, damaging the shuttle's TPS. While *Columbia* was still in orbit, some engineers suspected damage, but NASA managers limited the investigation on the grounds that little could be done even if problems were found.

NASA's shuttle safety regulations stated that external tank foam shedding and subsequent debris strikes upon the shuttle itself were safety issues that needed to be resolved before a launch was cleared, but launches were often given the go-ahead as engineers unsuccessfully studied the foam shedding problem. The majority of shuttle launches recorded such foam strikes and thermal tile scarring in violation of safety regulations. During re-entry of STS-107, the damaged area allowed the hot gases to penetrate and destroy the internal wing structure of the shuttle. This eventually caused the in-flight breakup of the vehicle. A massive ground search in parts of Texas, Louisiana, and Arkansas recovered crew remains and many shuttle fragments.

TECHNICAL AND ORGANIZATIONAL ISSUES

In what appears to be reminiscent of the Space Shuttle *Challenger* Accident, the *Columbia* Accident Investigation Board's recommendations addressed both technical and organizational issues. The Space Shuttle program was set back over 2 years by the disaster, a delay comparable only to that resulting from the *Challenger* disaster. Concurrently, construction of the International Space Station was put on hold, and the station relied entirely on the Russian Federal Space Agency for resupply and crew rotation.

SHUTTLE FLIGHT RISK MANAGEMENT

Investigations concluded that, in a risk-management scenario similar to the *Challenger* disaster, NASA management failed to recognize the relevance of engineering concerns for safety. Two examples were failures to honor engineers' requests for imaging to inspect possible damage, and failure to respond to engineers' requests about the status of astronaut inspection of the left wing. Engineering made three separate requests for Department of Defense (DOD) imaging of the shuttle in orbit to more precisely determine damage. While the images were not guaranteed to show the damage, the capability existed for imaging of sufficient resolution to provide meaningful examination. In fact, the investigation Board recommended that subsequent shuttle flights be imaged while in orbit using ground-based or space-based DOD technologies. NASA management did not honor the requests and in some cases intervened to stop the DOD from assisting.

NASA's chief engineer for the TPS was concerned about left wing damage and asked NASA management whether an astronaut would visually inspect it. NASA managers never responded.

Throughout the risk assessment process, senior NASA managers were influenced by their belief that nothing could be done even if damage was detected. This affected their stance on investigation urgency, thoroughness, and possible contingency actions. They decided to conduct a parametric "what-if" scenario study more suited to determine risk probabilities of future events, instead of inspecting and assessing the actual damage. The investigation report in particular singled out NASA manager Linda Ham for exhibiting this attitude.

Much of the risk assessment hinged on damage predictions to the TPS. These fall into two categories: damage to the silica tile on the wing lower surface and damage to the reinforced carbon–carbon (RCC) leading-edge panels.

Damage-prediction software, known as "Crater," was used to evaluate possible tile and RCC damage. The software predicted severe penetration of multiple tiles by the impact, but engineers downplayed this, believing that results showing that the software overstated damage from small projectiles meant that the same would be true of larger Spray-On Foam Insulation (SOFI) impacts. The program used to predict RCC damage was based on small ice impacts the size of cigarette butts, not larger SOFI impacts. With several prediction trials for SOFI paths of impact, the software predicted an ice impact would completely penetrate the RCC panel. Engineers downplayed this, also believing that impacts of the less-dense SOFI material would result in less damage than ice impacts. In an e-mail exchange, NASA managers questioned whether the density of the SOFI could be used as justification for reducing predicted damage. Despite engineering concerns about the energy imparted by the SOFI material, NASA managers ultimately accepted the rationale to reduce predicted damage of the RCC panels from complete penetration to slight damage to the panel's thin coating.

NASA managers felt a rescue or repair was impossible, so there was no point in trying to inspect the vehicle for damage while in orbit. However, the Board determined either a rescue mission or on-orbit repair, though risky, might have been possible had NASA verified the severity of the damage within 5 days into the mission.

Ultimately the NASA Mission Management Team felt there was insufficient evidence to indicate that the strike was an unsafe situation, so they declared the debris strike a "turnaround" issue, which was deemed not of the highest importance.

CONCLUDING REMARKS

As in the case of Space Shuttle *Challenger*, the same managerial laxity seemed to have occurred in the Space Shuttle *Columbia* accident. Are we really learning from project failures? In many respects, yes we do learn from project failures. But what often happens is that we become complacent about knowing what went wrong and how to fix it that we really don't implement structural project management to preempt a repeat. It is hoped that a guiding step-by-step approach, as advocated by the STEP methodology of this book, will enforce structure and order in integrating all aspects of a complex project, such as space missions. Technically based accidents can occur; but we should not fuel it with managerial ineptitude.

Many other developments had surfaced since the two shuttle accidents, some positive and some negative. Of course, as of 2008, the shuttle program is still running and is expected to continue until 2010. NASA is going strong and doing better and better with each mission. New science, technology, and engineering developments are helping the organization to respond to prevailing challenges while preparing for the future.

SATELLITE PROJECT FAILURE: ANOTHER STEP CASE EXAMPLE

Another recent case example of a large project failure is the demise of an ambitious satellite project (Taubman, 2007). The story, as recounted by Philip Taubman, cited the death of a spy satellite program that was doomed by an ineffective process. The case example shows gross flaws in the overall management process of the satellite program. Subsequent government investigations revealed the need for more rigorous processes for project cost estimation, systems engineering, and engineering management. This bears out the fact that highly technical projects often suffer not from the technical aspects, but from a lack of integrative project management. For example, the National Reconnaissance Office (NRO), which designs highly sophisticated spy satellites, has been faulted for managing its accounting books with surprisingly glum managerial skills. It somehow lost track of more than $2 billion in its own budget. This astonishing accounting snafu is attributed to lax management. A step-by-step STEP project management process can help to preempt similar failures.

DEATH OF A SPY SATELLITE

Reprinted from "In Death of Spy Satellite Program, Lofty Plans and Unrealistic Bids" by Philip Taubman, *The New York Times*, November 11, 2007 (Reprinted with permission, *The New York Times*).

According to Philip Taubman's investigative report, by May 2002, the government's effort to build a technologically audacious new generation of spy satellites was foundering. The contractor building the satellites, Boeing, was still giving Washington reassuring progress reports. But the program was threatening to outstrip its $5 billion budget and pivotal parts of the design seemed increasingly unworkable. Peter B. Teets, the new head of the nation's spy satellite agency, appointed a panel of experts to examine the secret project, telling them, according to one member, "Find out what's going on, find the terrible truth I suspect is out there."

The panel reported that the project, called Future Imagery Architecture (F.I.A), was far behind schedule and would most likely cost $2 billion to $3 billion more than planned, according to records from the satellite agency, the NRO.

Even so, the experts recommended pressing on. Just months after the September 11 terrorist attacks, and with the new satellites promising improved, more frequent images of foreign threats like terrorist training camps, nuclear weapons plants, and enemy military maneuvers, they advised Mr. Teets to seek an infusion of $700 million.

It took two more years, several more review panels and billions more dollars before the government finally killed the project—perhaps the most spectacular and expensive failure in the 50-year history of American spy satellite projects. The story behind that failure has remained largely hidden, like much of the workings of the nation's intelligence establishment.

But an investigation by *The New York Times* found that the collapse of the project, at a loss of at least $4 billion, was all but inevitable—the result of a troubled partnership between a government seeking to maintain the supremacy of its intelligence technology, but on a constrained budget, and a contractor all too willing to make promises it ultimately could not keep.

"The train wreck was predetermined on Day 1," said A. Thomas Young, a former aerospace executive who led a panel that examined the project.

The Future Imagery project is one of several satellite programs to break down in recent years, leaving the United States with outdated imaging technology. But perhaps more striking is that the multiple failures that led to the program's demise reveal weaknesses in the government's ability to manage complex contracts at a time when military and intelligence contracting is soaring.

The Times's examination found that the satellite agency put the Future Imagery contract out for bid in 1998 despite an internal assessment that questioned whether its lofty technological goals were attainable given the tight budget and schedule.

Boeing had never built the kind of spy satellites the government was seeking. Yet when Boeing said it could live within the stringent spending caps imposed by Congress and the satellite agency, the government accepted the company's optimistic projections, a Panglossian compact that set the stage for many of the travails that followed. Despite its relative inexperience, Boeing was given responsibility for monitoring its own work, under a new government policy of shifting control of big military projects to contractors. At the same time, the satellite agency, hobbled by budget cuts and the loss of seasoned staff members, lacked the expertise to make sound engineering evaluations of its own.

The satellites were loaded with intelligence collection requirements, as numerous intelligence and military services competed to influence their design. Boeing's initial design for the optical system that was the heart of one of the two new satellite systems was so elaborate that optical engineers working on the project said it could not be built. Engineers constructing a radar-imaging unit at the core of the other satellite could not initially produce the unusually strong radar signal that was planned.

A torrent of defective parts, like gyroscopes and electric cables, repeatedly stalled work. Even an elementary rule of spacecraft construction—never use tin because it deforms in space and can short-circuit electronic components—was violated by parts suppliers.

By the time the project, known by its initials, F.I.A., was killed in September 2005—a year after the first satellite was originally to have been delivered—cost estimates ran as high as $18 billion.

"The F.I.A. contract was technically flawed and unexecutable the day it was signed," said Robert J. Hermann, who ran the NRO from 1979 to 1981 and in 1996 led the panel that first recommended creation of a new satellite system. "Some top official should have thrown his badge on the table and screamed, 'We can't do this system at this price.' No one did."

Boeing's point man on the job was Ed Nowinski, an engineer who had become a top government spy satellite expert during 28 years at the Central Intelligence Agency (C.I.A.). "It was a perfect storm," Mr. Nowinski said ruefully. But he acknowledged that Boeing frequently provided the government with positive reports on the troubled project.

"Look, we did report problems," Mr. Nowinski said, "but it was certainly in my best interests to be very optimistic about what we could do."

Boeing, which fired Mr. Nowinski as the project fell apart, declined to comment. A spokeswoman, Diana Ball, said Boeing could not discuss classified programs.

The Times's examination was based on interviews with more than 30 government and industry officials involved with the project, many discussing it publicly for the first time. Some agreed to be interviewed on the condition that they not be identified because many aspects of the project remained classified. They said they were willing to talk because they hoped an airing of its history would help prevent similar misadventures in the future.

Asked about the recent problems with F.I.A. and other satellite programs, Senator Christopher S. Bond, Republican of Missouri and vice chairman of the Intelligence Committee, said, "It's fair to say we have lost double-digit billions on satellite programs that weren't effectively managed by the government."

This year, a stealth satellite program was killed by Mike McConnell, the director of national intelligence. Also, a new generation of infrared satellites for detecting missile launches has barely survived cost overruns and technical setbacks.

Taken together, these episodes represent a stark reversal for a satellite program born in the most perilous years of the cold war, when American technology answered the call of national defense by taking spying into space.

Today, space technology has lost its luster for young engineers, who are drawn increasingly to companies like Google and Apple. Defense experts say the entire acquisition system for space-based imagery technologies is in danger of breaking down. And the nation, at least for now, has been left without advanced new systems to replace a dwindling number of reconnaissance satellites first designed in the 1970s and updated in the 1990s.

Even though reconnaissance satellites are less useful in spying on terrorist groups than on more traditional threats like foreign military forces, they remain integral to intelligence and military operations, including monitoring nuclear and missile installations in Iran and North Korea. They are also critical to Pentagon mapmaking and the targeting of precision-guided weapons like cruise missiles.

"There is not a gap in the coverage we are providing, but our constellation is fragile," said Alden V. Munson Jr., deputy director of national intelligence for acquisition.

Since the F.I.A. debacle, the NRO has banned Boeing from bidding on new spy satellite contracts. But all the news was not bad for Boeing. The company received a $430 million kill fee for the optical satellite system. And, despite the ban, the radar-imaging satellite remained in Boeing's hands.

RESPONSE TO SOVIET THREAT

The first generation of photo reconnaissance satellites was developed in the waning months of the Eisenhower administration, in a frantic effort to measure the Soviet threat.

The satellite system, code-named Corona, was the product of an inspired partnership of government, science and industry. The C.I.A. set broad goals and then let the Lockheed Corporation, with help from the Air Force, figure out how to build the satellites, get them into orbit, and return the film canisters to earth without burning up as they plunged through the atmosphere.

In the mid-1970s, the same partnership developed systems that electronically captured and transmitted pictures moments after they were recorded. These electro-optical satellites were among the first devices to use the technology now common in digital cameras.

They were followed in the 1980s by radar-imaging satellites, which can see through clouds and operate in darkness, bouncing radar signals off the earth to plot terrain and paint images of objects on the ground.

By the 1990s, though, the threats to national security—and the world of satellite intelligence—were undergoing convulsive change.

Familiar targets like Soviet air bases and missile factories were being supplanted by the more varied and elusive threats of the post-cold-war world. At the same time, the armed services, eager for increased tactical intelligence after the 1991 Persian Gulf war, were demanding satellites that could stream battlefield data instantly to commanders around the globe.

In 1996, a commission created by the director of Central Intelligence recommended building a fleet of light, small, relatively inexpensive satellites that, according to a declassified version of the panel's report, could together be at least as effective as the Lockheed behemoths then in orbit. (They cost about $1 billion apiece, weighed 30,000 lb and were the size of a bus.)

Having more satellites in orbit, the theory went, would increase "revisit time," the number of times a day satellites pass above target sites. That would help combat increasing efforts to camouflage such sites.

Lighter satellites would require cheaper and less powerful rockets than the Titan IV's then in use, which could cost $450 million per launching. The panel also envisioned saving money and time by taking advantage of technologies and parts developed by commercial satellite companies.

But as the concept took shape, several powerful forces were bearing down, turning the satellite procurement system to quicksand, military experts said.

One was the new policy, cousin to the Clinton administration's effort to downsize government, of transferring control of big military projects to contractors, on the theory that they could best manage engineering work and control costs.

Another factor was a decline of American expertise in systems engineering, the science and art of managing complex engineering projects to weigh risks, gauge feasibility, test components, and ensure that the pieces come together smoothly.

Finally, troubled by the free-spending habits of the satellite agency, Congress demanded rigid spending guidelines for the satellite project.

The first concerns about the project's formula—high-concept technology on a fast schedule with a tightly managed budget—came from the satellite agency itself.

In early 1997, as the project began to move from conceptual thinking to concrete planning, the agency's acquisition board, which reviewed programs at an early stage, questioned the feasibility of the new approach, given the expected $5 billion budget cap for its first 5 years. As Dennis D. Fitzgerald, the agency's principal deputy director from August 2001 until last April, recalled, the board's

STEP Case Study: Space Shuttle *Challenger* 329

review "had the most reds and yellows"—agency parlance for cautionary notes—he had ever seen.

Even so, in January 1997, the agency invited military companies to a classified briefing about the project now called Future Imagery Architecture.

A COMPANY TRYING TO DIVERSIFY

Albert D. Wheelon, who founded the Directorate of Science and Technology at the C.I.A. in 1963 and played a leading role in the early development of spy satellites, said in an interview, "Writing winning proposals is different from building winning hardware."

This could be an apt epitaph for Boeing's handling of F.I.A.

Boeing, famous for making airplanes, had never built an electro-optical or radar-imaging spy satellite. But with the European Airbus consortium threatening its commercial airliner business, the company was trying to diversify.

By contrast, the other invited bidder, Lockheed, saw the contract almost as an entitlement, military and government officials said.

Lockheed all but owned the imagery-satellite franchise. Over four decades, as the company built successive generations of satellites, the government had, in effect, invested more than $30 billion in its operations. What is more, Lockheed had recently acquired the traditional builder of radar-imaging satellites, Martin Marietta (and with it a new name, Lockheed Martin).

"Lockheed believed it had this program in the bag," said Leslie Lewis, a military analyst who reviewed the project for a Rand Corporation study.

As Boeing mobilized, Ed Nowinski seemed the perfect man to pursue the prize.

Mr. Nowinski, 63, was familiar with the concept of smaller satellites from his years at the C.I.A. He had joined the agency in 1967 as an electrical engineer and worked on the first electro-optical systems. Eventually, he became the head of the agency's satellite development programs and of imagery operations at the NRO and received several medals for distinguished service.

Former coworkers describe Mr. Nowinski as a fine engineer and an easy colleague, an unassuming man who took pride in working on secret projects that enhanced American security. They also said he could be insufficiently demanding, a potential weakness for someone running a multibillion-dollar project. Mr. Nowinski did not contest the description in an interview.

His government career had ended abruptly in October 1995, when the C.I.A. fired him for using a government car for personal travel. Mr. Nowinski said he was trying to make the most efficient use of his time when he was swamped with work and had to travel frequently between his home and several government offices in the Washington area.

In 1998, a former C.I.A. colleague, Robert J. Kohler, invited Mr. Nowinski to help Boeing put together its satellite proposal. He was soon living in a rented apartment near the Boeing defense systems offices in Seal Beach, on the outskirts of Los Angeles, working 12 h a day, 7 days a week on Team 377, the company's secret planning group.

"I never imagined they would recompete the business," Mr. Nowinski said. "When Lockheed didn't call, Bob and I figured we'd go with the underdog."

Mr. Kohler recalled that Team 377 requested $100 million just to draft the proposal; he said Harry Stonecipher, Boeing's president at the time, gave his approval the next day. Before long, more than 300 engineers and other specialists were at work in Seal Beach.

If they looked like underdogs, they had history on their side. Mr. Fitzgerald, the reconnaissance office's former deputy director, said the government had traditionally found it hard to resist new bidders on space programs, with their allure of new ideas and lower costs. Indeed, of 18 government space programs reopened for competitive bidding between 1977 and 2002, all but two ended up changing hands, he said.

Mr. Fitzgerald explained the dynamic this way: "You as the incumbent are probably going to write a realistic proposal because you know what's involved and propose pretty much what you've been doing, since it has been successful. Your competitor, out of ignorance or guile, is going to write probably a more imaginative, creative proposal for which there is almost no backing."

He added, "It's a little like a divorce, and running off with another woman."

The leaders of Team 377 realized that the best hope of impressing the satellite agency was to design a system that was cheaper and better—more technologically daring—than anything Lockheed might propose. Having worked closely with Lockheed while at the C.I.A., Mr. Kohler said: "I knew what Lockheed Martin was going to do. We would do things 180° differently."

Multiple Design Challenges

Designing and building a precision-pointing, high-resolution electro-optical satellite—roughly the equivalent of the Hubble Space Telescope—requires melding many engineering disciplines.

The satellite must withstand the explosive force of being rocketed into orbit, then operate flawlessly for years in the unforgiving environment of space.

To position itself for picture taking, it requires delicately tuned attitude control and propulsion systems.

The electro-optical system presented an especially formidable challenge. The large, heavy satellites of the past had been effective at limiting the movement and vibrations that might mar picture taking, just as a tripod can eliminate blurred images with hand-held cameras.

"If you vibrate, you're looking at Jupiter," one satellite expert said.

Boeing, in effect, sought to replace the tripod with a system that would automatically adjust the image to compensate for any vibration, much as a camcorder does, but on a far grander, more exacting scale.

The team also wanted an optical system that could take wide-angle images, showing large areas on the ground, as well as tightly focused detailed pictures of small objects. The goal, to use an oversimplified analogy, was a revolutionary zoom lens.

As for the radar-imaging satellite, Boeing designed a relatively simple system with one major exception: to improve image quality, it would produce a far stronger radar signal than any previous satellite had.

Pulling off such complex new technology typically requires extensive testing and work on multiple solutions to especially difficult problems. There is no margin for error—once in orbit, a broken satellite cannot be easily fixed.

Yet the budget for F.I.A. was limited and not very elastic, unlike those for many earlier projects.

"Some programs are slightly underfunded, some are significantly underfunded," said Mr. Young, chairman of one of the panels that examined the project and a former Martin Marietta executive. "F.I.A. was grossly underfunded."

Congress had set a cap of $5 billion for the first 5 years, with spending limited to $1 billion a year. (It also budgeted $5 billion more for the life of the project, including multiple satellites.) While the prime contractor could seek additional financing for unanticipated costs, the contract would discourage overruns or delays with financial penalties.

Also, the satellite agency, under pressure from Congress to control costs, would no longer have a reserve fund. "From 1961 to 1995, the NRO had never delivered a program that I'm aware of on cost or on schedule," Mr. Fitzgerald said, adding, "But we always had this margin that would allow us to buy our way out of problems."

To underscore the importance of the budget cap, the agency changed its system for scoring contract bids. Previously, price had rarely accounted for more than 25% of a company's score. Now it would account for 50%.

As Boeing was putting the finishing touches on its proposal, Mr. Kohler said he warned the company that a $5 billion bid was unrealistic.

"I did a simple calculation," he recalled. "I took what it had cost to build a comparably complex system before, figured in inflation, and realized the project would cost $4 billion more than the government had planned and Boeing was proposing."

"I said, 'We can't submit that bid.'"

Mr. Nowinski rejects the idea that the bid was off base. "We were very meticulous in putting together the proposal," he said. Still, he acknowledged, "It's true there was little if any margin to work with."

Mr. Fitzgerald compared the bidding to liar's poker, a game based on the serial numbers on dollar bills that relies heavily on bluffing and gamesmanship.

"There's a lot of money on the table, and no one wants to say that they can't do it," he said. The ethic, he added, is "win the program at any cost and sort it out later. Correct the government's sins and my sins with overruns."

This time around, that would prove impossible.

Winning Bid Is Announced

The NRO announced its decision on September 3, 1999, after studying the bids for nearly a year. The top brass at Seal Beach gathered in shirt sleeves at 9 a.m., in the office of Roger Roberts, head of Boeing's satellite operations. Over the speakerphone, an agency official read a brief statement awarding both satellites to Boeing.

"The room was momentarily silent," Mr. Nowinski recalled. "We hadn't really expected to win the whole project. We figured we'd be lucky to get the radar system. I was stunned."

They threw open the door and informed a crowd of colleagues waiting in the outer office. The room erupted in cheers.

The final decision had been made by Keith R. Hall, who became the satellite agency's director in 1997 after serving as a senior intelligence official and deputy staff director of the Senate Intelligence Committee.

Mr. Hall, now a vice president of the consulting firm Booz Allen Hamilton, recalled in an interview that though both bids claimed to fall within the spending cap, an agency evaluation team had calculated that only Boeing's actually would. Its plan was also deemed the more technologically innovative.

Even a former Lockheed Martin executive vice president, Albert E. Smith, acknowledged that "Boeing wrote a better, cheaper proposal than we did."

The upshot, Mr. Hall said, was that there was really "no choice in the source selection." He added that he considered the Boeing proposal executable, if moderately risky.

The award announcement had barely been completed, though, when dissenting grenades started landing at the satellite agency.

Lockheed Martin, infuriated by the decision, filed a protest, which froze the project for several months as the agency reviewed its decision.

Eventually, Lockheed Martin withdrew its protest. Dennis R. Boxx, the company's senior vice president for corporate communications, said he could not comment on classified projects. But government and industry officials said the company stood down after the agency awarded it a consolation prize, a relatively small piece of the project.

Within a few months, two cost-estimating groups, one operated by the Pentagon, the other by the office that coordinates work among intelligence agencies, determined that the Boeing plan would bust the budget caps.

By then, Mr. Hall said, it was too late to reopen the bidding.

Nor did the cheering last long at Seal Beach. As Boeing moved from writing its proposal to building the hardware, assembling a workforce of thousands, outside engineers questioned the photo satellite's intricate optical system.

"There were a lot of bright young people involved in developing the concept, but they hadn't been involved in manufacturing sophisticated optical systems," said one military industry executive familiar with the project. "It soon became clear the system could not be built."

The design was eventually supplanted by a more conventional approach, partly to accommodate added intelligence collection requirements from Washington, Mr. Nowinski said.

Expectations about relying on the commercial satellite industry for parts and know-how proved wrong, since those companies curtailed production and laid off experienced technicians after the dot-com collapse.

Soon, defective parts began showing up in critical components, forcing costly delays at Boeing and some subcontractors.

"The No. 1 problem that killed us on this project was substandard parts," Mr. Nowinski said.

One of the electro-optical satellite's most important components—a set of oversize gyroscopes that help adjust the spacecraft's attitude for precision picture-taking—was flawed, said engineers involved in the project. The problem was traced to a subcontractor that had changed its manufacturing process for a crucial part, inadvertently producing a subtle but disabling alteration in the metallic structure that went undetected until Boeing discovered it, three years into the project.

Several kinds of integrated circuits for the electro-optical satellite also proved defective. Even rudimentary parts like electric cabling were unfit for use, several engineers said. Customized wiring did not conform to the orders and in some cases was contaminated by dirt.

As for the sister satellite, Mr. Nowinski said, "We thought the radar system would be a piece of cake."

But the plans were impeded by unexpected difficulties in increasing the strength of the radar signals that would be bounced off the earth. The problem, among other things, involved a vacuum-tube device called a traveling wave tube assembly. Perhaps most surprising was the appearance of parts containing tin, forbidden because it tends to sprout tiny irregularities, known as "tin whiskers," in space. One military industry executive said he was astounded when, several years into the project, he got a form letter from Boeing telling suppliers not to use tin.

"That told me there had been a total breakdown in discipline and systems engineering on the project," he said, "and that the company was operating on cruise control."

SIGNS OF A PROJECT IN TROUBLE

The tight schedule called for the radar-imaging satellite to be delivered in 2004 and its sister spacecraft the next year. Three years before that first deadline, government and industry officials say, it was becoming clear that the project was in trouble.

As costs escalated, Boeing cut back on testing and efforts to work several potential solutions to difficult technical problems. If a component failed, Boeing, lacking a backup approach, had to return to square one, forcing new delays.

Yet the company hesitated to report setbacks and ask for additional financing.

"When you've got a flawed program, or a flawed contract, you really have an obligation to go the customer and tell them," Mr. Young said. "Boeing wasn't doing that."

The reason, according to an internal reconnaissance office post-mortem was the budget cap, and the steep financial penalties for exceeding it. "The cost of an overrun was so ruinous that the strongest incentive it provided to the contractor was to prove they were on cost," the post-mortem found.

It did not help that the government ordered two major and several minor design changes that added $1 billion to cost projections. The changes, government and industry officials said, were intended to give the electro-optical satellite the flexibility to perform additional functions.

It was against this backdrop that Mr. Teets, the satellite agency's new director, formed the review group in May 2002 that recommended pressing on and seeking new financing.

The next year, the government ordered up another look at the project, as part of a broader examination of failing military space programs. The study, led by Mr. Young, reported that F.I.A. was "significantly underfunded and technically flawed" and "was not executable."

By this time, the government had approved an additional $3.6 billion. Still, rather than recommending cancellation, the Young panel said the program could be salvaged with even more financing and changes in the program and schedule.

In an interview, Mr. Young said the panel genuinely thought the project could be saved. Several members, though, said the group should have called for ending the program but stopped short because of its powerful supporters in Congress and the Bush administration. Among the most influential was Representative Jane Harman, the ranking Democrat on the House Intelligence Committee, whose Southern California district includes the Boeing complex where the satellites were being assembled.

The death sentence for F.I.A. was finally written in 2005. Another review board pronounced the program deeply flawed and said propping it up would require another $5 billion—raising the ante to $18 billion—and five more years. And even with that life support, Mr. Fitzgerald recalled, the panel was not confident that Boeing could come through.

That September, the director of National Intelligence, John D. Negroponte, killed the electro-optical program on the recommendation of the reconnaissance office's new director, Donald M. Kerr. Lockheed was engaged to reopen its production line and build an updated model of its old photo satellite.

Government officials say the delivery date for that model has slipped to 2009. Late last year, a Lockheed satellite carrying experimental imagery equipment failed to communicate with ground controllers after reaching orbit, rendering it useless.

Boeing calculated that its revenue losses from the cancellation would total about $1.7 billion for 2005 and 2006, less than 2% of forecast revenues. Having kept the radar-satellite contract, the company is expected to deliver the first one in 2008 or early 2009, at least 4 years behind the original schedule.

SEARCH FOR LESSONS

The satellite agency and military experts are still sifting through the wreckage, looking for lessons—beyond the budget issues—that would prevent a similar meltdown in the future.

In an interview in September, Mr. Kerr, who last month became the principal deputy director of National Intelligence, said a pivotal factor was selecting a company with no experience building imagery spy satellites, especially when contractors were being given greater responsibility for monitoring their own work. Boeing, he said, was "in a way exquisitely unprepared to exercise judgment in certain areas because it wasn't within their own experience."

The satellite office's oversight is faulted, too. Jimmie D. Hill, a former deputy director, said transferring management authority to military contractors was a morale killer for officers who worked on Air Force satellite projects, many of whom had been recruited to be midlevel managers at the NRO.

"Most of the best and the brightest young captains and majors said, 'To hell with it, there's nothing for me to do here, I'm going to go do something that's interesting,'" Mr. Hill said. "And so you have a void in capability right now."

There is wide agreement among military experts that F.I.A. sunk under a surfeit of data demands and technological risks.

"There's a good rule on projects like this," said Representative Heather A. Wilson, a New Mexico Republican on the Intelligence Committee. "Aim for only one miracle per program."

STEP Case Study: Space Shuttle *Challenger* 335

The government has taken remedial steps. While still at the satellite agency, Mr. Kerr said he was working to attract and keep experienced engineers and to improve cost-estimating and systems-engineering expertise. At his invitation, Virginia Tech University is offering a master's program in engineering management at agency headquarters outside Washington.

Mr. Munson, the deputy national intelligence director for acquisition, said competitive bidding for space programs would be initiated only among companies deemed qualified. And the intelligence office has formed an independent cost-estimating group to review project proposals and set budgets. "We are not going to start programs we can't afford," he said.

Keith Hall, the man who chose Boeing to build F.I.A., said the cost caps distorted the entire enterprise.

"If I had to do it over again, I should have decided at the time the cost cap was levied that we would just keep building what we had been building," he said, referring to the Lockheed satellites. "I shouldn't have allowed it to go forward."

In the dying days of F.I.A., Boeing fired Ed Nowinski. He returned to his retirement home in Florida, where he keeps a hand in the space business as a consultant.

He blames himself for some of the tribulations.

"You know, I might have been exactly the wrong guy for this project," he said. "After 25 to 30 years in the government, I think too much like a government guy. I was too sympathetic to the government, tried too hard to make their jobs easier."

He also faults himself for failing to assemble a stronger team at Boeing. "I should have been more brutal with the government and with my people," he said.

Mr. Nowinski remains convinced that with adequate time and money, Boeing could have built the electro-optical satellite. "We had solved most of the problems," he said.

But, he added, "When they say, 'We're turning the lights out, the game is over,' you might as well go home."

REFERENCES

Badiru, A, *Project Management in Manufacturing and High Technology Operations*, John Wiley & Sons, New York, 1996.

Taubman, P, "In death of Spy Satellite Program, lofty plans and unrealistic bids," *The New York Times*, November 11, 2007.

Appendix A: Engineering Code of Ethics*

Ethics have a lot to do with how STEPs are executed. Work ethics relate to how a person dedicates himself or herself to the task at hand. It is quite possible to have a highly qualified and competent individual with very low work ethics. In that case, a project many not be able to get much work out of the individual. The communication and cooperation processes of Triple C can help mitigate the adverse effects of low work ethics. This is done through statement of clear objectives, explanation of the importance of each person's role, assignment to team expectations, and empowerment to get things done. Ethical standards relate to an individual's sense of integrity and credibility. This infers a person's conscious and deliberate commitment to play by the rules. Even when rules have to be bent to get things done, it should be done within the sphere of honesty, honor, and justifiable reasoning. Figure A.1 shows work ethics and ethical standards as components of personal ethics. The code of ethics for engineers, as presented by the National Society of Professional Engineers (NSPE), is presented in this appendix.

PREAMBLE TO CODE OF ETHICS FOR ENGINEERS

Engineering is an important and learned profession. As members of this profession, engineers are expected to exhibit the highest standards of honesty and integrity. Engineering has a direct and vital impact on the quality of life for all people. Accordingly, the services provided by engineers require honesty, impartiality, fairness, and equity, and must be dedicated to the protection of the public health, safety, and welfare. Engineers must perform under a standard of professional behavior that requires adherence to the highest principles of ethical conduct.

FUNDAMENTAL CANONS

Engineers, in the fulfillment of their professional duties, shall

1. Hold paramount the safety, health, and welfare of the public
2. Perform services only in areas of their competence
3. Issue public statements only in an objective and truthful manner
4. Act for each employer or client as faithful agents or trustees
5. Avoid deceptive acts
6. Conduct themselves honorably, responsibly, ethically, and lawfully so as to enhance the honor, reputation, and usefulness of the profession

---

* Adapted from NSPE® Engineering Code of Ethics; NSPE® is a registered trademark of the National Society of Professional Engineers

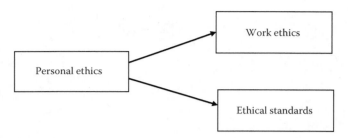

FIGURE A.1 Components of ethics.

Rules of Practice

1. Engineers shall hold paramount the safety, health, and welfare of the public.
    a. If engineers' judgment is overruled under circumstances that endanger life or property, they shall notify their employer or client and such other authority as may be appropriate.
    b. Engineers shall approve only those engineering documents that are in conformity with applicable standards.
    c. Engineers shall not reveal facts, data, or information without the prior consent of the client or employer except as authorized or required by law or this Code.
    d. Engineers shall not permit the use of their name or associate in business ventures with any person or firm that they believe is engaged in fraudulent or dishonest enterprise.
    e. Engineers shall not aid or abet the unlawful practice of engineering by a person or firm.
    f. Engineers having knowledge of any alleged violation of this Code shall report thereon to appropriate professional bodies and, when relevant, also to public authorities, and cooperate with the proper authorities in furnishing such information or assistance as may be required.
2. Engineers shall perform services only in the areas of their competence.
    a. Engineers shall undertake assignments only when qualified by education or experience in the specific technical fields involved.
    b. Engineers shall not affix their signatures to any plans or documents dealing with subject matter in which they lack competence, nor to any plan or document not prepared under their direction and control.
    c. Engineers may accept assignments and assume responsibility for coordination of an entire project and sign and seal the engineering documents for the entire project, provided that each technical segment is signed and sealed only by the qualified engineers who prepared the segment.
3. Engineers shall issue public statements only in an objective and truthful manner.

Appendix A

    a. Engineers shall be objective and truthful in professional reports, statements, or testimony. They shall include all relevant and pertinent information in such reports, statements, or testimony, which should bear the date indicating when it was current.
    b. Engineers may express publicly technical opinions that are founded upon knowledge of the facts and competence in the subject matter.
    c. Engineers shall issue no statements, criticisms, or arguments on technical matters that are inspired or paid for by interested parties, unless they have prefaced their comments by explicitly identifying the interested parties on whose behalf they are speaking and by revealing the existence of any interest the engineers may have in the matters.

4. Engineers shall act for each employer or client as faithful agents or trustees.
    a. Engineers shall disclose all known or potential conflicts of interest that could influence or appear to influence their judgment or the quality of their services.
    b. Engineers shall not accept compensation, financial or otherwise, from more than one party for services on the same project, or for services pertaining to the same project, unless the circumstances are fully disclosed and agreed to by all interested parties.
    c. Engineers shall not solicit or accept financial or other valuable consideration, directly or indirectly, from outside agents in connection with the work for which they are responsible.
    d. Engineers in public service as members, advisors, or employees of a governmental or quasi-governmental body or department shall not participate in decisions with respect to services solicited or provided by them or their organizations in private or public engineering practice.
    e. Engineers shall not solicit or accept a contract from a governmental body on which a principal or officer of their organization serves as a member.

5. Engineers shall avoid deceptive acts.
    a. Engineers shall not falsify their qualifications or permit misrepresentation of their or their associates' qualifications. They shall not misrepresent or exaggerate their responsibility in or for the subject matter of prior assignments. Brochures or other presentations incident to the solicitation of employment shall not misrepresent pertinent facts concerning employers, employees, associates, joint venturers, or past accomplishments.
    b. Engineers shall not offer, give, solicit, or receive, either directly or indirectly, any contribution to influence the award of a contract by public authority, or which may be reasonably construed by the public as having the effect or intent of influencing the awarding of a contract. They shall not offer any gift or other valuable consideration in order to secure work. They shall not pay a commission, percentage, or brokerage fee in order to secure work, except to a bona fide employee or bona fide established commercial or marketing agencies retained by them.

PROFESSIONAL OBLIGATIONS

1. Engineers shall be guided in all their relations by the highest standards of honesty and integrity.
   a. Engineers shall acknowledge their errors and shall not distort or alter the facts.
   b. Engineers shall advise their clients or employers when they believe a project will not be successful.
   c. Engineers shall not accept outside employment to the detriment of their regular work or interest. Before accepting any outside engineering employment, they will notify their employers.
   d. Engineers shall not attempt to attract an engineer from another employer by false or misleading pretenses.
   e. Engineers shall not promote their own interest at the expense of the dignity and integrity of the profession.
2. Engineers shall at all times strive to serve the public interest.
   a. Engineers are encouraged to participate in civic affairs, career guidance for youths, and work for the advancement of the safety, health, and well-being of their community.
   b. Engineers shall not complete, sign, or seal plans and specifications that are not in conformity with applicable engineering standards. If the client or employer insists on such unprofessional conduct, they shall notify the proper authorities and withdraw from further service on the project.
   c. Engineers are encouraged to extend public knowledge and appreciation of engineering and its achievements.
   d. Engineers are encouraged to adhere to the principles of sustainable development* in order to protect the environment for future generations.
3. Engineers shall avoid all conduct or practice that deceives the public.
   a. Engineers shall avoid the use of statements containing a material misrepresentation of fact or omitting a material fact.
   b. Consistent with the foregoing, engineers may advertise for recruitment of personnel.
   c. Consistent with the foregoing, engineers may prepare articles for the lay or technical press, but such articles shall not imply credit to the author for work performed by others.
4. Engineers shall not disclose, without consent, confidential information concerning the business affairs or technical processes of any present or former client or employer, or public body on which they serve.
   a. Engineers shall not, without the consent of all interested parties, promote or arrange for new employment or practice in connection with a specific project for which the engineer has gained particular and specialized knowledge.

---

* "Sustainable development" is the challenge of meeting human needs for natural resources, industrial products, energy, food, transportation, shelter, and effective waste management while conserving and protecting environmental quality and the natural resource base essential for future development.

b. Engineers shall not, without the consent of all interested parties, participate in or represent an adversary interest in connection with a specific project or proceeding in which the engineer has gained particular specialized knowledge on behalf of a former client or employer.
5. Engineers shall not be influenced in their professional duties by conflicting interests.
   a. Engineers shall not accept financial or other considerations, including free engineering designs, from material or equipment suppliers for specifying their product.
   b. Engineers shall not accept commissions or allowances, directly or indirectly, from contractors or other parties dealing with clients or employers of the engineer in connection with work for which the engineer is responsible.
6. Engineers shall not attempt to obtain employment or advancement or professional engagements by untruthfully criticizing other engineers, or by other improper or questionable methods.
   a. Engineers shall not request, propose, or accept a commission on a contingent basis under circumstances in which their judgment may be compromised.
   b. Engineers in salaried positions shall accept part-time engineering work only to the extent consistent with policies of the employer and in accordance with ethical considerations.
   c. Engineers shall not, without consent, use equipment, supplies, laboratory, or office facilities of an employer to carry on outside private practice.
7. Engineers shall not attempt to injure, maliciously or falsely, directly or indirectly, the professional reputation, prospects, practice, or employment of other engineers. Engineers who believe others are guilty of unethical or illegal practice shall present such information to the proper authority for action.
   a. Engineers in private practice shall not review the work of another engineer for the same client, except with the knowledge of such engineer, or unless the connection of such engineer with the work has been terminated.
   b. Engineers in governmental, industrial, or educational employ are entitled to review and evaluate the work of other engineers when so required by their employment duties.
   c. Engineers in sales or industrial employ are entitled to make engineering comparisons of represented products with products of other suppliers.
8. Engineers shall accept personal responsibility for their professional activities, provided, however, that engineers may seek indemnification for services arising out of their practice for other than gross negligence, where the engineer's interests cannot otherwise be protected.
   a. Engineers shall conform with state registration laws in the practice of engineering.
   b. Engineers shall not use association with a nonengineer, a corporation, or partnership as a "cloak" for unethical acts.

9. Engineers shall give credit for engineering work to those to whom credit is due, and will recognize the proprietary interests of others.
    a. Engineers shall, whenever possible, name the person or persons who may be individually responsible for designs, inventions, writings, or other accomplishments.
    b. Engineers using designs supplied by a client recognize that the designs remain the property of the client and may not be duplicated by the engineer for others without express permission.
    c. Engineers, before undertaking work for others in connection with which the engineer may make improvements, plans, designs, inventions, or other records that may justify copyrights or patents, should enter into a positive agreement regarding ownership.
    d. Engineers' designs, data, records, and notes referring exclusively to an employer's work are the employer's property. The employer should indemnify the engineer for use of the information for any purpose other than the original purpose.
    e. Engineers shall continue their professional development throughout their careers and should keep current in their specialty fields by engaging in professional practice, participating in continuing education courses, reading in the technical literature, and attending professional meetings and seminars.

Appendix B: Project Management Code of Ethics and Professional Conduct*

VISION AND APPLICABILITY

VISION AND PURPOSE

As *practitioners* of project management, we are committed to doing what is right and honorable. We set high standards for ourselves and we aspire to meet these standards in all aspects of our lives—at work, at home, and in service to our profession.

This Code of Ethics and Professional Conduct describes the expectations that we have of ourselves and our fellow practitioners in the global project management community. It articulates the ideals to which we aspire as well as the behaviors that are mandatory in our professional and volunteer roles.

The purpose of this Code is to instill confidence in the project management profession and to help an individual become a better practitioner. We do this by establishing a profession-wide understanding of appropriate behavior. We believe that the credibility and reputation of the project management profession is shaped by the collective conduct of individual practitioners.

We believe that we can advance our profession, both individually and collectively, by embracing this Code of Ethics and Professional Conduct. We also believe that this Code will assist us in making wise decisions, particularly when faced with difficult situations where we may be asked to compromise our integrity or our values.

Our hope that this Code of Ethics and Professional Conduct will serve as a catalyst for others to study, deliberate, and write about ethics and values. Further, we hope that this Code will ultimately be used to build upon and evolve our profession.

PERSONS TO WHOM THE CODE APPLIES

The Code of Ethics and Professional Conduct applies to
- All *PMI members*
- Individuals who are not members of PMI but meet one or more of the following criteria:

---

* Adapted from PMI® code of Ethics; PMI® is a registered trademark of the Project Management Institute

1. Nonmembers who hold a PMI certification
2. Nonmembers who apply to commence a PMI certification process
3. Nonmembers who serve PMI in a volunteer capacity

STRUCTURE OF THE CODE

The Code of Ethics and Professional Conduct is divided into sections that contain standards of conduct, which are aligned with the four values that were identified as most important to the project management community. A glossary can be found at the end of the standard. The glossary defines words and phrases used in the Code. For convenience, those terms defined in the glossary are italicized in the text of the Code.

VALUES THAT SUPPORT THIS CODE

Practitioners from the global project management community were asked to identify the values that formed the basis of their decision making and guided their actions. The values that the global project management community defined as most important were responsibility, respect, fairness, and honesty. This Code affirms these four values as its foundation.

ASPIRATIONAL AND MANDATORY CONDUCT

Each section of the Code of Ethics and Professional Conduct includes both aspirational standards and mandatory standards. The aspirational standards describe the conduct that we strive to uphold as *practitioners*. Although adherence to the aspirational standards is not easily measured, conducting ourselves in accordance with these is an expectation that we have of ourselves as professionals—it is not optional.

The mandatory standards establish firm requirements, and in some cases, limit or prohibit practitioner behavior. Practitioners who do not conduct themselves in accordance with these standards will be subject to disciplinary procedures before PMI's Ethics Review Committee.

RESPONSIBILITY

DESCRIPTION OF RESPONSIBILITY

Responsibility is our duty to take ownership for the decisions we make or fail to make, the actions we take or fail to take, and the consequences that result.

RESPONSIBILITY: ASPIRATIONAL STANDARDS

As *practitioners* in the global project management community

1. We make decisions and take actions based on the best interests of society, public safety, and the environment.
2. We accept only those assignments that are consistent with our background, experience, skills, and qualifications.
3. We fulfill the commitments that we undertake—we do what we say we will do.

Appendix B

4. When we make errors or omissions, we take ownership and make corrections promptly. When we discover errors or omissions caused by others, we communicate them to the appropriate body as soon as they are discovered. We accept accountability for any issues resulting from our errors or omissions and any resulting consequences.
5. We protect proprietary or confidential information that has been entrusted to us.
6. We uphold this Code and hold each other accountable to it.

Responsibility: Mandatory Standards

As *practitioners* in the global project management community, we require the following of ourselves and our fellow practitioners:

1. Regulations and legal requirements
   a. We inform ourselves and uphold the policies, rules, regulations, and laws that govern our work, professional, and volunteer activities.
   b. We report unethical or illegal conduct to appropriate management and, if necessary, to those affected by the conduct.
2. Ethics complaints
   a. We bring violations of this Code to the attention of the appropriate body for resolution.
   b. We only file ethics complaints when they are substantiated by facts.
   c. We pursue disciplinary action against an individual who retaliates against a person raising ethics concerns.

RESPECT

Description of Respect

Respect is our duty to show a high regard for ourselves, others, and the resources entrusted to us. Resources entrusted to us may include people, money, reputation, the safety of others, and natural or environmental resources.

An environment of respect engenders trust, confidence, and performance excellence by fostering mutual cooperation—an environment where diverse perspectives and views are encouraged and valued.

Respect: Aspirational Standards

As *practitioners* in the global project management community:

1. We inform ourselves about the norms and customs of others and avoid engaging in behaviors they might consider disrespectful.
2. We listen to others' points of view, seeking to understand them.
3. We approach directly those persons with whom we have a conflict or disagreement.
4. We conduct ourselves in a professional manner, even when it is not reciprocated.

RESPECT: MANDATORY STANDARDS

As *practitioners* in the global project management community, we require the following of ourselves and our fellow practitioners:

1. We negotiate in good faith.
2. We do not exercise the power of our expertise or position to influence the decisions or actions of others in order to benefit personally at their expense.
3. We do not act in an *abusive manner* toward others.
4. We respect the property rights of others.

FAIRNESS

DESCRIPTION OF FAIRNESS

Fairness is our duty to make decisions and act impartially and objectively. Our conduct must be free from competing self interest, prejudice, and favoritism.

FAIRNESS: ASPIRATIONAL STANDARDS

As *practitioners* in the global project management community

1. We demonstrate transparency in our decision-making process.
2. We constantly reexamine our impartiality and objectivity, taking corrective action as appropriate.
3. We provide equal access to information to those who are authorized to have that information.
4. We make opportunities equally available to qualified candidates.

FAIRNESS: MANDATORY STANDARDS

As practitioners in the global project management community, we require the following of ourselves and our fellow practitioners:

1. *Conflict-of-interest* situations
   a. We proactively and fully disclose any real or potential conflicts of interest to the appropriate stakeholders.
   b. When we realize that we have a real or potential *conflict of interest*, we refrain from engaging in the decision-making process or otherwise attempting to influence outcomes unless or until we have made full disclosure to the affected stakeholders, we have an approved mitigation plan, and we have obtained the consent of the stakeholders to proceed.
2. Favoritism and discrimination
   a. We do not hire or fire, reward or punish, or award or deny contracts based on personal considerations, including, but not limited to, favoritism, nepotism, or bribery.

b. We do not discriminate against others based on, but not limited to, gender, race, age, religion, disability, nationality, or sexual orientation.
c. We apply the rules of the organization (employer, *Project Management Institute*, or other group) without favoritism or prejudice.

HONESTY

DESCRIPTION OF HONESTY

Honesty is our duty to understand the truth and act in a truthful manner both in our communications and in our conduct.

HONESTY: ASPIRATIONAL STANDARDS

As *practitioners* in the global project management community

1. We earnestly seek to understand the truth
2. We are truthful in our communications and in our conduct
3. We provide accurate information in a timely manner
4. We make commitments and promises, implied or explicit, in good faith
5. We strive to create an environment in which others feel safe to tell the truth

HONESTY: MANDATORY STANDARDS

As *practitioners* in the global project management community, we require the following of ourselves and our fellow practitioners:

1. We do not engage in or condone behavior that is designed to deceive others, including, but not limited to, making misleading or false statements, stating half-truths, providing information out of context or withholding information that, if known, would render our statements as misleading or incomplete.
2. We do not engage in dishonest behavior with the intention of personal gain or at the expense of others.

Ethics Glossary

Abusive manner. Conduct that results in physical harm or creates intense feelings of fear, humiliation, manipulation, or exploitation in another person.

Conflict of interest. A situation that arises when a practitioner of project management is faced with making a decision or doing some act that will benefit the practitioner or another person or organization to which the practitioner owes a *duty of loyalty* and at the same time will harm another person or organization to which the practitioner owes a similar *duty of loyalty*. The only way practitioners can resolve conflicting duties is to disclose the conflict to those affected and allow them to make the decision about how the practitioner should proceed.

Duty of loyalty. A person's responsibility, legal or moral, to promote the best interest of an organization or other person with whom they are affiliated.

Project Management Institute (PMI). The totality of the Project Management Institute, including its committees, groups, and chartered components such as chapters, colleges, and specific interest groups.

PMI member. A person who has joined the PMI as a member.

PMI-sponsored activities. Activities that include, but are not limited to, participation on a PMI member advisory group, PMI standard development team, or another PMI working group or committee. This also includes activities engaged in under the auspices of a chartered PMI component organization—whether it is in a leadership role in the component or another type of component educational activity or event.

Practitioner. A person engaged in an activity that contributes to the management of a project, portfolio, or program, as part of the project management profession.

PMI volunteer. A person who participates in *PMI-sponsored activities*, whether a member of the PMI or not.

APPENDIX C: Project Acronyms and Glossary

A&E	Architecture and Engineering
AACE	American Association of Cost Engineers
ABC	Activity-based costing
ACO	Administrative Contracting Officer
ACV	At-completion variance
ACWP	Actual cost of work performed
ADM	Arrow diagramming method
ADP	Automated data processing
ADPE	Automated data processing equipment
ADR	Arrow diagramming method
AF	Award Fee
AFR	Air Force Regulation
AGE	Auxiliary Ground Equipment
AHP	Analytical Hierarchy Process
AIS	Automated Information System
ANSI	American National Standards Institute
AOA	Activity on arrow
AON	Activity on node
APR	Acquisition plan review
AQL	Acceptable quality level
AR	Acceptance review
ARB	Acquisition Review Board
ARC	Appraisal requirements for CMMI
ARO	After Receipt of Order
ARO	Army Research Office
ASAPM	American Society for the Advancement of Project Management
ASCR	Annual System Certification Review
AT	Acceptance test
ATE	Automatic test equipment
ATP	Acceptance test procedure
AUW	Authorized unpriced work
B&P	Bid and proposal funds
BAA	Broad agency announcement
BAC	Budget at completion
BAFO	Best and final offer
BCE	Baseline cost estimate
BCWP	Budgeted cost of work performed
BCWS	Budgeted cost of work scheduled
BIT	Built-in test

BITE	Built-in test equipment
BNB	Bid/no bid
BOA	Basic ordering agreement
BOE	Basis of estimate
BOM	Bill of material
BPA	Blanket purchase agreement
BTW	By the way
BY (1)	Base year
BY (2)	Budget year
C/SCSC	Cost/schedule control system criteria
C/SSR	Cost/schedule status report
CA	Contract administrator
CAAS	Contracted advisory and assistance services
CAC	Cost at completion
CAD	Computer-aided design
CADM	Computer-aided document management
CAIV	Cost as an independent variable
CAM (1)	Computer-aided manufacturing
CAM (2)	Cost account manager
CAR	Contract acceptance review
CAS	Cost accounting standards
CASE (1)	Computer aided software engineering
CASE (2)	Computer aided systems engineering
CAT	Computer-aided testing
CBD	Commerce Business Daily
CBJ	Congressional budget justification
CBJR	Congressional budget justification review
CCA	Change control authority
CCB (1)	Change control board
CCB (2)	Configuration control board
CCN	Contract change notice
CCO	Contract change order
CCP	Contract change proposal
CDCG	Contract data classification guide
CDD	Concept definition document
CDR	Critical design review
CDRL (1)	Contract data requirements lists
CDRL (2)	Contract documentation requirements list
CEO	Chief executive officer
CET	Cost evaluation team
CFE	Contractor furnished equipment
CFSR	Contract funds status report
CI (1)	Configuration item
CI (2)	Continuous improvement
CIAR	Configuration Item Acceptance Review
CICA	Competition in Contracting Act of 1984

Appendix C

CID	Commercial item description
CIR (1)	Contract implementation review
CIR (2)	Contract inspection report
CIT	Component integration and test
CITRR	Configuration item test readiness
CLIN	Contract line item numbers
CM	Configuration management
CMM	Capability maturity model
CMMI	Capability maturity model integration
CMO	Configuration Management Officer
CMSP	Contractor Management Systems evaluation program
CO (1)	Change order
CO (2)	Contracting officer
COCOMO	Constructive cost model
CONOPS	Concept of operations
COR	Contracting officer's representative
COTR	Contracting officer's technical representative
COTS	Commercial off-the-shelf
COW	Cards on the wall planning
CPA	Certified Public Accountant
CPAF	Cost plus award fixed
CPC	Computer program component
CPCI	Computer program configuration item
CPFF	Cost plus fixed fee
CPI (1)	Continuous Process Improvement
CPI (2)	Cost Performance Index
CPIF	Cost plus incentive fee
CPO	Contractor project office
CPR	Cost performance report
CPU	Central processing unit
CPVR	Construction performance verification review
CQI	Continuous quality improvement
CRADA	Cooperative Research and Development Agreement
CRD	Critical resource diagramming
CRWG	Computer resources working group
CSC	Computer software component
CSCI	Computer software configuration item
CSE	Chief systems engineer
CSOM	Computer system operators manual
CSSR	Contract system status report
CSU	Computer software unit
CTC (1)	Collaborate to consensus
CTC (2)	Contract target cost
CTC (3)	Cost to complete
CTP	Contract target price
CV_t	Cost variance

CWBS	Contract work breakdown structure
CY	Calendar year
DA&R	Decomposition analysis and resolution
DAR	Deactivation approval review
DARPA	Defense Advanced Research Projects Agency
DCAA	Defense Contract Audit Agency
DCAS	Defense Contract Administration Service
DCN	Documentation change notice
DCR	Design concept review
DD 250	DD 250
DDT&E	Design, development, test, and evaluation
DID	Data item description
DLA	Defense Logistics Agency
DMAIC	Define, Measure, Analyze, Improve, Control
DMO	Documentation management officer
DP	Data processing
DPAS	Defense Priorities and Allocation System
DPRO	Defense Plant Representative Office
DR	Discrepancy report
DRD	Documentation requirements description
DRR	Development readiness review
DSMC	Defense Systems Management College
DT&E	Development test and evaluation
DTC	Design-to-cost
DTS	Design-to-schedule
DVR	Documentation verification review
EAC	Estimate at completion
ECCM	Electronic counter-countermeasures
ECD	Estimated completion date
ECM	Electronic countermeasures
ECN	Engineering change notice
ECP	Engineering change proposal
ECR	Engineering change request
EDM	Engineering development model
EI	End item
EMC	Electromagnetic compatibility
EMI	Electromagnetic interference
EO	Engineering order
ERB	Engineering review broad
ESS	Environmental stress screening
ETC	Estimate to complete
ETR	Estimated time to repair
EV	Earned value
EVT	Earned value technique
EW	Electronic warfare
FA	First article

Appendix C

FAR	Federal Acquisition Regulations
FARA	Federal Acquisition Reform Act
FASA	Federal Acquisition Streamlining Act
FAT (1)	Factory acceptance test
FAT (2)	First article test
FCA	Functional configuration audit
FCCM	Facilities capital cost of money
FCR (1)	Final contract review
FCR (2)	Facility contract review
FDR	Final design review
FFBD	Functional flow block diagram
FFP	Firm fixed price contract
FFRDC	Federally Funded Research and Development Center
FMEA	Failure mode and effects analysis
FMECA	Failure mode, effects, and criticality analysis
FOC	Full operational capability
FOIA	Freedom of Information Act
FOM	Figure of merit
FP	Fixed Price Contract
FPAF	Fixed Price Award Fee
FPIF	Fixed Price Incentive Fee
FPR (1)	Final proposal review
FPR (2)	Fixed price redeterminable
FPVR	Facility performance verification review
FQR	Formal qualification review
FQT	Formal qualification testing
FRB	Failure review board
FRR	Facility readiness review
FSOW	Facility scope of work
FTRR	Facility test readiness review
FY	Fiscal year
G&A	General and administrative costs
GAO	General Accounting Office
GAS	General accounting system
GFE	Government furnished equipment
GFF	Government furnished facilities
GFI	Government furnished information
GFM	Government furnished material
GFP	Government furnished property
GLS	Global logistics support
GOCO	Government owned, contractor operated
GOGO	Government owned, government operated
GOTS	Government off-the-shelf
GPO	Government project office
GSA	General Services Administration
GSC	Global supply chain

GSE	Ground support equipment
HAC	House Appropriations Committee
HCI	Human–computer interface
HQ	Headquarters
HW	Hardware
HWCI	Hardware configuration item
IAW	In accordance with
ICD	Interface control document
ICP	Interface control plan
ICWG	Interface control working group
ID (1)	Identifier
ID (2)	Independent development
ID (3)	Indefinite delivery contract
IDD	Interface design document
IDEAL	IDEAL
IDEFO	Integrated definition for functional modeling
IE	Information engineering
IEEE	Institute of Electrical and Electronics Engineers
IFB	Invitation for bid
IG	Inspector General
IGCE	Independent government cost estimate
ILS	Integrated logistics support
ILS	Integrated logistics support
INCOSE	International Council on Systems Engineering
INI	Interest/no interest
IOC	Initial operational capability
IPT (1)	Integrated product teams
IPT (2)	Integrated project teams
IQ	Indefinite quantity
IR&D	Independent research and development
IRR	Internal rate of return
IRS	Interface requirements specification
IS	Interface specification
ISCO	Integrated schedule commitment
ISO	International Organization for Standardization
IV&V (1)	Independent verification and validation
IV&V (2)	Integration verification and validation
L/H	Labor hour contract
LAN	Local area network
LCC	Life cycle cost
LOB	Line of business
LOC (1)	Lines of code
LOC (2)	Logistics operations center
LOE	Level of effort
LOI	Letter of intent
LRIP	Low rate initial production

MBO	Management by objectives
MBWA	Management by walking around
MDT	Mean down time
MIL-SPEC	Military specification
MIL-STD	Military standard
MIPR	Military Interdepartmental Purchase Request
MIS	Management information system
MIV	Management information center
MOA	Memorandum of agreement
MOE	Measure of effectiveness
MOP	Measure of performance
MOU	Memorandum of understanding
MPS	Master project schedule
MR	Management reverse
MRB	Material review board
MRP (1)	Manufacturing resource planning
MRP (2)	Material resource planning
MTBF	Mean time between failures
MTTR	Mean time to repair
MYP	Multiyear procurement
N/A	Not applicable
NBV	Net book value
NC	Numerical control
NCR	Nonconformance report
NDI	Nondevelopment item
NIH	Not invented here
NLT	No later than
NMT	Not more than
NPV	Net present value
NTE	Not to exceed
O&M	Operations and maintenance
OAR	Operational acceptance review
OBS	Organizational breakdown structure
ODC	Other direct costs
OFR	Option of first refusal
OGA	One Generation Ahead
OGC	Office of General Council
OH	Overhead
OJT	On-the-job training
OMB	Office of Management and Budget
OOA	Object-oriented analysis
OOD	Object-oriented design
ORC	Operational readiness certificate
ORD	Operational requirements document
ORR	Operational readiness review
OSHA	Occupational Safety and Health Administration

OT&E	Operational test and evaluation
OVR	Operational validation review
PA	Product assurance
PAR	Product acceptance review
PBL	Performance-based logistics
PBS	Product breakdown structure
PC (1)	Personal computer
PC (2)	Project cycle
PCA	Physical configuration audit
PCCB	Project configuration control board
PCO (1)	Procuring contracting officer
PCO (2)	Principal contracting officer
PCR	Project completion review
PDM	Precedence diagramming method
PDP	Previously developed products
PDR	Preliminary design review
PERT	Project evaluation review technique
PET	Proposal evaluation teams
PIP	Product improvement plan
PIR (1)	Project implementation review
PIR (2)	Project initiation review
PL	Public law
PM (1)	Program manager
PM (2)	Project manager
PM&P	Parts, material, and processes
PMB	Performance measurement baseline
PMBOK	Project Management Body of Knowledge
PMI	Project Management Institute
PMP	Project Management Professional
PMS	Performance measurement system
PNP	Pursue/no pursue
POC	Point of contact
POM	Program management memorandum
PPI	Proposal preparation instructions
PPL	Project products list
PPLFS	Projects products list fact sheets
PPPI	Preplanned product improvement
PPR	Project plans review
PRB	Project review board
PRICE	Program review information for costing and estimating
PRR	Production readiness review
PSR	Project specification review
PWAA	Project Work Authorizing Agreement
PY	Prior year
QA	Quality assurance
QAR	Qualification acceptance review

Appendix C

QC	Quality control
QFD	Quality function deployment
QRC	Quick reaction capability
R&D	Research and development
RAD	Rapid application development
RAM	Random access memory
RDT&E	Research, development, test, and evaluation
RFC	Request for change
RFI	Request for information
RFP	Request for proposal
RFQ	Request for quotation
RIF	Reduction in force
ROI	Return on investment
ROM (1)	Rough order on magnitude
ROM (2)	Read-only memory
RTM	Requirements traceability matrix
RTVM	Requirement traceability and verification matrix
RVM	Requirements verification matrix
S/C	Subcontract
SAP	System acquisition plan
SAR	System acceptance review
SBA	Small business administration
SCA	Subcontract administrator
SCAMPI	Standard CMMI appraisal method for process improvement
SCE	Software capability evaluation
SCN	Specification change notice
SCR	System concept review
SDD	Software design document
SDF	Software development file
SDL	Software development library
SDP	Software development plan
SDR	System design review
SDRL	Subcontract documentation requirements list
SEB	Source evaluation board
SEI	Software Engineering Institute
SEI&T	Systems engineering, integration, and test
SEMP	Systems Engineering Management Plan
SETA	Systems engineering and technical assistance
SI	System integrator
SIPOC	Suppliers, Inputs, Process, Outputs, Customers
SMAP	Software management and assurance program
SMT	Subcontract management team
SOP	Standard operating procedure
SOW	Statement of work
SPI	Schedule performance index
SPM	Software programmers manual

SPO	System project office
SPR	Software problem report
SPS	Software product specification
SQA	Software quality assurance
SRD	System requirements document
SRR	System requirements review
SRS	Software requirements specification
SSA	Source selection authority
SSAC	Source selection advisory council
SSAR	Source selection authorization review
SSDD	System/segment design document
SSE	Software support environment
SSEB	Source selection evaluation board
SSIR	Source selection initiation review
SSM	Software sizing model
SSO	Source selection official
SSP	Source selection plan
SSR	Software specification review
SSS	System/segment specification
STE	Special test equipment
STP	System test plan
STR	Software test report
STRR	System test readiness review
SUM	Software users manual
SV	Schedule variance
SW	Software
SWAG	Scientific wild anatomical guess
T&E	Test and evaluation
T&M	Time and materials contract
TAAF	Test, analyze, and fix
TBD	To be determined
TBR	To be resolved
TBS	To be supplied
TCPI	To complete performance index
TD	Test director
TEM	Technical exchange meeting
TET	Technical evaluation team
TIM	Technical interchange meeting
TM	Technical manual
TOC	Theory of constraints
TOC	Total ownership cost
TP	Test procedures
TPM	Technical performance measurement
TQM	Total quality management
TR (1)	Time remaining

Appendix C

TR (2)	Test report
TRR	Test readiness review
TTC	Time to complete
UAR	User acceptance review
UB	Undistributed budget
URR	User readiness review
VA&R	Verification analysis and resolution
VAC	Variance at completion
VDD	Version description document
VE	Value engineering
VECP	Value engineering change proposal
VRIC	Vendor request for information or change
VV&T	Verification, validation, and test
WAN	Wide area network
WBS	Work breakdown structure
W-Mgt	W theory management
WO/WA	Work order/work authorization
WP	Work packages
WR	Work remaining
X-Mgmt	X theory (or authoritative) management
Y-Mgmt	Y theory (or supportive) management
Z-Mgmt	Z theory (or participative) management

Ab initio (Latin for "from the beginning"): Project contract or agreement executed right from the start of a project.

ABC (activity based costing): Bottom-up estimating and summation based on material and labor required for activities making up a project.

Accept: The act of formally receiving or acknowledging a deliverable and regarding it as being true, sound, suitable, or complete.

Acceptance: The act of formally signifying satisfaction with an outcome or a deliverable.

Acceptance criteria: Those criteria, including performance requirements and essential conditions, which must be met before project deliverables are accepted.

Acquire project team: The process of obtaining the human resources needed to complete the project.

Activity: A component of work performed during the course of a project. See also schedule activity.

Activity attributes: Multiple attributes associated with each schedule activity that can be included within the activity list. Activity attributes include activity codes, predecessors, successors, logical relationships, leads and lags, resource requirements, imposed dates, constraints, and assumptions.

Activity based costing: See ABC.

Activity based management: The achievement of strategic objectives and customer satisfaction by managing value-added activities.

Activity code: One or more numerical or text values that identify characteristics of the work or in some way categorize the schedule activity that allows filtering and ordering of activities within reports.

Activity definition: The process of identifying the specific schedule activities that need to be performed to produce the various project deliverables.

Activity description: A short phrase or label for each schedule activity used in conjunction with an activity identifier to differentiate that project schedule activity from other schedule activities. The activity description normally describes the scope of work of the schedule activity.

Activity duration: The time in calendar units between the start and finish of a schedule activity. See also actual duration, original duration, and remaining duration.

Activity duration estimating: The process of estimating the number of work periods that will be needed to complete individual schedule activities.

Activity identifier: A short unique numeric or text identification assigned to each schedule activity to differentiate that project activity from other activities. Typically unique within any one project schedule network diagram.

Activity list: A documented tabulation of schedule activities that shows the activity description, activity identifier, and a sufficiently detailed scope of work description so project team members understand what work is to be performed.

Activity-on-arrow (AOA): A project network diagramming technique in which activities are represented by lines (or arrows) and nodes represent starting and ending points. See arrow diagramming method. Activity-on-node (AON). See precedence diagramming method.

Activity-on-node (AON): A project network technique in which nodes represent activities and lines (or arrows) represent precedence relationships.

Activity resource estimating: The process of estimating the types and quantities of resources required to perform each schedule activity.

Activity sequencing: The process of identifying and documenting dependencies among schedule activities.

Actual cost (AC): Total costs actually incurred and recorded in accomplishing work performed during a given time period for a schedule activity or WBS component. Actual cost can sometimes be direct labor hours alone, direct costs alone, or all costs including indirect costs. It is also referred to as the actual cost of work performed (ACWP). See also earned value management and earned value technique.

Actual cost of work performed (ACWP): See actual cost (AC).

Actual duration: The time in calendar units between the actual start date of the schedule activity and either the data date of the project schedule if the schedule activity is in progress or the actual finish date if the schedule activity is complete.

Actual finish: The point in time that work actually ended on a schedule activity. (Note: In some application areas, the schedule activity is considered "finished" when work is "substantially complete.")

Appendix C

Actual start: The point in time that work actually started on a schedule activity.

Affinity diagram: A pictorial clustering of items into similar (or related) categories.

Analogous estimating: An estimating technique that uses the values of parameters, such as scope, cost, budget, and duration or measures of scale such as size, weight, and complexity from a previous, similar activity as the basis for estimating the same parameter or measure for a future activity. It is frequently used to estimate a parameter when there is a limited amount of detailed information about the project (e.g., in the early phases). Analogous estimating is a form of expert judgment. Analogous estimating is most reliable when the previous activities are similar in fact and not just in appearance, and the project team members preparing the estimates have the needed expertise.

Application area: A category of projects that have common components significant in such projects, but are not needed or present in all projects. Application areas are usually defined in terms of either the product (i.e., by similar technologies or production methods) or the type of customer (i.e., internal versus external, government versus commercial) or industry sector (i.e., utilities, automotive, aerospace, information technologies). Application areas can overlap.

Apportioned effort: Effort applied to project work that is not readily divisible into discrete efforts for that work, but which is related in direct proportion to measurable discrete work efforts. Contrast with discrete effort.

Approve/approval: The act of formally confirming, sanctioning, ratifying, or agreeing to something.

Approved change request: A change request that has been processed through the integrated change control process and approved. Contrast with requested change.

Arrow: The graphic presentation of a schedule activity in the arrow diagramming method or a logical relationship between schedule activities in the precedence diagramming method.

Arrow diagramming method (ADM): A schedule network diagramming technique in which schedule activities are represented by arrows. The tail of the arrow represents the start and the head represents the finish of the schedule activity. (The length of the arrow does not represent the expected duration of the schedule activity.) Schedule activities are connected at points called nodes (usually drawn as small circles) to illustrate the sequence in which the schedule activities are expected to be performed. See also precedence diagramming method.

As-of date: See data date.

Assumptions: Assumptions are factors that, for planning purposes, are considered to be true, real, or certain without proof or demonstration. Assumptions affect all aspects of project planning, and are part of the progressive elaboration of the project. Project teams frequently identify, document, and validate assumptions as part of their planning process. Assumptions generally involve a degree of risk.

Assumptions analysis: A technique that explores the accuracy of assumptions and identifies risks to the project from inaccuracy, inconsistency, or incompleteness of assumptions.

Authority: The right to apply project resources, expend funds, make decisions, or give approvals.

Backward pass: The calculation of late finish dates and late start dates for the uncompleted portions of all schedule activities; determined by working backwards through the schedule network logic from the project's end date. The end date may be calculated in a forward pass or set by the customer or sponsor. See also schedule network analysis.

Bar chart: A graphic display of schedule-related information, in the typical bar chart. Schedule activities or work breakdown structure components are listed down the left side of the chart, dates are shown across the top, and activity durations are shown as date-placed horizontal bars; also called a Gantt chart.

Baseline: The approved time-phased plan (for a project, a work breakdown structure component, a work package, or a schedule activity), plus or minus approved project scope, cost, schedule, and technical changes. Generally refers to the current baseline, but may refer to the original or some other baseline. Usually used with a modifier (e.g., cost baseline, schedule baseline, performance measurement baseline, technical baseline). See also performance measurement baseline.

Baseline finish date: The finish date of a schedule activity in the approved schedule baseline. See also scheduled finish date.

Best practices: Processes, procedures, and techniques that have consistently demonstrated achievement of expectations and that are documented for the purposes of sharing, repetition, replication, adaptation, and refinement.

Baseline start date: The start date of a schedule activity in the approved schedule baseline. See also scheduled start date.

Bill of materials (BOM): A documented formal hierarchical tabulation of the physical assemblies, subassemblies, and components needed to fabricate a product.

Bottom-up estimating: A method of estimating a component of work. The work is decomposed into more detail. An estimate is prepared of what is needed to meet the requirements of each of the lower, more detailed pieces of work, and these estimates are then aggregated into a total quantity for the component of work. The accuracy of bottom-up estimating is driven by the size and complexity of the work identified at the lower levels. Generally smaller work scopes increase the accuracy of the estimates.

Brainstorming: A general data gathering and creativity technique that can be used to identify risks, ideas, or solutions to issues by using a group of team members or subject-matter experts. Typically, a brainstorming session is structured so that each participant's ideas are recorded for later analysis.

Budget: The approved estimate for the project or any work breakdown structure component or any schedule activity. See also estimate.

Appendix C

Budget at completion: The sum of all the budget values established for the work to be performed on a project or a work breakdown structure component or a schedule activity. The total planned value for the project.

Budgeted cost of work performed (BCWP): See earned value (EV); budgeted cost of work scheduled (BCWS); planned value (PV).

Buffer: See reserve.

Buyer: The acquirer of products, services, or results for an organization.

Calendar unit: The smallest unit of time used in scheduling the project. Calendar units are generally in hours, days, or weeks, but can also be in quarter years, months, shifts, or even in minutes.

Change control: Identifying, documenting, approving or rejecting, and controlling changes to the project baselines.

Change Control Board: A formally constituted group of stakeholders responsible for reviewing, evaluating, approving, delaying, or rejecting changes to the project, with all decisions and recommendations being recorded.

Change control system: A collection of formal documented procedures that define how project deliverables and documentation will be controlled, changed, and approved. In most application areas the change control system is a subset of the configuration management system.

Change request: Requests to expand or reduce the project scope, modify policies, processes, plans, or procedures, modify costs or budgets, or revise schedules. Requests for a change can be direct or indirect, externally or internally initiated, and legally or contractually mandated or optional. Only formally documented requested changes are processed and only approved change requests are implemented.

Chart of accounts: Any numbering system used to monitor project costs by category (e.g., labor, supplies, materials, and equipment). The project chart of accounts is usually based upon the corporate chart of accounts of the primary performing organization. Contrast with code of accounts.

Charter: See project charter.

Checklist: Items listed together for convenience of comparison or to ensure the actions associated with them are managed appropriately and not forgotten.

An example is a list of items to be inspected that is created during quality planning and applied during quality control.

Claim: A request, demand, or assertion of rights by a seller against a buyer, or vice versa, for consideration, compensation, or payment under the terms of a legally binding contract, such as for a disputed change.

Close project: The process of finalizing all activities across all of the project process groups to formally close the project or phase.

Closing processes: Those processes performed to formally terminate all activities of a project or phase and transfer the completed product to others or close a cancelled project.

Code of accounts: Any numbering system used to uniquely identify each component of the work breakdown structure. Contrast with chart of accounts.

Co-location: An organizational placement strategy where the project team members are physically located close to one another in order to improve communication, working relationships, and productivity.

Common cause: A source of variation that is inherent in the system and predictable. On a control chart, it appears as part of the random process variation (i.e., variation from a process that would be considered normal or not unusual), and is indicated by a random pattern of points within the control limits. Also referred to as random cause. Contrast with special cause.

Communication: A process through which information is exchanged among persons using a common system of symbols, signs, or behaviors.

Communication management plan: The document that describes the communications needs and expectations for the project; how and in what format information will be communicated; when and where each communication will be made; and who is responsible for providing each type of communication. A communication management plan can be formal or informal, highly detailed or broadly framed, based on the requirements of the project stakeholders. The communication management plan is contained in, or is a subsidiary plan of, the project management plan.

Communications planning: The process of determining the information and communications needs of the project stakeholders: who they are, their levels of interest and influence on the project, who needs what information, when will they need it, and how it will be given to them.

Compensation: Something given or received, a payment or recompense, usually something monetary or in kind for products, services, or results provided or received.

Component: A constituent part, element, or piece of a complex whole.

Configuration management system: A subsystem of the overall project management system. It is a collection of formal documented procedures used to apply technical and administrative direction and surveillance to identify and document the functional and physical characteristics of a product, result, service, or component; control any changes to such characteristics; record and report each change and its implementation status; and support the audit of the products, results, or components to verify conformance to requirements. It includes the documentation, tracking systems, and defined approval levels necessary for authorizing and controlling changes. In most application areas, the configuration management system includes the change control system.

Constraint: The state, quality, or sense of being restricted to a given course of action or inaction. An applicable restriction or limitation, either internal or external to the project, that will affect the performance of the project or a process. For example, a schedule constraint is any limitation or restraint placed on the project schedule that affects when a schedule activity can be scheduled and is usually in the form of fixed imposed dates. A cost constraint is any limitation or restraint placed on the project budget such as funds available over time. A project resource constraint is any limitation or restraint placed on resource usage, such as what resource

Appendix C   365

skills or disciplines are available and the amount of a given resource available during a specified time frame.

Contingency: See reserve.

Contingency allowance: See reserve.

Contingency reserve: The amount of funds, budget, or time needed above the estimate to reduce the risk of overruns of project objectives to a level acceptable to the organization.

Contract: A contract is a mutually binding agreement that obligates the seller to provide the specified product or service or result and obligates the buyer to pay for it.

Contract administration: The process of managing the contract and the relationship between the buyer and seller, reviewing and documenting how a seller is performing or has performed to establish required corrective actions and provide a basis for future relationships with the seller, managing contract-related changes and, when appropriate, managing the contractual relationship with the outside buyer of the project.

Contract closure: The process of completing and settling the contract, including resolution of any open items and closing each contract.

Contract management plan: The document that describes how a specific contract will be administered and can include items such as required documentation delivery and performance requirements. A contract management plan can be formal or informal, highly detailed or broadly framed, based on the requirements in the contract. Each contract management plan is a subsidiary plan of the project management plan.

Contract statement of work (SOW): A narrative description of products, services, or results to be supplied under contract.

Contract work breakdown structure (CWBS): A portion of the work breakdown structure for the project developed and maintained by a seller contracting to provide a subproject or project component.

Control: Comparing actual performance with planned performance, analyzing variances, assessing trends to effect process improvements, evaluating possible alternatives, and recommending appropriate corrective action as needed.

Control account (CA): A management control point where the integration of scope, budget, actual cost, and schedule takes place, and where the measurement of performance will occur. Control accounts are placed at selected management points (specific components at selected levels) of the work breakdown structure. Each control account may include one or more work packages, but each work package may be associated with only one control account. Each control account is associated with a specific single organizational component in the organizational breakdown structure (OBS). Previously called a cost account. See also work package.

Control account plan (CAP): A plan for all the work and effort to be performed in a control account. Each CAP has a definitive statement of work, schedule, and time-phased budget. Previously called a cost account plan.

Control chart: A graphic display of process data over time and against established control limits, and that has a centerline that assists in detecting a trend of plotted values toward either control limit.

Control limits: The area composed of three standard deviations on either side of the centerline, or mean, of a normal distribution of data plotted on a control chart that reflects the expected variation in the data. See also specification limits.

Controlling: See control.

Corrective action: Documented direction for executing the project work to bring expected future performance of the project work in line with the project management plan.

Cost: The monetary value or price of a project activity or component that includes the monetary worth of the resources required to perform and complete the activity or component, or to produce the component. A specific cost can be composed of a combination of cost components including direct labor hours, other direct costs, indirect labor hours, other indirect costs, and purchased price. (However, in the earned value management methodology, in some instances, the term cost can represent only labor hours without conversion to monetary worth.) See also actual cost and estimate.

Cost baseline: See baseline.

Cost budgeting: The process of aggregating the estimated costs of individual activities or work packages to establish a cost baseline.

Cost control: The process of influencing the factors that create variances, and controlling changes to the project budget.

Cost estimating: The process of developing an approximation of the cost of the resources needed to complete project activities.

Cost management plan: The document that sets out the format and establishes the activities and criteria for planning, structuring, and controlling the project costs. A cost management plan can be formal or informal, highly detailed or broadly framed, based on the requirements of the project stakeholders. The cost management plan is contained in, or is a subsidiary plan, of the project management plan.

Cost of quality (COQ): Determining the costs incurred to ensure quality. Prevention and appraisal costs (cost of conformance) include costs for quality planning, quality control (QC), and quality assurance to ensure compliance to requirements (i.e., training, QC systems, etc.). Failure costs (cost of nonconformance) include costs to rework products, components, or processes that are noncompliant, costs of warranty work and waste, and loss of reputation.

Cost performance index (CPI): A measure of cost efficiency on project. It is the ratio of earned value (EV) to actual costs (AC). CPI = EV/AC. A CPI value equal to or greater than 1 indicates a favorable condition and a value less than 1 indicates an unfavorable condition.

Cost-plus-fee (CPF): A type of cost-reimbursable contract where the buyer reimburses the seller for the seller's allowable costs for performing the contract work and the seller also receives a fee calculated as an agreed upon percentage of the costs. The fee varies with the actual cost.

Cost-plus-fixed-fee (CPFF) contract: A type of cost-reimbursable contract where the buyer reimburses the seller for the seller's allowable costs (allowable costs are defined by the contract) plus a fixed amount of profit (fee).

Cost-plus-incentive-fee (CPIF) contract: A type of cost-reimbursable contract where the buyer reimburses the seller for the seller's allowable costs (allowable costs are defined by the contract), and the seller earns its profit if it meets defined performance criteria.

Cost-plus-percentage of cost (CPPC): See cost-plus-fee.

Cost-reimbursable contract: A type of contract involving payment (reimbursement) by the buyer to the seller for the seller's actual costs, plus a fee typically representing seller's profit. Costs are usually classified as direct costs or indirect costs. Direct costs are costs incurred for the exclusive benefit of the project, such as salaries of full-time project staff. Indirect costs, also called overhead and general and administrative cost, are costs allocated to the project by the performing organization as a cost of doing business, such as salaries of management indirectly involved in the project, and cost of electric utilities for the office. Indirect costs are usually calculated as a percentage of direct costs. Cost-reimbursable contracts often include incentive clauses where, if the seller meets or exceeds selected project objectives, such as schedule targets or total cost, then the seller receives from the buyer an incentive or bonus payment.

Cost variance (CV): A measure of cost performance on a project. It is the algebraic difference between earned value (EV) and actual cost (AC). CV = EV − AC. A positive value indicates a favorable condition and a negative value indicates an unfavorable condition.

Crashing: A specific type of project schedule compression technique performed by taking action to decrease the total project schedule duration after analyzing a number of alternatives to determine how to get the maximum schedule duration compression for the least additional cost. Typical approaches for crashing a schedule include reducing schedule activity durations and increasing the assignment of resources on schedule activities. See schedule compression and see also fast tracking.

Create WBS (work breakdown structure): The process of subdividing the major project deliverables and project work into smaller, more manageable components.

Criteria: Standards, rules, or tests on which a judgment or decision can be based, or by which a product, service, result, or process can be evaluated.

Critical activity: Any schedule activity on a critical path in a project schedule. Most commonly determined by using the critical path method. Although some activities are "critical," in the dictionary sense, without being on the critical path, this meaning is seldom used in the project context.

Critical chain method: A schedule network, analysis technique that modifies the project schedule to account for limited resources. The critical chain method mixes deterministic and probabilistic approaches to schedule network analysis.

Critical path: Generally, but not always, the sequence of schedule activities that determines the duration of the project. Generally, it is the longest path through the project. However, a critical path can end, as an example, on a schedule milestone that is in the middle of the project schedule and that has a finish-no-later-than imposed date schedule constraint. See also critical path method.

Critical path method (CPM): A schedule network analysis technique used to determine the amount of scheduling flexibility (the amount of float) on various logical network paths in the project schedule network, and to determine the minimum total project duration. Early start and finish dates are calculated by means of forward pass and backward pass computations using a specified start date. Late start and finish dates are calculated by means of a backward pass, starting from a specified completion date, which sometimes is the project early finish date determined during the forward pass calculation.

Current finish date: The current estimate of the point in time when a schedule activity will be completed, where the estimate reflects any reported work progress. See also scheduled finish date and baseline finish date.

Current start date: The current estimate of the point in time when a schedule activity will begin, where the estimate reflects any reported work progress. See also scheduled start date and baseline start date.

Customer: The person or organization that will use the project's product or service or result. (See also user.)

Data date (DD): The date up to or through which the project's reporting system has provided actual status and accomplishments. In some reporting systems, the status information for the data date is included in the past and in some systems the status information is in the future. Also called as-of-date and time-now date.

Date: A term representing the day, month, and year of a calendar, and, in some instances, the time of day.

Decision Tree Analysis: The decision tree is a diagram that describes a decision under consideration and the implications of choosing one or another of the available alternatives. It is used when some future scenarios or outcomes of actions are uncertain. It incorporates probabilities and the costs or rewards of each logical path of events and future decisions, and uses expected monetary value analysis to help the organization identify the relative values of alternate actions. See also expected monetary value analysis.

Decompose: See decomposition.

Decomposition: A planning technique that subdivides the project scope and project deliverables into smaller, more manageable components, until the project work associated with accomplishing the project scope and providing the deliverables is defined in sufficient detail to support executing, monitoring, and controlling the work.

Defect: An imperfection or deficiency in a project component where that component does not meet its requirements or specifications and needs to be either repaired or replaced.

Defect repair: Formally documented identification of a defect in a project component with a recommendation to either repair the defect or completely replace the component.

Deliverable: Any unique and verifiable product, result, or capability to perform a service that must be produced to complete a process, phase, or project. Often used more narrowly in reference to an external deliverable, which is a deliverable that is subject to approval by the project sponsor or customer. See also product, service, and result.

Delphi Technique: An information gathering technique used as a way to reach a consensus of experts on a subject. Experts on the subject participate in this technique anonymously. A facilitator uses a questionnaire to solicit ideas about the important project points related to the subject. The responses are summarized and are then recirculated to the experts for further comment. Consensus may be reached in a few rounds of this process. The Delphi technique helps reduce bias in the data and keeps any one person from having undue influence on the outcome.

Dependency: See logical relationship.

Design review: A management technique used for evaluating a proposed design to ensure that the design of the system or product meets the customer requirements or to assure that the design will perform successfully, can be produced, and can be maintained.

Develop project charter: The process of developing the project charter that formally authorizes a project.

Develop project management plan: The process of documenting the actions necessary to define, prepare, integrate, and coordinate all subsidiary plans into a project management plan.

Develop project scope statement (preliminary): The process of developing the preliminary project scope statement that provides a high-level scope narrative.

Develop project team: The process of improving the competencies and interaction of team members to enhance project performance.

Direct and manage project execution: The process of executing the work defined in the project management plan to achieve the project's requirements defined in the project scope statement.

Discipline: A field of work requiring specific knowledge and that has a set of rules governing work conduct (e.g., mechanical engineering, computer programming, cost estimating, etc.).

Discrete effort: Work effort that is directly identifiable to the completion of specific work breakdown structure components and deliverables, and that can be directly planned and measured. Contrast with apportioned effort.

Document: A medium and the information recorded thereon, that generally has permanence and can be read by a person or a machine. Examples include project management plans, specifications, procedures, studies, and manuals.

Documented procedure: A formalized written description of how to carry out an activity, process, technique, or methodology.

Dummy activity: A schedule activity of zero duration used to show a logical relationship in the arrow diagramming method. Dummy activities are used when logical relationships cannot be completely or correctly described with schedule activity arrows. Dummy activities are generally shown graphically as a dashed line headed by an arrow.

Duration: The total number of work periods (not including holidays or other nonworking periods) required to complete a schedule activity or work breakdown structure component. Usually expressed as workdays or workweeks. Sometimes incorrectly equated with elapsed time. Contrast with effort. See also original duration, remaining duration, and actual duration.

Early finish date (EF): In the critical path method, the earliest possible point in time on which the uncompleted portions of a schedule activity (or the project) can finish, based on the schedule network, logic, the data date, and any schedule constraints. Early finish dates can change as the project progresses and as changes are made to the project management plan.

Early start date (ES): In the critical path method, the earliest possible point in time on which the uncompleted portions of a schedule activity (or the project) can start, based on the schedule network logic, the data date, and any schedule constraints. Early start dates can change as the project progresses and as changes are made to the project management plan.

Earned value (EV): The value of completed work expressed in terms of the approved budget assigned to that work for a schedule activity or work breakdown structure component. Also referred to as the budgeted cost of work performed (BCWP).

Earned value management (EVM): A management methodology for integrating scope, schedule, and resources, and for objectively measuring project performance and progress. Performance is measured by determining the BCWP (i.e., earned value) and comparing it to the ACWP (i.e., actual cost). Progress is measured by comparing the earned value to the planned value.

Earned value technique (EVT): A specific technique for measuring the performance of work for a work breakdown structure component, control account, or project. Also referred to as the earning rules and crediting method.

Effort: The number of labor units required to complete a schedule activity or work breakdown structure component. Usually expressed as staff hours, staff days, or staff weeks. Contrast with duration.

Enterprise: A company, business, firm, partnership, corporation, or governmental agency.

Enterprise environmental factors: Any or all external environmental factors and internal organizational environmental factors that surround or influence the project's success. These factors are from any or all of the enterprises involved in the project, and include organizational culture and structure, infrastructure, existing resources, commercial databases, market conditions, and project management software.

Estimate: A quantitative assessment of the likely amount or outcome. Usually applied to project costs, resources, effort, and durations and is usually preceded by a modifier (i.e., preliminary, conceptual, feasibility, order-of-magnitude, definitive). It should always include some indication of accuracy (e.g., plus or minus percent).

Estimate at completion (EAC): The expected total cost of a schedule activity, a work breakdown structure component, or the project when the defined scope of work will be completed. EAC is equal to the actual cost (AC) plus the estimate to complete (ETC) for all of the remaining work. EAC = AC + ETC. The EAC may be calculated based on performance to date or estimated by the project team based on other factors, in which case it is often referred to as the latest revised estimate. See also earned value technique and estimate to complete.

Estimate to complete (ETC): The expected cost needed to complete all the remaining work for a schedule activity, work breakdown structure component, or the project. See also earned value technique and estimate at completion.

Event: Something that happens, an occurrence, an outcome.

Exception report: Document that includes only major variations from the plan (rather than all variations).

Execute: Directing, managing, performing, and accomplishing the project work, providing the deliverables, and providing work performance information.

Executing: See execute.

Executing processes: Those processes performed to complete the work defined in the project management plan to accomplish the project's objectives defined in the project scope statement.

Execution: See execute.

Expected monetary value (EMV) analysis: A statistical technique that calculates the average outcome when the future includes scenarios that may or may not happen. A common use of this technique is within decision tree analysis. Modeling and simulation are recommended for cost and schedule risk analysis because it is more powerful and less subject to misapplication than expected monetary value analysis.

Expert judgment: Judgment provided based upon expertise in an application area, knowledge area, discipline, industry, etc., as appropriate for the activity being performed. Such expertise may be provided by any group or person with specialized education, knowledge, skill, experience, or training, and is available from many sources, including other units within the performing organization; consultants; stakeholders, including customers, professional, and technical associations; and industry groups.

Failure mode and effect analysis (FMEA): An analytical procedure, in which each potential failure mode in every component of a product is analyzed to determine its effect on the reliability of that component and, by itself or in combination with other possible failure modes, on the reliability of the product or system and on the required function of the component; or the examination of a product (at the system and/or lower levels) for all ways that a failure may occur. For each potential failure, an estimate is made of its effect on the total system and of its impact. In addition, a review is undertaken of the action planned to minimize the probability of failure and to minimize its effects.

Fast tracking: A specific project schedule compression technique that changes network logic to overlap phases that would normally be done in sequence, such as the design phase and construction phase, or to perform schedule activities in parallel. See schedule compression and see also crashing.

Finish date: A point in time associated with a schedule activity's completion. Usually qualified by one of the following: actual, planned, estimated, scheduled, early, late, baseline, target, or current.

Finish-to-finish (FF): The logical relationship where completion of work of the successor activity cannot finish until the completion of work of the predecessor activity. See also logical relationship.

Finish-to-start (FS): The logical relationship where initiation of work of the successor activity depends upon the completion of work of the predecessor activity. See also logical relationship.

Firm-fixed-price (FFP) contract: A type of fixed price contract where the buyer pays the seller a set amount (as defined by the contract} regardless of the seller's costs.

Fixed-price-incentive-fee (FPIF) contract: A type of contract where the buyer pays the seller a set amount (as defined by the contract) and the seller can earn an additional amount if the seller meets defined performance criteria.

Fixed-price or lump-sum contract: A type of contract involving a fixed total price for a well-defined product. Fixed price contracts may also include incentives for meeting or exceeding selected project objectives such as schedule targets. The simplest form of a fixed price contract is a purchase order.

Float: Also called slack. See total float and see also free float.

Flowcharting: The depiction in a diagram format of the inputs, process actions, and outputs of one or more processes within a system.

Forecasts: Estimates or predictions of conditions and events in the project's future based on information and knowledge available at the time of the forecast. Forecasts are updated and reissued based on work performance information provided as the project is executed. The information is based on the project's past performance and expected future performance, and includes information that could impact the project in the future such as estimate at completion and estimate to complete.

Forward pass: The calculation of the early start and early finish dates for the uncompleted portions of all network activities. See also schedule network analysis and backward pass.

Free float (FF): The amount of time that a schedule activity can be delayed without delaying the early start of any immediately following schedule activities. See also total float.

Functional manager: Someone with management authority over an organizational unit within a functional organization. The manager of any group that actually makes a product or performs a service. Sometimes called a line manager.

Functional organization: A hierarchical organization where each employee has one clear superior, staff are grouped by areas of specialization, and managed by a person with expertise in that area.
Funds: A supply of money or pecuniary resources immediately available.
Gantt chart: See bar chart.
Goods: Commodities, wares, merchandise.
Grade: A category or rank used to distinguish items that have the same functional use, but do not share the same requirements for quality.
Ground rules: A list of acceptable and unacceptable behaviors adopted by a project team to improve working relationships, effectiveness, and communication.
Hammock activity: See summary activity.
Historical information: Documents and data on prior projects including project files, records, correspondence, closed contracts, and closed projects.
Human Resource Planning: The process of identifying and documenting project roles, responsibilities and reporting relationships as well as creating the staffing management plan.
Imposed date: A fixed date imposed on a schedule activity or schedule milestone, usually in the form of a "start no earlier than" and "finish no later than" date.
Influence diagram: Graphical representation of situations showing causal influences, time ordering of events, and other relationships among variables and outcomes.
Influencer: Persons or groups that are not directly related to the acquisition or use of the projects and products, but due to their position in the customer organization, can influence, positively or negatively, the course of the project.
Information distribution: The process of making needed information available to project stakeholders in a timely manner.
Initiating processes: Those processes performed to authorize and define the scope of a new phase or project or that can result in the continuation of halted project work. A large number of the initiating processes are typically done outside the project's scope of control by the organization, program, or portfolio processes and those processes provide input to the project's initiating processes group.
Initiator: A person or organization that has both the ability and authority to start a project.
Input: Any item, whether internal or external to the project that is required by a process before that process proceeds. An input may be an output from a predecessor process.
Inspection: Examining or measuring to verify whether an activity, component, product, result, or service conforms to specified requirements.
Integral: Essential to completeness, requisite, constituent with, formed as a unit with another component.
Integrated: Interrelated, interconnected, interlocked, or enmeshed components blended and unified into a functioning or unified whole.

Integrated change control: The process of reviewing all change requests, approving changes, and controlling changes to deliverables and organizational process assets.

Invitation for bid (IFB): Generally, this term is equivalent to request for proposal. However, in some application areas, it may have a narrower or more specific meaning.

Issue: A point or matter in question or in dispute, or a point or matter that is not settled and is under discussion or over which there are opposing views or disagreements.

Knowledge: Knowing something with the familiarity gained through experience, education, observation, or investigation.

Knowledge area process: An identifiable project management process within a knowledge area.

Knowledge area, project management: See project management knowledge area.

Lag: A modification of a logical relationship that directs a delay in the successor activity. For example, in a finish-to-start dependency with a 10-day lag, the successor activity cannot start until 10 days after the predecessor activity has finished. See also lead.

Late finish date (LF): In the critical path method, the latest possible point in time that a schedule activity may be completed based upon the schedule network logic, the project completion date, and any constraints assigned to the schedule activities without violating a schedule constraint or delaying the project completion date. The late finish dates are determined during the backward pass calculation of the project schedule network.

Late start date (LS): In the critical path method, the latest possible point in time that a schedule activity may begin based upon the schedule network logic, the project completion date, and any constraints assigned to the schedule activities without violating a schedule constraint or delaying the project completion date. The late start dates are determined during the backward pass calculation of the project schedule network.

Latest revised estimate: See estimate at completion.

Lead: A modification of a logical relationship that allows an acceleration of the successor activity. For example, in a finish-to-start dependency with a ten-day lead, the successor activity can start 10 days before the predecessor activity has finished. See also lag. A negative lead is equivalent to a positive lag.

Lessons learned: The learning gained from the process of performing the project. Lessons learned may be identified at any point. Also considered a project record, to be included in the lessons learned knowledge base.

Lessons learned knowledge base: A store of historical information and lessons learned about both the outcomes of previous project selection decisions and previous project performance.

Level of effort (LOE): Support-type activity (e.g., seller or customer liaison, project cost accounting, project management, etc.) that does not readily lend itself to measurement of discrete accomplishment. It is generally characterized by a uniform rate of work, performance over a period of time determined by the activities supported.

Leveling: See resource leveling.
Life cycle: See project life cycle.
Log: A document used to record and describe or denote selected items identified during execution of a process or activity. Usually used with a modifier, such as issue, quality control, action, or defect.
Logic: See network logic.
Logic diagram: See project schedule network diagram.
Logical relationship: A dependency between two project schedule activities, or between a project schedule activity and a schedule milestone. See also precedence relationship. The four possible types of logical relationships are: Finish-to-start; Finish-to-finish; Start-to-start; and Start-to-Finish.
Manage project team: The process of tracking team member performance, providing feedback, resolving issues, and coordinating changes to enhance project performance.
Manage stakeholders: The process of managing communications to satisfy the requirements of, and resolve issues with, project stakeholders.
Master schedule: A summary level project schedule that identifies the major deliverable and work breakdown structure components and key schedule milestones. See also milestone schedule.
Materiel: The aggregate of things used by an organization in any undertaking, such as equipment, apparatus, tools, machinery, gear, material, and supplies.
Matrix organization: Any organizational structure in which the project manager shares responsibility with the functional managers for assigning priorities and for directing the work of persons assigned to the project.
Methodology: A system of practices, techniques, procedures, and rules used by those who work in a discipline.
Milestone: A significant point or event in the project. See also schedule milestone.
Milestone schedule: A summary-level schedule that identifies the major schedule milestones. See also master schedule.
Monitor: Collect project performance data with respect to a plan, produce performance measures, and report and disseminate performance information.
Monitor and control project work: The process of monitoring and controlling the processes required to initiate, plan, execute, and close a project to meet the performance objectives defined in the project management plan and project scope statement.
Monitoring: See monitor.
Monitoring and controlling processes: Those processes performed to measure and monitor project execution so that corrective action can be taken when necessary to control the execution of the phase or project.
Monte Carlo analysis: A technique that computes or iterates the project cost or project schedule many times using input values selected at random from probability distributions of possible costs or durations, to calculate a distribution of possible total project cost or completion dates.
Near-critical activity: A schedule activity that has low total float. The concept of near-critical is equally applicable to a schedule activity or schedule

network path. The limit below which total float is considered near-critical is subject to expert judgment and varies from project to project.

Network: See project schedule network diagram.

Network analysis: See schedule network analysis.

Network logic: The collection of schedule activity dependencies that makes up a project schedule network diagram.

Network loop: A schedule network path that passes the same node twice. Network loops cannot be analyzed using traditional schedule network analysis techniques such as critical path method.

Network open end: A schedule activity without any predecessor activities or successor activities creating an unintended break in a schedule network path. Network open ends are usually caused by missing logical relationships.

Network path: Any continuous series of schedule activities connected with logical relationships in a project schedule network diagram.

Networking: Developing relationships with persons who may be able to assist in the achievement of objectives and responsibilities.

Node: One of the defining points of a schedule network; a junction point joined to some or all of the other dependency lines. See also arrow diagramming method and precedence diagramming method.

Objective: Something toward which work is to be directed, a strategic position to be attained, or a purpose to be achieved, a result to be obtained, a product to be produced, or a service to be performed.

Operations: An organizational function performing the ongoing execution of activities that produce the same product or provide a repetitive service. Examples are production operations, manufacturing operations, and accounting operations.

Opportunity: A condition or situation favorable to the project, a positive set of circumstances, a positive set of events, a risk that will have a positive impact on project objectives, or a possibility for positive changes. Contrast with threat.

Organization: Group of persons organized for some purpose or to perform some type of work within an enterprise.

Organization chart: A method for depicting interrelationships among a group of persons working together toward a common objective.

Organizational breakdown structure (OBS): A hierarchically organized depiction of the project organization arranged so as to relate the work packages to the performing organizational units. (Sometimes OBS is written as Organization Breakdown Structure with the same definition.)

Organizational process assets: Any or all process-related assets from any or all of the organizations involved in the project that are or can be used to influence the project's success. These process assets include formal and informal plans, policies, procedures, and guidelines. The process assets also include the organization's knowledge bases such as lessons learned and historical information.

Original duration (OD): The activity duration originally assigned to a schedule activity and not updated as progress is reported on the activity. Typically

Appendix C

used for comparison with actual duration and remaining duration when reporting schedule progress.

Output: A product, result, or service generated by a process. May be an input to a successor process.

Parametric estimating: An estimating technique that uses a statistical relationship between historical data and other variables (e.g., square footage in construction, lines of code in software development) to calculate an estimate for activity parameters, such as scope, cost, budget, and duration. This technique can produce higher levels of accuracy depending upon the sophistication and the underlying data built into the model. An example for the cost parameter is multiplying the planned quantity of work to be performed by the historical cost per unit to obtain the estimated cost.

Pareto chart: A histogram, ordered by frequency of occurrence, that shows how many results were generated by each identified cause.

Path convergence: The merging or joining of parallel schedule network paths into the same node in a project schedule network diagram. Path convergence is characterized by a schedule activity with more than one predecessor activity.

Path divergence: Extending or generating parallel schedule network paths from the same node in a project schedule network diagram. Path divergence is characterized by a schedule activity with more than one successor activity.

Percent complete (PC or PCT): An estimate, expressed as a percent, of the amount of work that has been completed on an activity or a work breakdown structure component.

Perform quality assurance (QA): The process of applying the planned, systematic quality activities (such as audits or peer reviews) to ensure that the project employs all processes needed to meet requirements.

Perform quality control (QC): The process of monitoring specific project results to determine whether they comply with relevant quality standards and identifying ways to eliminate causes of unsatisfactory performance.

Performance measurement baseline: An approved plan for the project work against which project execution is compared and deviations are measured for management control. The performance measurement baseline typically integrates scope, schedule, and cost parameters of a project, but may also include technical and quality parameters.

Performance reporting: The process of collecting and distributing performance information. This includes status reporting, progress measurement, and forecasting.

Performance reports: Documents and presentations that provide organized and summarized work performance information, earned value management parameters and calculations, and analyses of project work progress and status. Common formats for performance reports include bar charts, S-curves, histograms, tables, and project schedule network diagram showing current schedule status.

Performing organization: The enterprise whose personnel are most directly involved in doing the work of the project.

Phase: See project phase.

Plan contracting: The process of documenting the products, services, and results requirements and identifying potential sellers.

Plan purchases and acquisitions: The process of determining what to purchase or acquire, and determining when and how to do so.

Planned finish date: See schedule finish date.

Planned start date: See scheduled start date.

Planned value: The authorized budget assigned to the scheduled work to be accomplished for a schedule activity or work breakdown structure component. Also referred to as the budgeted cost of work scheduled (BCWS).

Planning package: A WBS component below the control account with known work, content but without detailed schedule activities. See also control account.

Planning processes: Those processes performed to define and mature the project scope, develop the project management plan, and identify and schedule the project activities that occur within the project.

Portfolio: A collection of projects or programs and other work that are grouped together to facilitate effective management of that work to meet strategic business objectives. The projects or programs of the portfolio may not necessarily be interdependent or directly related.

Portfolio Management: The centralized management of one or more portfolios, which includes identifying, prioritizing, authorizing, managing, and controlling projects, programs, and other related work, to achieve specific strategic business objectives.

Position description: An explanation of a project team member's roles and responsibilities.

Practice: A specific type of professional or management activity that contributes to the execution of a process and that may employ one or more techniques and tools.

Precedence diagramming method (PDM): A schedule network diagramming technique in which schedule activities are represented by boxes (or nodes). Schedule activities are graphically linked by one or more logical relationships to show the sequence in which the activities are to be performed.

Precedence relationship: The term used in the precedence diagramming method for a logical relationship. In current usage, however, precedence relationship, logical relationship, and dependency are widely used interchangeably regardless of the diagramming method used.

Predecessor activity: The schedule activity that determines when the logical successor activity can begin or end.

Preventive action: Documented direction to perform an activity that can reduce the probability of negative consequences associated with project risks.

Probability and impact matrix: A common way to determine whether a risk is considered low, moderate, or high by combining the two dimensions of a risk: Its probability of occurrence, and its impact on objectives if it occurs.

Procedure: A series of steps followed in a regular definitive order to accomplish something.

Process: A set of interrelated actions and activities performed to achieve a specified set of products, results, or services.

Process Group: See Project Management Process Groups.

Procurement documents: Those documents utilized in bid and proposal activities, which include buyer's invitation for bid, invitation for negotiations, request for information, request for quotation, request for proposal and seller's responses.

Procurement management plan: The document that describes how procurement processes from developing procurement documentation through contract closure will be managed.

Product: An artifact that is produced, is quantifiable, and can be either an end item in itself or a component item. Additional words for products are materiel and goods. Contrast with result and service. See also deliverable.

Product life cycle: A collection of generally sequential, nonoverlapping product phases whose name and number are determined by the manufacturing and control needs of the organization. The last product life cycle phase for a product is generally the product's deterioration and death. Generally, a project life cycle is contained within one or more product life cycles.

Product scope: The features and functions that characterize a product, service, or result.

Product scope description: The documented narrative description of the product scope.

Program: A group of related projects managed in a coordinated way to obtain benefits and control not available from managing them individually. Programs may include elements of related work outside of the scope of the discrete projects in the program.

Program management: The centralized coordinated management of a program to achieve the program's strategic objectives and benefits.

Program Management Office (PMO): The centralized management of a particular program or programs such that corporate benefit is realized by the sharing of resources, methodologies, tools and techniques, and related high-level project management focus. See also project management office.

Progressive elaboration: Continuously improving and detailing a plan as more detailed and specific information and more accurate estimates become available as the project progresses, and thereby producing more accurate and complete plans that result from the successive iterations of the planning process.

Project: A temporary endeavor undertaken to create a unique product, service, or result.

Project calendar: A calendar of working days or shifts that establishes those dates on which schedule activities are worked and nonworking days that determine those dates on which schedule activities are idle. Typically defines holidays, weekends, and shift hours. See also resource calendar.

Project charter: A document issued by the project initiator or sponsor that formally authorizes the existence of a project, and provides the project manager with the authority to apply organizational resources to project activities.

Project initiation: Launching a process that can result in the authorization and scope definition of a new project.

Project life cycle: A collection of generally sequential project phases whose name and number are determined by the control needs of the organization or organizations involved in the project. A life cycle can be documented with a methodology.

Project Management (PM): The application of knowledge, skills, tools, and techniques to project activities to meet the project requirements.

Project Management Body of Knowledge (PMBOK): An inclusive term that describes the sum of knowledge within the profession of project management. As with other professions such as law, medicine, and accounting, the body of knowledge rests with the practitioners and academics that apply and advance it. The complete *PMBOK* includes proven traditional practices that are widely applied and innovative practices that are emerging in the project management profession.

Project Management Information System (PMIS): An information system consisting of the tools and techniques used to gather, integrate, and disseminate the outputs of project management processes. It is used to support all aspects of the project from initiating through closing, and can include both manual and automated systems.

Project management knowledge area: An identified area of project management defined by its knowledge requirements and described in terms of its component processes, practices, inputs, outputs, tools, and techniques.

Project Management Office (PMO): An organizational body or entity assigned various responsibilities related to the centralized and coordinated management of those projects under its domain. The responsibilities of a PMO can range from providing project management support functions to actually being responsible for the direct management of a project. See also program management office.

Project management plan: A formal, approved document that defines how the projected is executed, monitored, and controlled. It may be summary or detailed and may be composed of one or more subsidiary management plans and other planning documents.

Project management process: One of the processes, unique to project management and described in the *PMBOK* guide.

Project management process group: A logical grouping of the project management processes described in the *PMBOK* guide. The project management process groups include initiating processes, planning processes, executing processes, monitoring and controlling processes, and closing processes. Collectively, these five groups are required for any project, have clear internal dependencies, and must be performed in the same sequence on each project, independent of the application area or the specifics of the

Appendix C   381

applied project life cycle. Project management process groups are not project phases.

Project Management Professional (PMP): A person certified as a PMP by the Project Management Institute (PMI).

Project management software: A class of computer software applications specifically designed to aid the project management team with planning, monitoring, and controlling the project, including cost estimating, scheduling, communications, collaboration, configuration management, document control, records management, and risk analysis.

Project management system: The aggregation of the processes, tools, techniques, methodologies, resources, and procedures to manage a project. The system is documented in the project management plan and its content will vary depending upon the application area, organizational influence, complexity of the project, and the availability of existing systems. A project management system, which can be formal or informal, aids a project manager in effectively guiding a project to completion. A project management system is a set of processes and the related monitoring and control functions that are consolidated and combined into a functioning, unified whole.

Project management team: The members of the project team who are directly involved in project management activities. On some smaller projects, the project management team may include virtually all of the project team members.

Project manager (PM): The person assigned by the performing organization to achieve the project objectives.

Project organization chart: A document that graphically depicts the project team members and their interrelationships for a specific project.

Project phase: A collection of logically related project activities, usually culminating in the completion of a major deliverable. Project phases (also called phases) are mainly completed sequentially, but can overlap in some project situations. Phases can be subdivided into subphases and then components; this hierarchy, if the project or portions of the project are divided into phases, is contained in the work breakdown structure. A project phase is a component of a project life cycle. A project phase is not a project management process group.

Project process groups: The five process groups required for any project that have clear dependencies and that are required to be performed in the same sequence on each project, independent of the application area or the specifics of the applied project life cycle. The process groups are initiating, planning, executing, monitoring and controlling, and closing.

Project schedule: The planned dates for performing schedule activities and the planned dates for meeting schedule milestones.

Project schedule network diagram: Any schematic display of the logical relationships among the project schedule activities. Always drawn from left to right to reflect project work chronology.

Project scope: The work that must be performed to deliver and product, service, or result with the specified features and functions.

Project scope management plan: The document that describes how the project scope will be defined, developed, and verified and how the work breakdown structure will be created and defined, and that provides guidance on how the project scope will be managed and controlled by the project management team. It is contained in or is a subsidiary plan of the project management plan. The project scope management plan can be informal and broadly framed, or formal and highly detailed, based on the needs of the project.

Project scope statement: The narrative description of the project scope, including major deliverables, project objectives, project assumptions, project constraints, and a statement of work that provides a documented basis for making future project decisions and for confirming or developing a common understanding of project scope among the stakeholders. A statement of what needs to be accomplished.

Project summary work breakdown structure (PSWBS): A work breakdown structure for the project that is only developed down to the subproject level of detail within some legs of the WBS, and where the detail of those subprojects are provided by use of contract work breakdown structures.

Project team: All the project team members, including the project management team, the project manager and, for some projects, the project sponsor.

Project team directory: A documented list of project team members, their project roles, and communication information.

Project team members: The persons who report either directly or indirectly to the project manager, and who are responsible for performing project work, as a regular part of their assigned duties.

Project organization: Any organizational structure in which the project manager has full authority to assign priorities, apply resources, and direct the work of persons assigned to the project.

Qualitative risk analysis: The process of prioritizing risks for subsequent further analysis or action by assessing and combining their probability of occurrence and impact.

Quality: The degree to which a set of inherent characteristics fulfills requirements.

Quality management plan: The quality management plan describes how the project management team will implement the performing organization's quality policy. The quality management plan is a component or a subsidiary plan of the project management plan. The quality management plan may be formal or informal, highly detailed, or broadly framed, based on the requirements of the project.

Quality planning: The process of identifying which quality standards are relevant to the project and determining how to satisfy them.

Quantitative risk analysis: The process of numerically analyzing the effect on overall project objectives of identified risks.

Regulation: Requirements imposed by a governmental body. These requirements can establish product, process, or service characteristics—including applicable administrative provisions—that have government-mandated compliance.

Appendix C

Reliability: The probability of a product performing its intended function under specific conditions for a given period of time.

Remaining duration: The time in calendar units between the data date of the project schedule and the finish date of a schedule activity that has an actual start date. This represents the time needed to complete a schedule activity where the work is in progress.

Request for information: A type of procurement document whereby the buyer requests a potential seller to provide various pieces of information related to a product or service or seller capability.

Request for proposal (RFP): A type of procurement document used to request proposals from prospective sellers of products or services. In some application areas, it may have a narrower or more specific meaning.

Request for quotation (RFQ): A type of procurement document used to request price quotations from prospective sellers of common or standard products or services. Sometimes used in place of request for proposal and in some application areas, it may have a narrower or more specific meaning.

Request seller responses: The process of obtaining information, quotations, bids, offers, or proposals, as appropriate.

Requested change: A formally documented change request that is submitted for approval to the integrated change control process. Contrast with approved change request.

Requirement: A condition or capability that must be met or possessed by a system, product, service, result, or component to satisfy a contract, standard, specification, or other formally imposed documents. Requirements include the quantified and documented needs, wants, and expectations of the sponsor, customer, and other stakeholders.

Reserve: A provision in the project management plan to mitigate cost and schedule risk. Often used with a modifier (e.g., management reserve, contingency reserve) to provide further detail on what types of risk are meant to be mitigated.

Reserve analysis: An analytical technique to determine the essential features and relationships of components in the project management plan to establish a reserve for the schedule duration, budget, estimated cost, or funds for a project.

Residual risk: A risk that remains after risk responses have been implemented.

Resource: Skilled human resources (specific disciplines either individually or in crews or teams), equipment, services, supplies, commodities, materiel, budgets, or funds.

Resource breakdown structure (RBS): A hierarchical structure of resources by resource category and resource type used in resource leveling schedules and to develop resource-limited schedules, and which may be used to identify and analyze project human resource assignments.

Resource calendar: A calendar of working days and nonworking days that determines those dates on which each specific resource is idle or can be active. Typically defines resource specific holidays and resource availability periods. See also project calendar.

Resource-constrained schedule: See resource-limited schedule.

Resource histogram: A bar chart showing the amount of time that a resource is scheduled to work over a series of time periods. Resource availability may be depicted as a line for comparison purposes. Contrasting bars may show actual amounts of resource used as the project progresses.

Resource leveling: Any form of schedule network analysis in which scheduling decisions (start and finish dates) are driven by resource constraints (e.g., limited resource availability or difficult-to-manage changes in resource availability levels).

Resource-limited schedule: A project schedule whose schedule activity, scheduled start dates, and scheduled finish dates reflect expected resource availability. A resource-limited schedule does not have any early or late start or finish dates. The resource-limited schedule total float is determined by calculating the difference between the critical path method late finish date and the resource-limited scheduled finish date. Sometimes called resource-constrained schedule. See also resource leveling.

Resource planning: See activity resource estimating.

Responsibility matrix: A structure that relates the project organizational breakdown structure to the work breakdown structure to help ensure that each component of the project's scope of work is assigned to a responsible person.

Result: An output from performing project management processes and activities. Results include outcomes (e.g., integrated systems, revised process, restructured organization, tests, trained personnel, etc.) and documents (e.g., policies, plans, studies, procedures, specifications, reports, etc.).

Retainage: A portion of a contract payment that is withheld until contract completion to ensure full performance of the contract terms.

Rework: Action taken to bring a defective or nonconforming component into compliance with requirements or specifications.

Risk: An uncertain event or condition that, if it occurs, has a positive or negative effect on a project's objectives. See also risk category and risk breakdown structure.

Risk acceptance: A risk response planning technique that indicates that the project team has decided not to change the project management plan to deal with a risk, or is unable to identify any other suitable response strategy.

Risk avoidance: A risk response planning technique for a threat that creates changes to the project management plan that are meant to either eliminate the risk or to protect the project objectives from its impact. Generally, risk avoidance involves relaxing the time, cost, scope, or quality objectives.

Risk breakdown structure (RBS): A hierarchically organized depiction of the identified project risks arranged by risk category and subcategory that identifies the various areas and causes of potential risks. The risk breakdown structure is often tailored to specific project types.

Risk category: A group of potential causes of risk. Risk causes may be grouped into categories such as technical, external, organizational, environmental, or project management. A category may include subcategories such as technical maturity, weather, or aggressive estimating. See also risk breakdown structure.

Risk database: A repository that provides for collection, maintenance, and analysis of data gathered and used in the risk management processes.

Risk identification: The process of determining which risks might affect the project and documenting their characteristics.

Risk management plan: The document describing how project risk management will be structured and performed on the project. It is contained in or is a subsidiary plan of the project management plan. The risk management plan can be informal and broadly framed, or formal and highly detailed, based on the needs of the project. Information in the risk management plan varies by application area and project size. The risk management plan is different from the risk register that contains the list of project risks, the results of risk analysis, and the risk responses.

Risk management planning: The process of deciding how to approach, plan, and execute risk management activities for a project.

Risk mitigation: A risk response planning technique associated with threats that reduces the probability of occurrence or impact of a risk to below an acceptable threshold.

Risk monitoring and control: The process of tracking identified risks, monitoring residual risks, identifying new risks, executing risk response plans, and evaluating their effectiveness throughout the project life cycle.

Risk register: The document containing the results of the qualitative risk analysis, quantitative risk analysis, and risk response planning. The risk register details all identified risks, including description, category, cause, probability of occurring, impact(s) on objectives, proposed responses, owners, and current status. The risk register is a component of the project management plan.

Risk response planning: The process of developing options and actions to enhance opportunities and to reduce threats to project objectives.

Risk transference: A risk response planning technique that shifts the impact of a threat to a third party, together with ownership of the response.

Role: A defined function to be performed by a project team member, such as testing, filing, inspecting, coding.

Rolling wave planning: A form of progressive elaboration planning where the work to be accomplished in the near term is planned in detail at a low level of the work breakdown structure, while the work far in the future is planned at a relatively high level of the work breakdown structure, but the detailed planning of the work to be performed within another one or two periods in the near future is done as work is being completed during the current period.

Root cause analysis: An analytical technique used to determine the basic underlying reason that causes a variance or a defect or a risk. A root cause may underlie more than one variance or defect or risk.

Schedule: See project schedule and see also schedule model.

Schedule activity: A discrete scheduled component of work performed during the course of a project. A schedule activity normally has an estimated duration, an estimated cost, and estimated resource requirements. Schedule activities are connected to other schedule activities or schedule milestones with logical relationships, and are decomposed from work packages.

Schedule analysis: See schedule network analysis.

Schedule compression: Shortening the project schedule duration without reducing the project scope. See also crashing and fast tracking.

Schedule control: The process of controlling changes to the project schedule.

Schedule development: The process of analyzing schedule activity sequences, schedule activity durations, resource requirements, and schedule constraints to create the project schedule.

Schedule management plan: The document that establishes criteria and the activities for developing and controlling the project schedule. It is contained in or is a subsidiary of the project management plan. The schedule management plan may be formal or informal, highly detailed or broadly framed, based on the needs of the project.

Schedule milestone: A significant event in the project schedule, such as an event restraining future work or marking the completion of a major deliverable. A schedule milestone has zero duration. Sometimes called a milestone activity. See also milestone.

Schedule model: A model used in conjunction with manual methods or project management software to perform schedule network analysis to generate the project schedule for use in managing the execution of a project. See also project schedule.

Schedule network analysis: The technique of identifying early and late start dates, as well as early and late finish dates, for the uncompleted portions of project schedule activities. See also critical path method, critical chain method, what-if analysis, and resource leveling.

Schedule performance index (SPI): A measure of schedule efficiency on a project. It is the ratio of EV to PV. SPI = EV/ PV. An SPI value equal to or greater than one indicates a favorable condition; and a value of less than one indicates an unfavorable condition. See also earned value management.

Schedule variance (SV): A measure of schedule performance on a project. It is the algebraic difference between the EV and the PV. SV = EV − PV. See also earned value management.

Scheduled finish date: The point in time that work was scheduled to finish on a schedule activity. The scheduled finish date is normally within the range of dates delimited by the early finish date and the late finish date. It may reflect resource leveling of scarce resources. Sometimes called planned finish date.

Scheduled start date: The point in time that work was scheduled to start on a schedule activity. The scheduled start date is normally within the range of dates delimited by the early start date and the late start date. It may reflect resource leveling of scarce resources. Sometimes called planned start date.

Appendix C 387

Scope: The sum of the products, services, and results to be provided as a project. See also project scope and product scope.
Scope baseline: See baseline.
Scope change: Any change to the project scope. A scope change almost always requires an adjustment to the project cost or schedule.
Scope control: The process of controlling changes to the project scope.
Scope creep: Adding features and functionality (project scope) without addressing the effects on time, costs, and resources, or without customer approval.
Scope definition: The process of developing a detailed project scope statement as the basis for future project decisions.
Scope verification: The process of formalizing acceptance of the completed project deliverable.
S-curve: Graphic display of cumulative costs, labor hours, percentage of work, or other quantities, plotted against time. The name derives from the S-like shape of the curve (flatter at the beginning and end, steeper in the middle) produced on a project that starts slowly, accelerates, and then tails off. Also a term for the cumulative likelihood distribution that is a result of a simulation, a tool of quantitative risk analysis.
Secondary risk: A risk that arises as a direct result of implementing a risk response.
Select sellers: The process of reviewing offers, choosing from among potential sellers and negotiating a written contract with a seller.
Seller: A provider or supplier of products, services, or results to an organization.
Sensitivity analysis: A quantitative risk analysis and modeling technique used to help determine which risks have the most potential impact on the project. It examines the extent to which the uncertainty of each project element affects the objective being examined when all other uncertain elements are held at their baseline values. The typical display of results is in the form of a tornado diagram.
Service: Useful work performed that does not produce a tangible product or result, such as performing any of the business functions supporting production or distribution. Contrast with product and result. See also deliverable.
Simulation: A simulation uses a project model that translates the uncertainties specified at a detailed level into their potential impact on objectives that are expressed at the level of the total project. Project simulations use computer models and estimates of risk usually expressed as a probability distribution of possible costs or durations at a detailed work level and are typically performed using Monte Carlo analysis.
Skill: Ability to use knowledge, a developed aptitude, and a capability to effectively and readily execute or perform an activity.
Slack: See total float and free float.
Special cause: A source of variation that is not inherent in the system, is not predictable, and is intermittent. It can be assigned to a defect in the system. On a control chart, points beyond the control limits, or nonrandom patterns within the control limits, indicate it. Also referred to as assignable cause. Contrast with common cause.

Specification: A document that specifies, in a complete, precise, verifiable manner, the requirements, design, behavior, or other characteristics of a system, component, product, result, or service and, often, the procedures for determining whether these provisions have been satisfied. Examples are requirement specification, design specification, product specification, and test specification.

Specification limits: The area, on either side of the centerline, or mean, of data plotted on a control chart that meets the customer's requirements for a product or service. This area may be greater than or less than the area defined by the control limits. See also control limits.

Sponsor: The person or group that provides the financial resources, in cash or in kind, for the project.

Staffing management plan: The document that describes when and how human resource requirements will be met. It is contained in, or is a subsidiary plan of, the project management plan. The staffing management plan can be informal and broadly framed, or formal and highly detailed, based on the needs of the project. Information in the staffing management plan varies by application area and project size.

Stakeholder: Persons and organizations such as customers, sponsors, performing organization and the public, who are actively involved in the project, or whose interests may be positively or negatively affected by execution or completion of the project. They may also exert influence over the project and its deliverable.

Standard: A document established by consensus and approved by a recognized body that provides for common and repeated use, rules, guidelines, or characteristics for activities or their results, aimed at the achievement of the optimum degree of order in a given context.

Start date: A point in time associated with a schedule activity's start usually qualified by one of the following: actual, planned, estimated, scheduled, early, late, target, baseline, or current.

Start-to-Finish: The logical relationship where completion of the successor schedule activity is dependent upon the initiation of the predecessor schedule activity. See also logical relationship.

Start-to-Start: The logical relationship where initiation of the work of the successor schedule activity depends upon the initiation of the work of the predecessor schedule activity. See also logical relationship.

Statement of work (SOW): A narrative description of products, services, or results to be supplied.

Subnetwork: A subdivision (fragment) of a project schedule network diagram usually representing a subproject or a work package. Often used to illustrate or study some potential or proposed schedule condition, such as changes in preferential schedule logic or project scope.

Subphase: A subdivision of a phase.

Subproject: A smaller portion of the overall project created when a project is subdivided into more manageable components or pieces. Subprojects are usually represented in the work breakdown structure. A subproject can be referred to as a project, managed as a project, and acquired from a seller. May be referred to as a subnetwork in a project schedule network diagram.

Appendix C

Successor: See successor activity.

Successor activity: The schedule activity that follows a predecessor activity, as determined by their logical relationship.

Summary activity: A group of related schedule activities aggregated at some summary level, and displayed/reported as a single activity at that summary level. See also subproject and subnetwork.

SWOT (Strengths, Weaknesses, Opportunities, and Threats analysis): This information gathering technique examines the project from the perspective of each project's strengths, weaknesses, opportunities, and threats to increase the breadth of the risks considered by risk management.

System: An integrated set of regularly interacting or interdependent components created to accomplish a defined objective, with defined and maintained relationships among its components, and the whole producing or operating better than the simple sum of its components. Systems may be either physically process based or management process based or more commonly a combination of both. Systems for project management are composed of project management processes, techniques, methodologies, and tools operated by the project management team.

Target completion date: An imposed date that constrains or otherwise modifies the schedule network analysis.

Target finish date: The date that work is planned (targeted) to finish on a schedule activity.

Target schedule: A schedule adopted for comparison purposes during schedule network analysis, which can be different from the baseline schedule. See also baseline.

Target start date: The date that work is planned (targeted) to start on a schedule activity.

Task: A term for work, whose meaning and placement within a structured plan for project work varies by the application area, industry, and brand of project management software.

Technical performance measurement: A performance measurement technique that compares technical accomplishments during project execution to the project management plan's schedule of planned technical achievements. It may use key technical parameters of the product produced by the project as a quality metric. The achieved metric values are part of the work performance information.

Template: A partially complete document in a predefined format that provides a defined structure for collecting, organizing, and presenting information and data. Templates are often based upon documents created during prior projects. Templates can reduce the effort needed to perform work and increase the consistency of results.

Threat: A condition or situation unfavorable to the project, a negative set of circumstances, a negative set of events, a risk that will have a negative impact on a project objective if it occurs, or a possibility for negative changes. Contrast with opportunity.

Three-point estimate: An analytical technique that uses three cost or duration estimates to represent the optimistic, most likely, and pessimistic scenarios.

This technique is applied to improve the accuracy of the estimates of cost or duration when the underlying activity or cost component is uncertain.

Threshold: A cost, time, quality, technical, or resource value used as a parameter, and which may be included in product specifications. Crossing the threshold should trigger some action, such as generating an exception report.

Time and material (T&M) contract: A type of contract that is a hybrid contractual arrangement containing aspects of both cost-reimbursable and fixed-price contracts. Time and material contracts resemble cost-reimbursable type arrangements in that they have no definitive end because the full value of the arrangement is not defined at the time of the award. Thus, time and material contracts can grow in contract value as if they were cost-reimbursable-type arrangements. Conversely, time and material arrangements can also resemble fixed price arrangements. For example, the unit rates are preset by the buyer and seller, when both parties agree on the rates for the category of senior engineers.

Time-now date: See data date.

Time-scaled schedule network diagram: Any project schedule network diagram drawn in such a way that the positioning and length of the schedule activity represents its duration. Essentially, it is a bar chart that includes schedule network logic.

Total float: The total amount of time that a schedule activity may be delayed from its early start date without delaying the project finish date, or violating a schedule constraint. Calculated using the critical path method technique and determining the difference between the early finish dates and late finish dates. See also free float.

Total quality management (TQM): A common approach to implementing a quality improvement program within an organization.

Trend analysis: An analytical technique that uses mathematical models to forecast future outcomes based on historical results. It is a method of determining the variance from a baseline of a budget, cost, schedule, or scope parameter by using prior progress reporting periods' data and projecting how much that parameter's variance from baseline might be at some future point in the project if no changes are made in executing the project.

Triggers: Indications that a risk has occurred or is about to occur. Triggers may be discovered in the risk identification process and watched in the risk monitoring and control process. Triggers are sometimes called risk symptoms or warning signs.

Triple constraint: A framework for evaluating competing demands. The triple constraint is often depicted as a triangle where one of the sides or one of the corners represents one of the parameters being managed by the project team.

User: The person or organization that will use the project's product or service. See also customer.

Validation: The technique of evaluating a component or product during or at the end of a phase or project to ensure it complies with the specified requirements. Contrast with verification.

Appendix C

Value engineering (VE): A creative approach used to optimize project life cycle costs, save time, increase profits, improve quality, expand market share, solve problems, and use resources more effectively.

Variance: A quantifiable deviation, departure, or divergence away from a known baseline or expected value.

Variance analysis: A method for resolving the total variance in the set of scope, cost, and schedule variables into specific component variances that are associated with defined factors affecting the scope, cost, and schedule variables.

Verification: The technique of evaluating a component or product at the end of a phase or project to assure or confirm it satisfies the conditions imposed. Contrast with validation.

Virtual team: A group of persons with a shared objective who fulfill their roles with little or no time spent meeting face to face. Various forms of technology are often used to facilitate communication among team members. Virtual teams can be comprised of persons separated by great distances.

Voice of the customer: A planning technique used to provide products, services, and results that truly reflect customer requirements by translating those customer requirements into the appropriate technical requirements for each phase of project product development.

War room: A room used for project conferences and planning, often displaying charts of cost, schedule status, and other key project data.

Work: Sustained physical or mental effort, exertion, or exercise of skill to overcome obstacles and achieve an objective.

Work authorization: A permission and direction, typically written, to begin work on a specific schedule activity or work package or control account. It is a method for sanctioning project work to ensure that the work is done by the identified organization, at the right time, and in the proper sequence.

Work authorization system: A subsystem of the overall project management system. It is a collection of formal documented procedures that defines how project work will be authorized (committed) to ensure that the work is done by the identified organization, at the right time, and in the proper sequence. It includes the steps, documents, tracking system, and defined approval levels needed to issue work authorizations.

Work breakdown structure (WBS): A deliverable-oriented hierarchical decomposition of the work, to be executed by the project team to accomplish the project objectives and create the required deliverables. It organizes and defines the total scope of the project. Each descending level represents an increasingly detailed definition of the project work. The WBS is decomposed into work packages. The deliverable orientation of the hierarchy includes both internal and external deliverables. See also work package, control account, contract work breakdown structure, and project summary work breakdown structure.

Work breakdown structure component: An entry in the work breakdown structure that can be at any level.

Work breakdown structure dictionary: A document that describes each component in the WBS. For each WBS component, the WBS dictionary includes a brief definition of the scope or statement of work, defined deliverable(s), a list of associated activities, and a list of milestones. Other information may include responsible organization, start and end dates, resources required, an estimate of cost, charge number, contract information, quality requirements, and technical references to facilitate performance of the work.

Work item: See activity and schedule activity.

Work package: A deliverable or project work component at the lowest level of each branch of the work breakdown structure. The work package includes the schedule activities and schedule milestones required to complete the work package deliverable or project work component. See also control account.

Work performance information: Information and data, on the status of the project schedule activities being performed to accomplish the project work, collected as part of the direct and manage project execution processes. Information includes status of deliverable, implementation status for change requests, corrective actions, preventive actions and defect repairs, forecasted estimates to complete, reported percent of work physically completed; achieved value of technical performance measures, start and finish dates of schedule activities.

Work-around: A response to a negative risk that has occurred. Distinguished from contingency plan in that a workaround is not planned in advance of the occurrence of the risk event.

Index

A

Activity-based costing (ABC) methodology
   activity-based management (ABM), 154
   cost allocation, 153
   decomposition/breakdown, 154
   SMART principle, 154
Activity-on-arrow (AOA) approach, 95, 97–98
Activity-on-node (AON) network, 95, 97–98;
     see also Precedence diagramming
     method (PDM)
AOA, see Activity-on-arrow approach
AON, see Activity-on-node network
Arrow diagramming method (ADM), 95; see also
   Activity-on-arrow (AOA) approach
Attribute data, 163–164

B

Backward pass computation, 103–104
Benefit–cost ratio analysis
   cash flow analysis, 132–133
   commodity escalation rate, 133–134
   discounted payback period, 129–130
   double investment, 130–131
   hyperinflation, 135
   inflation concepts, 134–135
   interest rate, 132
   mild and moderate inflation, 135
   severe inflation, 135
   simple payback period, 129
Bottom-up budgeting, 83, 143–144
Breakeven analysis
   mathematical relationship, 136
   profit function, 138
   profit vs. area of loss, 139
   revenue and cost functions, 140
   single-point analysis, 141
   total revenue, 136
Budget distribution, 143–144

C

Capability maturity model integration (CMMI), 62–63
Capable process (C_p), 177–178
Carpé Futurum, 63
CMMI, see Capability maturity model integration
Communications management
   block diagram, 220
   implementation, 220
   multiperson complexity
     channels, 223
     combination formula, 222–223
     cooperation plots, 224
     permutation formula, 223–224
   step-by-step implementation, 219–221
   tools and techniques
     information distribution, 221
     performance reporting, 222
     planning, 221
     stakeholder management, 222
   Triple C model
     integrated and hierarchical
       processes, 224
     planning, scheduling, and control, 225
     types, 225–226
Community feasibility, 77
Contract feasibility analysis, 261–263
Control charts
   calculations
     control limits, 166–168
     inherent process variability, 171
     natural process limit, 170, 172
   data collection strategy, 164
   out-of-control patterns, 166
   plotting
     clutch smoothness, 169, 171
     control charts, moving range and
       individual, 169
     range and average chart, 168–169
   sampling frequency, 165
   stable process, 165–166
   subgroup sample size, 164–165
   trend analysis
     MINITAB software, 172
     test 1, 172–174
     test 2, 174
     test 3, 174–175
     test 4, 175
     test 5, 175–176
     test 6, 175–176
     test 7, 176–177
     test 8, 177
     zone control charting technique, 172
   types of data, 163–164
   X-bar and range charts, 164
Cost–control pie chart, 145–146
Cost-driven/cost-push inflation, 134
Cost management
   budgeting, 116–117

contemporary earned value technique (EVT)
  actual cost (AC) and cost variance (CV), 151
  estimate at completion (EAC), 152–153
  estimate to complete (ETC), 151–152
  planned value (PV), 150
  schedule variance (SV), 151
cost and schedule control systems criteria (C/SCSC)
  contract management, 146–147
  cost–schedule–performance relationship, 148–149
  development, work content base, 149–150
  integrated approach, 147–148
  risk management, 147
  R&M 2000 standard, 149–150
  work rate analysis technique, 149
cost control, 116, 118
inflation effects
  cash flow analysis, 132–133
  commodity escalation rate, 133–134
  concepts, 134–135
  definition, 131, 134
  hyperinflation, 135
  interest rate, 132
  mild and moderate inflation, 135
  severe inflation, 135
portfolio management
  project assessment approach, 118
  return on investment (ROI), 117
project cost estimation, 116–117
  ballpark figure, 141
  bottom-up budgeting, 143–144
  budgeting and risk allocation, 144–145
  cost monitoring, 145–146
  optimistic and pessimistic, 142
  project budget allocation, 142
  top-down budgeting, 142–143
project management body of knowledge (PMBOK), 115
Cost performance index (CPI), 145
Cost reimbursable contracts, 259
CPI, *see* Cost performance index
CPM, *see* Critical path method
Crashing ratios, 110
Critical path method (CPM)
  activity precedence relationships, 100–101
  analysis, 101–102
  backward pass computation, 103–104
  forward pass calculation, 102–103
  Gantt charts, 107–109
  network scheduling
    activity-on-arrow (AOA) approach, 95, 97–98
    critical resource diagramming (CRD), 95
    network components, 97–99
    network control, tracking, 97
    planning phase, 96–97
    precedence diagramming method (PDM), 95, 99
    project network diagram, 95–96
Cultural feasibility, 76, 262
Cumulative CPI (CPI^C), 151

D

Decision process, flaws
  flight readiness review hierarchy, 289–290
  Mission Management Team, 291
  shuttle program management structure, 290
  solid rocket booster (SRB), 291–292
Decision tree analysis
  decision table, 248
  event, 246–247
  probability tree, 249–252
  total number of paths, 248–249
  weather-dependent task durations, 251
Demand-driven/demand-pull inflation, 134
Department of Defense Architecture Framework (DODAF), 5–6
Divide and conquer concept, 22
DODAF, *see* Department of Defense Architecture Framework

E

EAC, *see* Estimate at completion
Economic feasibility, 76, 262
Economic order quantity (EOQ)
  formulation, notations, 274–275
  inventory pattern, 275
  replenishment quantity, 275
  total relevant costs (TRC), 276
Economies of scale, 119
Enterprise environmental factor, 53
EOQ, *see* Economic order quantity
Estimate at completion (EAC), 152–153
Estimate to complete (ETC), 151–152
Ethical codes
  engineers
    components, 337–338
    fundamental canons, 337
    professional obligations, 340–342
    rules of practice, 338–339
  professional conduct
    aspirational and mandatory conduct, 344
    fairness, 346–347
    honesty, 347
    respect, 345–346
    responsibility, 344–345
    structure, 344
    vision and purpose, 343
Expert judgment, 53–54

Index

F

Financial feasibility, 76, 262
Finite element analysis, 182
Fixed price/lump sum contracts, 259
Flight Readiness Review (FRR), 289–294
Formal organization structure
    functional organization, 25–26
    informal organizational structure, 24
    limitations, 25
    positive characteristics, 24–25
FRR, *see* Flight Readiness Review
FRR record, O-ring concern
    Aldrich's testimony, 313–314
    Hardy's testimony, 309–310
    Jerry Mason's testimony, 301
    Lucas' testimony, 311–312
    McDonald's testimony, 293–294, 299, 305
    Mulloy's testimony, 292–293, 300, 305–311
    Roger Boisjoly's testimony, 296–299, 301–304
    Thomas's testimony, 314
Full-duplex communication, 232

G

Gantt charts, 107–109
Government Accountability Office (GAO), 10

H

Half-duplex communication, 232
Hammersmith's project alert scale, 86
Human resource management
    aging workforce, 188–189
    conceptual approach, 210
    elements
        advantages, management approaches, 190–191
        auto-management skills, 191
        career goals and paths, 191–192
        characteristics, technical training programs, 195
        leadership qualities, 198
        management approaches, 190
        Maslow's hierarchy of needs, 193–194, 196–197
        motivation–hygiene factors, 195
        professional peers, 195–196
        professional preservation, 195
        project components, 190
        technical skills, 192–193
        Theories X and Y, 194
        work content, 195
        work simplification, 198
    employees standard competencies, 188
    extended training model
        characteristics and dependency types, 214
        RRP record, 215–216
        stage/phase 1, 210–214
        stage/phase 2, 216–217
        WBS task, 215
    knowledge workers, 189
    performance management
        LINK concept, 201–202
        unbiased leadership, 199, 201
    quantitative modeling, worker assignment
        assignment methods, 202
        assignment solution, 203–204
        cost matrix, 203
        transportation method, 202–203
    step-by-step implementation, 198
        block diagram and implementation, 199
        inputs, tools and techniques, and outputs, 200–201
    technical training
        characteristics, 209
        model, 208
        process, 208–209
    worker assignment problem
        cost matrix method, 203
        formulation, 202
        solution, 204
    work rate analysis
        data table, 205
        formulation, 206
        relationship, 204–205
        resource schedule chart, 207
        tabulation, 206–207
Hungarian method, 203
Hyperinflation, 135

I

Imposed precedence relationship, 100
Inflation effects
    cash flow analysis, 132–133
    commodity escalation rate, 133–134
    concepts, 134–135
    definition, 131, 134
    hyperinflation, 135
    interest rate, 132
    mild and moderate inflation, 135
    severe inflation, 135
Informal organizational structure, 24
Integrated change control
    components, 59
    configuration management system, 59–60
    inputs, 60
    outputs, 61
    tools and techniques, 60

Internal rate of return (IRR), 127–128
Invitation for bids (IFB), *see* Request for proposal (RFP)

K

Kaizen concept, 182–183

L

Lean task value rating system, 183–185
Limit theorem, 125

M

Managerial feasibility, 262
Material contracts, 259
Material requirement planning (MRP), 214
Matrix organization structure
    advantages, 28
    disadvantages, 29
    matrix blend, 28–29
    project characteristic levels, 30
MAUT, *see* Multiattribute utility theory
Minimum attractive rate of return (MARR), 121
MINITAB software, 172
MRP, *see* Material requirement planning
Multiattribute utility theory (MAUT), 268

N

Network scheduling, CPM
    activity-on-arrow (AOA) approach, 95, 97–98
    critical resource diagramming (CRD), 95
    network components, 97–99
    network control, tracking, 97
    planning phase, 96–97
    precedence diagramming method (PDM), 95, 99
    project network diagram, 95–96

O

Organizational process asset, 53
Orion space vehicle crew module, 281–282

P

PDM, *see* Precedence diagramming method
Performance measurement baseline (PMB), 119
PERT formula, 142
PMB, *see* Performance measurement baseline
PMBOK, *see* Project Management Body of Knowledge
PMIS, *see* Project management information system
PMM, *see* Project management methodology
PMS, *see* Project management system
Political feasibility, 77, 262
Portfolio management, 116–118
Precedence diagramming method (PDM), 95, 99
Probability tree diagram
    task selection, 249
    weather-dependent tasks, 250
Procedural precedence relationship, 100
Process capability analysis
    capability index (C_{pk})
        applications, 181–182
        characteristics, 179–181
        specification, 178–179
    capable process (C_p), 177–178
    definition, 177
    distribution, 177–178
Procurement invitation (PI), *see* Request for proposal (RFP)
Procurement management
    compromise programming (CP) approach, 273–274
    contract relationship
        reason for changes, 266
        Triple C linkage, 265–266
    contracts
        categories, 259
        completion and term, 260
        contractor statement, 261
        cost-reimbursable contract, 259–260
        risk levels, 260
    discount option evaluation
        cost curves, 277
        optimal order quantity, 277–278
        quantity discount, 278–279
        reduction strategies, 277
    economic order quantity (EOQ) model
        formulation, notations, 274–275
        inventory pattern, 275
        replenishment quantity, 275
        total relevant costs (TRC), 276
    goal programming approach, 271–272
    procurement cycle, 253–254
    project proposal
        management section, 264–265
        technical section, 264
    rating procedure
        requirements and computation steps, 267
        vendor rating matrix layout, 268
    selection techniques
        Wadhwa–Ravindran model, 269–271
        weighted objective model, 271
    sensitivity analysis, 279–280

Index

step-by-step implementation
  block diagram, 254–255
  contract types and categories, 259
  feasibility analysis, 261–263
  implementation, process groups, 255
  inputs, tools, techniques, and outputs, 256–258
  organization process assets, 261
total relevant cost calculation, 277
vendor rating system, 266–267
Procurement request (PR), *see* Request for proposal (RFP)
Program evaluation and review technique (PERT), 95, 113
Project balance technique
  activity-based costing (ABC) methodology
    activity-based management (ABM), 154
    cost allocation, 153
    decomposition/breakdown, 154
    SMART principle, 154
  contemporary earned value technique (EVT)
    actual cost (AC) and cost variance (CV), 151
    estimate at completion (EAC), 152–153
    estimate to complete (ETC), 151–152
    planned value (PV), 150
    schedule variance (SV), 151
  cost and schedule control systems criteria (C/SCSC)
    contract management, 146–147
    cost–schedule–performance relationship, 148–149
    development, work content base, 149–150
    integrated approach, 147–148
    risk management, 147
    R&M 2000 standard, 149–150
    work rate analysis technique, 149
  cost control elements, 150
Project cost elements
  benefit–cost ratio analysis
    cash flow analysis, 132–133
    commodity escalation rate, 133–134
    discounted payback period, 129–130
    double investment, 130–131
    hyperinflation, 135
    inflation concepts, 134–135
    interest rate, 132
    mild and moderate inflation, 135
    severe inflation, 135
    simple payback period, 129
  breakeven analysis
    mathematical relationship, 136
    profit function, 138
    profit *vs.* area of loss, 139
    revenue and cost functions, 140
    single-point analysis, 141
    total revenue, 136

cash flow analysis
  arithmetic gradient series, 126–127
  capitalized cost formula, 125
  compound amount factor, 121–122, 124
  internal rate of return (IRR), 127–128
  present value factor, 122–123
  time value of money calculations, 121
  uniform series capital recovery factor, 123–124
  uniform series sinking fund factor, 125
cost factors, 119–121
Project cost estimation
  bottom-up budgeting, 143–144
  budgeting and risk allocation, 144–145
  cost monitoring, 145–146
  optimistic and pessimistic estimates, 142
  project budget allocation, 142
  top-down budgeting, 142–143
Project integration management
  block diagram, 44
  CMMI and Carpé Futurum, 62–63
  conventional element, 47, 50
  input–process–output analysis, 45–49
  integrated change control
    components, 59
    configuration management system, 59–60
    inputs, 60
    outputs, 61
    tools and techniques, 60
  integration element, 45
  management plan development, 55–56
  optimism, pessimism, and realism (OPR), 64
  preliminary project scope statement, 54–55
  project charter development
    approach and organization, 52
    brevity and conciseness, 50
    business opportunities/organizational problems, 48, 50
    components, 50–51
    constraints and assumptions, 51
    financial summary, 52
    impact statement, 51
    output, 54
    owner and manager, 52
    scope, goals and objectives, 51
    tools and techniques, 53–54
    Triple C approach, 48
  project closure, 61–62
  project execution, 57–58
  project life cycle, 46, 49
  Project Management Body of Knowledge (PMBOK), 45, 47
  soft and hard skill interface, 43–44
  sustainability, 64–65
  work monitoring and control, 58–59
Projectized organization
  advantages, 27

disadvantages, 27–28
matrix organization structure, 28–30
personnel structure, 26
Project life cycle
  characteristics, 17, 19
  definition, 17
  implementation clusters, 16–17
  knowledge areas, 16, 18
  phases, 17, 19
Project management
  definition, 12
  emergence
    cross-functional application, 11–12
    general management skills, 11
    Government Accountability Office (GAO), 10
    intersection themes, 11
    safety violation, 11–12
  general applicability, 31–32
  knowledge areas
    characteristic/mnemonic symbols, 14–15
    components, 15–16
    IST-CQH-CRP, 14
  large project, 3
  midsize project, 3–4
  process
    corporate life cycle, 19
    integrated loop, 20
    product life cycle, 19
    project life cycle, 16–19
    two-dimensional matrix, 16
  Quick Quiz, 2–3
  small project, 4
  systems architecture, 5–6
Project Management Body of Knowledge (PMBOK)
  communications management, 219
  cost management, 115
  human resource management, 187
  integration management, 45, 47
  risk management, 239, 241
Project management information system (PMIS), 13, 53
Project management methodology (PMM), 12, 53
Project management system (PMS), 13
Project performance factors, 246
Project selection methods, 53

Q

Quality loss function (QLF), 160–161
Quality management
  foundations, 157, 160
  Kaizen concept, 182–183
  lean principles and applications, 182
  lean task value rating system, 183–185
  quality assurance, 157, 159
  quality control, 157, 159
  quality planning, 157–158
  Six Sigma approach
    capable process (C_p), 177–178
    control limits calculation, 166–168
    data collection strategy, 164
    defect sources, identification and elimination, 161–162
    lean, project management, 185
    out-of-control pattern, 166
    roles and responsibility, 162
    sampling frequency, 165
    stable process, 165–166
    statistical process control (SPC), 162–163
    subgroup sample size, 164–165
    X-bar and range charts, 164
  Taguchi loss function, 160–161
Quantity discount, 276, 278–279
Quick-and-dirty assessment, 129
Quick Quiz, 2–3

R

Request for bid (RFB), *see* Request for proposal (RFP)
Request for proposal (RFP), 79–81
Request for tender (RFT), *see* Request for proposal (RFP)
Resource requirement planning (RRP), 214
Return on investment (ROI), 117
Risk management
  block diagram, 241
  cost uncertainties, 244–245
  decision tree analysis
    decision table, 248
    event, definition, 246–247
    probability tree, 249–252
    total number of paths, 248–249
    weather-dependent task durations, 251
  definition, 239–241
  elements and possibilities, 240
  implementation, 242
  life cycle, 240
  performance uncertainties, 246
  project decisions, 242–244
  schedule uncertainties, 245–246
  step-by-step implementation
    across process groups, 242
    qualitative risk analysis, 243
    quantitative risk analysis, 244
    risk identification, 243
    risk planning, 242
    risk response planning, 244
  task selection
    consequences, 247
    decision table, 247
    decision tree, 248

Index

modified probability tree, 252
probability tree diagram, 249
project completion time, 251
tree representation, 248
R&M 2000 standard, 149–150
RRP, *see* Resource requirement planning
Rule of 72 approach, 130–131

S

Safety feasibility, 77, 262
Schedule performance index (SPI), 151
Scope management
  block diagram, 68–69
  budget planning
    bottom-up budgeting, 83
    specifications, 82–83
    top-down budgeting, 83
    zero-base budgeting, 84
  control, 72
  DMAIC application
    analyze and control stage, 74
    define stage, 72–73
    improve stage, 74
    measure stage, 73
    Six Sigma methodology, 72
  elements, 68–69
  feasibility
    analysis, 76–77
    dimensions, 77–78
    industry conversion, 78–79
    inputs–tools and techniques–output analysis, 69–71
    local resources and work force assessment, 79
    Pareto distribution, 67–68
  planning, 69
  product scope definition, 67
  project proposal
    preparation, 81–82
    request for proposal (RFP), 79–81
  project scope definition
    decision making, 71–72
    primary purpose, 69
    product and stakeholder analysis, 71
    trade-off analysis, 67–68
  suppliers, inputs, process, outputs, customers (SIPOC) diagram, 74–76
  trade-off analysis, 67–68
  verification, 72
  work breakdown structure (WBS)
    deliverables, 72
    Hammersmith's project alert scale, 86
    multidimensional project scope response surface, 87–88
    planning levels, 86
    product assurance activities, 86–87
  project review, 85
  project size, 86
  selection criteria, 84–85
  selection hierarchy, 85
Simplex communication, 231
Six Sigma approach
  capable process (C_p), 177–178
  control limits calculation, 166–168
  data collection strategy, 164
  DMAIC, 72, 237
  defect sources, identification and elimination, 161–162
  lean, project management, 185
  out-of-control pattern, 166
  roles and responsibility, 162
  sampling frequency, 165
  stable process, 165–166
  statistical process control (SPC), 162–163
  subgroup sample size, 164–165
  X-bar and range charts, 164
SMART, *see* Specific, measurable, aligned, realistic, timed technique
SMART communication
  modes, 231–232
  principles, 230
  Triple C matrix, 231
Social feasibility, 77, 262
Solid rocket booster (SRB), 291–292
SOW, *see* Statement of work
Space shuttle challenger
  additional commission findings
    ice problem, 320–321
    postinvestigation development, 321–322
  Challenger mission
    accident, 288–289
    flaws in decision process, 289–292
    investigation, 289
    launch delays, 287
    launch problems, 287–288
  Columbia disaster, 322–323
  design decisions, 284–285
  development chronology, 286–287
    Enterprise orbiter, 285
    operational phase, 286
    test flight series, 285–286
  development process, 285
  management decision ambiguity
    Aldrich's discussion, 319–320
    Cioffoletti's testimony, 317–318
    freeze protection plan, 315
    icing condition, 316
    orbiter thermal protection system (TPS), 317
    Peller's testimony, 318–319
    Petrone's testimony, 318
  managerial analysis, 283
  mission background and authorization, 283–284

next generation space vehicles, 281–282
presidential commission findings, 315
risk management, 323–324
satellite project failure, 325
spy satellite death, 325–327
    company diversification, 329–330
    lesson search, 334–335
    multiple design challenges, 330–331
    Soviet threat, 327–329
    trouble signs, 333–334
    winning bid announcement, 331–333
technical and organizational issues, 323
testimonies
    Aldrich's testimony, 313–314
    Boisjoly's testimony, 296–299, 301–304
    Hardy's testimony, 309–310
    Lucas' testimony, 311–312
    Mason's testimony, 301
    McDonald's testimony, 293–294, 299, 305
    Mulloy's testimony, 292–293, 300, 305–311
    Thomas's testimony, 314
SPC, *see* Statistical process control
Specific, measurable, aligned, realistic, timed (SMART) technique, 148, 154
SPI, *see* Schedule performance index
SRB, *see* Solid rocket booster
Stakeholders, 13
Standard deviation equation, 164
Statement of work (SOW), 23, 52–53, 261
Statistical process control (SPC), 162–163
STEP methodology
    building blocks, 7–8
    define, measure, analyze, improve, control (DMAIC) process, 32–33
    engineering challenges, 8
    formal organization structure
        functional organization, 25–26
        informal organizational structure, 24
        limitations, 25
        positive characteristics, 24–25
    integrated systems approach, 38–39
    management steps, 39–40
    elements
        engineering projects, 33–34
        outsourcing, 34
        Triple C approach, 35
    output category, 37
    plan, do, check, act (PDCA) loop, 32
    plan-execute-learn-close (PELC) quadrants, 32
    project definitions, 12–14
    project execution levels, 35–36
    projectized organization
        advantages, 27
        disadvantages, 27–28
        matrix organization structure, 28–30
        personnel structure, 26
    project management
        emergence, 10–12
        general applicability, 31–32
        knowledge areas, 14–16
        large project, 3
        midsize project, 3–4
        process, 16–20
        Quick Quiz, 2–3
        small project, 4
        systems architecture, 5–6
    project organization structure, 23
    project plan elements, 30–31
    project requirements, 40–42
    projects and operations, 20–21
    society advancement, 35
    STEM initiatives and project conceptions, 8–10
    success/failure factors, 21–22
    systems engineering and program management, 4–5
    unique aspects, 37–38
    work breakdown structure (WBS)
        divide and conquer concept, 22
        elements, 22–23
        hypothetical design project, 23–24
        interrelationship flowchart, 22
        preparation, 23

T

Taguchi loss function, 160–161
Technical feasibility, 261
Technical precedence relationship, 100
Thermal protection system (TPS), 317
Three dimensional (3-D) cube relationship, 94
Time contracts, 259
Time management
    Crashing
        comprehensive analysis, 111–112
        CPM network, 109–110
        task duration, 109
        time and cost data, 109–110
    critical chain analysis
        bottleneck operation, 113
        constraint management, 112
        CPM/PERT network analysis, 113–114
        project scheduling, 113
    critical path activity
        analysis, 105
        backward-pass analysis, 103, 105
        definition, 104
        slack/float types, 104–105
    Gantt charts, 107–109
    network templates, 107
    step-by-step implementation
        activity definition, 90–92
        block diagram, 89, 91

Index

project schedule control, 91, 94
project schedule development, 91, 93
resource estimation, 91–92
rolling wave planning, 93–94
sequencing, 91–92, 94
three dimensional (3-D) cube relationship, 94
work breakdown structure (WBS), 91
subcritical path analysis, 106–107
time–cost–quality trade-off axis, 89–90
Top-down budgeting, 83, 142–143
Total process improvement (TPM), 162
Toyota production system, 162
TPM, *see* Total process improvement
TPS, *see* Thermal protection system
Triple C model
  addressing questions, 224
  applications, 236–237
  communication
    barriers, 229
    factors influencing, 228–229
    recommendations, 229–230
    SMART, 230–232
    types, 226, 231–232
  conflict resolution
    guidelines, 236
    types, 235–236
  cooperation
    factors, 232
    security guidelines, 233–234
    types, 226, 232–233
  coordination
    responsibility matrix, 234–235
    types, 226
  DMAIC
    Six Sigma, 237
    sustainability, 237–238
  integrated and hierarchical processes, 224
  planning, scheduling, and control, 224–225
  project charter development, 48

U

Unstable process distribution, 166–167

V

Variable data, 163

W

Wadhwa–Ravindran model, 269–271
Wage-driven/wage-push inflation, 134–135
Work breakdown structure (WBS), 91, 214
  deliverables, 72
  divide and conquer concept, 22
  elements, 22–23
  Hammersmith's project alert scale, 86
  hypothetical design project, 23–24
  interrelationship flowchart, 22
  multidimensional project scope response surface, 87–88
  planning levels, 86
  preparation, 23
  product assurance activities, 86–87
  project review, 85
  project size, 86
  selection criteria, 84–85
  selection hierarchy, 85

Z

Zero-base budgeting, 84